8 *Years*
Previous Solved Papers

ESE 2025

UPSC ENGINEERING SERVICES EXAMINATION

Preliminary Examination

General Studies & Engineering Aptitude

Objective Solved Papers

- ✓ Topicwise presentation
- ✓ Analysis of previous years papers
- ✓ Detailed solutions with **additional information**
- ✓ Matched and verified with the official keys of UPSC

www.madeeasypublications.org

MADE EASY Publications Pvt. Ltd.

Corporate Office: 44-A/4, Kalu Sarai (Near Hauz Khas Metro Station), New Delhi-110016

Contact: 9021300500

E-mail: infomep@madeeasy.in

Visit us at: www.madeeasypublications.org

ESE-2025 : Preliminary Examination
General Studies and Engineering Aptitude
Topicwise Objective Solved Papers : (2017-2024)

1st Edition : 2023

2nd Edition : 2024

Preface

To get a thorough knowledge and to succeed in the growing competition in Engineering Services Examination reading just the theory will not suffice. To supersede other competitors, an aspirant needs a thorough practice of variety of questions. With the introduction of 9 Non-Technical subjects and Current Affairs in Paper-I of ESE Prelims, it has become mandatory to get well versed with these subjects by getting acquainted with every possible variety of questions. To help every aspirant to score high marks in the exam, MADE EASY has come up with **General Studies and Engineering Aptitude Previous Year Objective Solved Papers** with subject and topicwise bifurcations. This book contains questions of ESE-2017, ESE-2018, ESE-2019, ESE-2020, ESE-2021, ESE-2022 and ESE-2023.

B. Singh (Ex. IES)

MADE EASY team has put sincere efforts in framing and compilation of questions with accurate explanations, supplemented with relevant theory and illustrations of subjects namely:

- Engineering Aptitude covering Logical Reasoning and Analytical Ability
- Engineering Mathematics and Numerical Analysis
- General Principles of Design, Drawing, Importance of Safety
- Standards and Quality Practices in Production, Construction, Maintenance and Services
- Basics of Energy and Environment
- Basics of Project Management
- Basics of Material Science and Engineering
- Information and Communication Technologies (ICT)
- Ethics and Values in Engineering Profession

Note: For Current Affairs, students are advised to go through the MADE EASY Current Affairs Magazine Annual Edition.

It is impossible to acknowledge all the individuals who helped us, but would like to sincerely thank all authors, editors and reviewers for putting their painstaking efforts to publish this book.

With Best Wishes

B. Singh

CMD, MADE EASY

GENERAL STUDIES AND ENGINEERING APTITUDE

Objective Solved Papers

of UPSC Engineering Services Examination

Contents

Reasoning & Aptitude

CONTENTS

Reasoning & Aptitude

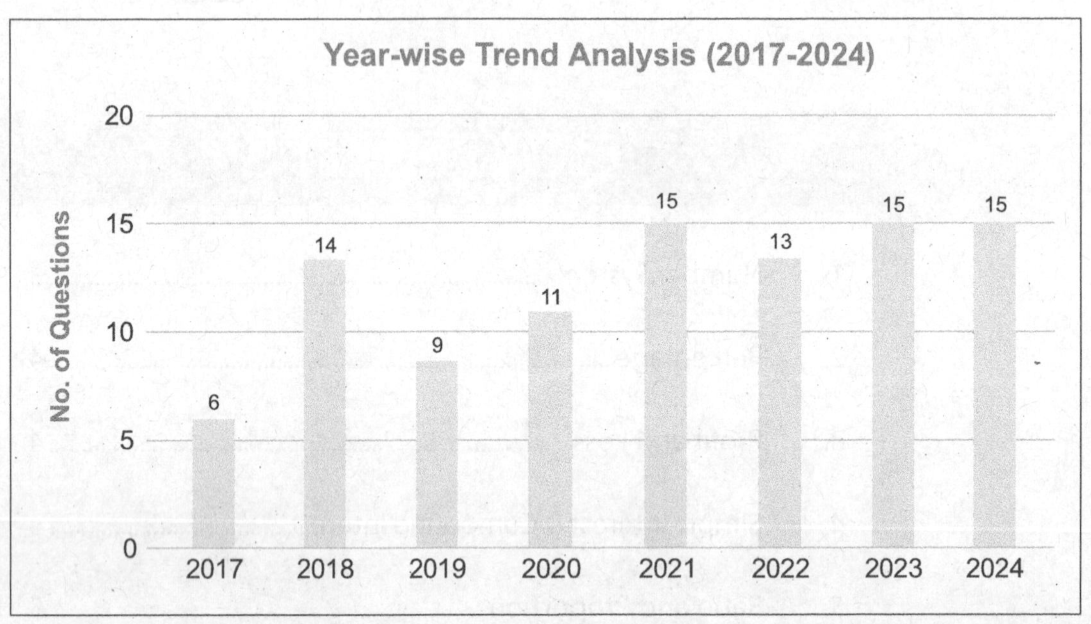

TOPIC-WISE TREND ANALYSIS (2017-2024)		
Sl. No.	Topics	Total Number of Questions
1.	Number System	12
2.	Percentage	3
3.	Profit and Loss	2
4.	Simple Interest & Compound Interest	2
5.	Ratio and Proportion	2
6.	Average	2
7.	Time and Work	6
8.	Time, Speed and Distance	3
9.	Progression	1
10.	Probability	6
11.	Set Theory	2
12.	Blood Relationship	1
13.	Directions	4
14.	Analytical Reasoning	32
15.	Principles of Equation	9
16.	Area and Volume	14

UNIT

I

Reasoning & Aptitude

1. NUMBER SYSTEM

1.1 The sum of squares of successive integers 8 to 16, both inclusive, will be
(a) 1126 (b) 1174
(c) 1292 (d) 1356
[ESE Prelims : 2018]

1.2 A number of friends decided to go on a picnic and planned to spend Rs. 96 on eatables. Four of them, however, did not turn up. As a consequence, the remaining ones had to contribute Rs. 4 each extra. The number of those friends who attended the picnic is
(a) 8 (b) 12
(c) 16 (d) 20
[ESE Prelims : 2020]

1.3 The sum of all the natural numbers between 1 and 101 which are divisible by 5 is
(a) 1000 (b) 1050
(c) 1500 (d) 2550
[ESE Prelims : 2020]

1.4 Sum of the series $2^2 + 4^2 + 6^2 + ... + 20^2$ is
(a) 1040 (b) 1540
(c) 2540 (d) 3080
[ESE Prelims : 2021]

1.5 21 mango trees, 42 apple trees and 56 orange trees have to be planted in rows such that each row contains the same number of trees of one variety only. Minimum number of rows in which the above trees may be planted is
(a) 9 (b) 12
(c) 14 (d) 17
[ESE Prelims : 2022]

1.6 A general wishes to draw up his 36562 soldiers in the form of a solid square. After arranging them, he found that some of them are left over. How many are left?
(a) 81 (b) 75
(c) 61 (d) 52
[ESE Prelims : 2022]

1.7 Two boys begin together to write out a booklet containing 817 lines. The first boy starts with first line, writing at the rate of 200 lines an hour and the second boy starts with the last line. He writes line 817 and so on backwards proceeding at the rate of 150 lines an hour. At what line will they meet?
(a) 469th (b) 467th
(c) 465th (d) 463th
[ESE Prelims : 2022]

1.8 Anuhya picks a number. She doubles the number, squares the result, divides the square by 3, subtracts 30 from the quotient, and gets 18. What are the possible numbers that Anuhya could have picked?
(a) 6 or –6 (b) 16 or –16
(c) 26 or –26 (d) 36 or –36
[ESE Prelims : 2024]

1.9 Nothing is known about the personal life of the ancient Green Mathematician Diophantus except for the information in the following:

"Diophantus passed $\frac{1}{6}$ of his life in childhood, $\frac{1}{12}$ in youth, and $\frac{1}{7}$ more as a bechelor. Five years after his marriage was born a son who died four years before his father, at $\frac{1}{2}$ his father's (final) age." How old was Diophantus when he died?
(a) 64 (b) 54
(c) 74 (d) 84
[ESE Prelims : 2024]

1.10 Select a two-digit number between 50 and 100. Add 83 to your number. From this number form a new number by adding the digit in the hundreds place to the number formed by the other two digits (the digits in the tens place and the ones place). Now subtract this newly formed number from your original number, to arrive at the final result. What is the final result?

(a) 16 (b) 26
(c) 36 (d) 46

[ESE Prelims : 2024]

1.11 The members of a club are going to elect a president from four nominees. In each first-place vote receives 4 points, each second-place vote receives 3 points, each third-place vote receives 2 points, and each last place vote receives 1 point. If the 100 members of the club mark their ballots as shown in the table below, who will be elected president?

	Rankings					
Avalon	2	2	2	2	3	2
Branson	1	4	4	4	2	1
Columbus	3	3	1	3	1	3
Dunkrirk	4	1	3	1	4	4
Number of voters	30	24	18	12	10	6

(a) Avalon (b) Branson
(c) Columbus (d) Dunkirk

[ESE Prelims : 2024]

1.12 A man has a certain number of small boxes to pack into parcels. If he packs 3, 4, 5 or 6 in a parcel, he is left one, if he packs 7 in a parcel, none is left over. What is the number of boxes, he may have to pack?

(a) 106 (b) 301
(c) 309 (d) 400

[ESE Prelims : 2024]

2. PERCENTAGE

2.1 Rajiv spends 40% of his salary on food, 20% on house rent, 10% on entertainment and 10% on conveyance. If his savings at the month end are ₹ 2,000, then his monthly salary is:

(a) ₹ 6,000 (b) ₹ 8,000
(c) ₹ 10,000 (d) ₹ 12,000

[ESE Prelims : 2017]

2.2 The population of a village is 5500. If the number of males increases by 11% and the number of females increases by 20% then the population becomes 6330. The population of female in the village is

(a) 2000 (b) 2500
(c) 3000 (d) 3500

[ESE Prelims : 2021]

2.3 Rohith spends 40% of his monthly income on food items and 50% of the remaining on clothes and conveyance. He saves one-third of the remaining amount after spending on food, clothes and conveyance. If he saves ₹19,200 every year, what is his monthly income?

(a) ₹32,000 (b) ₹16,000
(c) ₹12,000 (d) ₹6,000

[ESE Prelims : 2022]

3. PROFIT AND LOSS

3.1 A total of 324 notes comprising of ₹ 20 and ₹ 50 denominations make a sum of ₹ 12,450. The number of ₹ 20 notes is

(a) 200 (b) 144
(c) 125 (d) 110

[ESE Prelims : 2017]

3.2 A man sold a chair and a table together for Rs. 7,600, thereby making a profit of 25% on the chair and 10% on the table. By selling them together for Rs. 7,500 he would make a profit of 10% on the chair and 20% on the table. Then the cost price of chair and table will be

(a) Rs. 3000 and Rs. 4000
(b) Rs. 3500 and Rs. 4000
(c) Rs. 3000 and Rs. 3500
(d) Rs. 3500 and Rs. 3500

[ESE Prelims : 2019]

4. SIMPLE INTEREST & COMPOUND INTEREST

4.1 In how many years will a sum ₹800 at 10% per annum compounded semi-annually become ₹926.10?

(a) $1\frac{1}{3}$ years (b) $1\frac{1}{2}$ years
(c) $2\frac{1}{3}$ years (d) $2\frac{1}{2}$ years

[ESE Prelims : 2021]

4.2 The difference between simple interest and compound interest on a sum for 2 years at 8% when the interest is compounded annually is ₹16. It the interest were compounded half yearly the difference in two interests would be nearly

(a) ₹21.35 (b) ₹24.64
(c) ₹27.85 (d) ₹29.94

[ESE Prelims : 2022]

5. RATIO AND PROPORTION

5.1 The present ages of 3 brothers are in the proportion 3 : 4 : 5. After 10 years the sum of their ages will be 78. What are their ages now?

(a) 12, 16 and 20 (b) 15, 20 and 25
(c) 21, 28 and 35 (d) 24, 32 and 40

[ESE Prelims : 2017]

5.2 If $(2x + 3y) : (3x + 5y) = 18 : 29$, then the value of $x : y$ is

(a) 4 : 1 (b) 4 : 5
(c) 3 : 4 (d) 3 : 1

[ESE Prelims : 2021]

6. AVERAGE

6.1 A student purchases some books for Rs. 1600. If he had bought 8 more books for the same amount, each book would cost Rs. 10 less. The number of books he buys is

(a) 30 (b) 32
(c) 34 (d) 36

[ESE Prelims : 2019]

6.2 Ten years ago father was 12 times as old as his son and after 10 years father will be 2 times older than his son. The present ages of father and son respectively are

(a) 32 years and 14 years
(b) 34 years and 14 years
(c) 32 years and 12 years
(d) 34 years and 12 years

[ESE Prelims : 2020]

7. TIME AND WORK

7.1 Five Men can paint a building in 20 days, 8 Women can paint the same building in 25 days and 10 Boys can paint it in 30 days. If a team has 2 Men, 6 Women and 5 Boys, how long will it take to paint the building?

(a) 12 days (b) 13 days
(c) 14 days (d) 15 days

[ESE Prelims : 2017]

7.2 A group of workers estimate to finish a work in 10 days, but 5 workers could not join the work. If the rest of them finished the work in 12 days, the number of members present in the team originally is

(a) 50 (b) 45
(c) 35 (d) 30

[ESE Prelims : 2017]

7.3 A small production unit now works 6 days per week with $3\frac{1}{3}$ hours of first shift every one of the 6 days and 3 hours of second shift for each of the first 5 days. Wage negotiations led to an agreement to work on 5 days a week with both shifts together clocking $7\frac{1}{2}$ hours per day with an 8% increase in weekly wages. How much change in the hourly production would mean parity in the agreement for both management and employees?

(a) 3.68% (b) 2.15%
(c) 1.82% (d) 1.33%

[ESE Prelims : 2018]

7.4 A clock strikes once at 1 o'clock, twice at 2 o'clock, thrice at 3 o'clock and so on. The number of times it strikes in 24 hours will be

(a) 116 (b) 136
(c) 156 (d) 196

[ESE Prelims : 2020]

7.5 A is twice as good a workman as B and together, they finish a piece of work in 18 days. In how many days will A alone finish the work?

(a) 28 days (b) 30 days
(c) 27 days (d) 29 days

[ESE Prelims : 2021]

7.6 A tank can be filled by 20 buckets each of capacity 13.5 litres. If the capacity of each bucket be 9 litres, how many buckets will fill the same tank?

(a) 30 (b) 25
(c) 20 (d) 15

[ESE Prelims : 2022]

8. TIME, SPEED AND DISTANCE

8.1 Which of the following conditions hold good for a train which crosses the bridge of length l in time t_1 and crosses another bridge of length $\frac{l}{2}$ in time t_2?

1. $t_2 = \frac{t_1}{2}$
2. $2t_2 > t_1$
3. $t_2 < \frac{t_1}{2}$
4. Speed of train is $\frac{l}{10}$ if $t_1 - t_2 = 5$

Select the correct answer using the codes given below:

(a) 1 and 4 only (b) 2 and 4 only
(c) 1 and 3 only (d) 2 and 3 only

[ESE Prelims : 2019]

8.2 A tourist covers half of his journey by train at 60 km/h, half of the remainder by bus at 30 km/h and the rest by cycle at 10 km/h. Average speed of the tourist during the journey is

(a) 36 km/h (b) 33 km/h
(c) 24 km/h (d) 18 km/h

[ESE Prelims : 2019]

8.3 What is the angle between the hour hand and minute hand of a clock at 3 : 30?

(a) 105° (b) 180°
(c) 75° (d) 90°

[ESE Prelims : 2021]

9. PROGRESSION

9.1 Let the sum of the squares of successive integers 0, 1, 2, ..., $n-1$, n be denoted by S. Let the sum of the cubes of the same integers be denoted by C. It is desirable that $\frac{C}{S}$, as n increases in steps of 'unity' from 'zero', is given by the series:

$\frac{0}{1}, \frac{3}{3}, \frac{9}{5}, \frac{18}{7}, \frac{30}{9}, \ldots$ (for $n = 0, 1, 2, 3, 4 \ldots$). What will this ratio be for $n = 8$?

(a) $\frac{108}{17}$ (b) $\frac{103}{17}$
(c) $\frac{103}{15}$ (d) $\frac{100}{15}$

[ESE Prelims : 2018]

10. PROBABILITY

10.1 Four persons are chosen at random from a group of 3 men, 2 women and 4 children. The chance that exactly 2 of them are children, is

(a) $\frac{2}{9}$ (b) $\frac{4}{5}$
(c) $\frac{7}{12}$ (d) $\frac{10}{21}$

[ESE Prelims : 2021]

10.2 A library has two books each having three copies and three other books each having two copies. In how many ways can all these books be arranged in a shelf so that copies of the same book are not separated?

(a) 80 (b) 100
(c) 120 (d) 140

[ESE Prelims : 2022]

10.3 If six people greet each other at a meeting by shaking hands with one another, how many handshakes will take place?

(a) 14 (b) 16
(c) 15 (d) 18

[ESE Prelims : 2024]

11. SET THEORY

11.1 In a group of 1000 people, 750 speak Hindi and 400 speak English. The number of only Hindi speaking people is

(a) 150 (b) 350
(c) 600 (d) 750

[ESE Prelims : 2020]

11.2 An activities director for a cruise ship has surveyed 240 passengers. Of the 240 passengers, 135 like swimming, 80 like swimming and dancing, 150 like dancing, 40 like swimming and games, 65 like games, 25 like dancing and games, 15 like all three activities. How many passengers like exactly two of the three types of activities?

(a) 220 (b) 20
(c) 30 (d) 100

[ESE Prelims : 2024]

12. BLOOD RELATIONSHIP

12.1 Mary introduces Jack as the son of the only daughter of my father's wife. How is Jack related to Mary?
(a) Brother (b) Son
(c) Husband (d) Father
[ESE Prelims : 2021]

13. DIRECTIONS

13.1 A man walked 3 km towards East, then 5 km towards North-East, then 8 km towards South and finally 5 km towards North-East direction. The distance of his present location from the starting point will be
(a) 9 km (b) 11 km
(c) 15 km (d) 21 km
[ESE Prelims : 2020]

13.2 A man walks 1 km to East and then he turns to South and walks 5 km. Again, he turns to East and walks 2 km. After, he turns to North and walks 9 km. Now, how far he is from his station point?
(a) 3 km (b) 4 km
(c) 5 km (d) 7 km
[ESE Prelims : 2021]

13.3 Rohith went 15 km to the west from his house, then he turned left and walked 20 km. He then turned east and walked 25 km and finally turning left covered 20 km. How far is he from his house?
(a) 5 km (b) 10 km
(c) 40 km (d) 80 km
[ESE Prelims : 2023]

13.4 The following map shows the 10 bridges and 3 islands between the suburbs of North Bay and South Bay. During your morning workout, you decide to jog over each bridge exactly once. Which one of the following statements is correct?

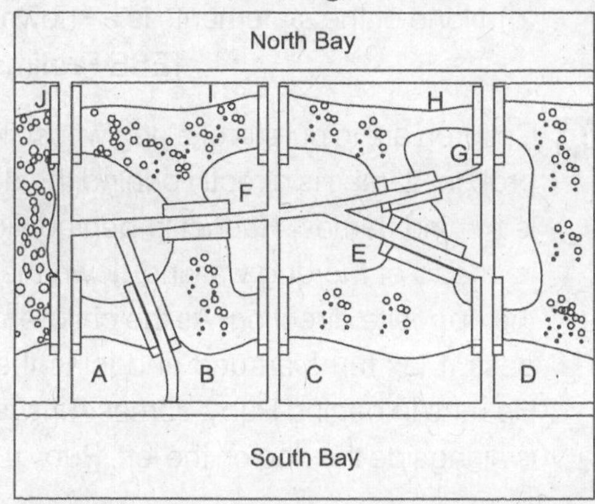

(a) You want to start from North Bay and that your workout concludes after you jog over the 'D' bridge.
(b) You want to start from North Bay and that your workout concludes after you jog over the 'E' bridge.
(c) You want to start from North Bay and that your workout concludes after you jog over the 'H' bridge.
(d) You want to start from North Bay and that your workout concludes after you jog over the 'G' bridge.
[ESE Prelims : 2024]

14. ANALYTICAL REASONING

14.1 The 12 digits on the face of a clock are to be represented employing contributions of only the number 9 as either 9 or $\sqrt{9}$. The other prescribed conditions are (i) the least number of uses alone are permitted; and (ii) when alternates are possible, use of 9 will be preferred over use of $\sqrt{9}$, which should be used minimally. How many times would $\sqrt{9}$ have to be used?
(a) 6 (b) 5
(c) 4 (d) 3
[ESE Prelims : 2018]

14.2 In a particular test, the marks scored by 4 candidates – *A*, *B*, C and *D* are as follows:
- Marks obtained by *A* and *B* add to 100;
- Marks obtained by *C* and *D* add up to those scored by *A*;
- B scores 4 times of *D*;
- D scores 10 marks less than *C*.

The marks obtained by *C* will be
(a) 30 (b) 15
(c) 20 (d) 25
[ESE Prelims : 2018]

14.3 Consider the following gold articles P, Q, R, S and T with different weights:
- P weighs twice as much as Q
- Q weighs four and a half times as much as R
- R weighs half as much as S
- S weighs half as much as T
- T weighs less than P but more than R

Article T will be lighter in weight than

(a) P and S (b) P and R
(c) P and Q (d) Q and R

[ESE Prelims : 2020]

14.4 Consider the following students in an examination:

- A scored more than B
- C scored as much as D
- E scored less than F
- B scored more than C
- F scored less than D

Who scored the lowest?

(a) E (b) C
(c) D (d) F

[ESE Prelims : 2020]

14.5 The number of triangles in the given figure is

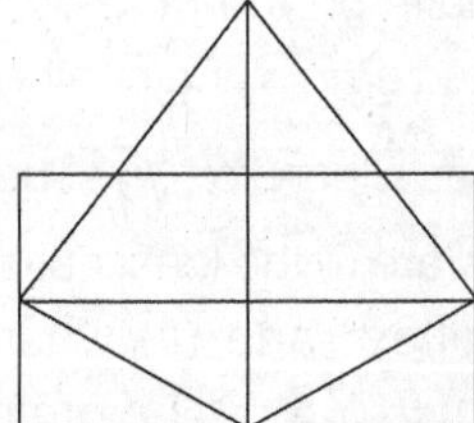

(a) 11 (b) 13
(c) 15 (d) 17

[ESE Prelims : 2021]

14.6 Statement 1 : A has more coins than B.
Statement 2 : B has fewer coins than C.
Statement 3 : C has fewer coins than A.
If the statement 1 is true and statement 2 is false, then the statement 3 is

(a) True (b) False
(c) Uncertain (d) Insufficient data

[ESE Prelims : 2021]

14.7 *A*, *B*, *C*, *D* and *E* are five different integers. When written in the ascending order of values, the difference between any two adjacent integers is 4. *D* is the greatest and *A* is the least. *B* is greater than *E* but less than *C*. The sum of the integers is equal to *E*. What is the positive difference between the lowest and the highest integers?

(a) 8 (b) 6
(c) 16 (d) 18

[ESE Prelims : 2021]

14.8 Some environmentalists question the prudence of exploiting features of the environment, arguing that there are no economic benefits to be gained from forests, mountains, or wetlands that no longer exist. Many environmentalists claim that because nature has intrinsic value it would be wrong to destroy such features of the environment, even if the economic costs of doing so were outweighed by the economic costs of not doing so.

Which one of the following can be logically inferred from the passage?

(a) It is economically imprudent to exploit features of the environment.
(b) Some environmentalists appeal to a noneconomic justification in questioning the defensibility of exploiting features of the environment.
(c) Most environmentalists appeal to economic reasons in questioning the defensibility of exploiting features of the environment.
(d) Many environmentalists provide only a noneconomic justification in questioning the defensibility of exploiting features of the environment.

[ESE Prelims : 2022]

14.9 Fact 1 : Jessica has four children.

Fact 2 : Two of the children have blue eyes and two of the children have brown eyes.

Fact 3 : Half of the children are girls.

If the first three statements are facts, which of the following statements must also be a fact?

I. At least one girl has blue eyes.
II. Two of the children are boys.
III. The boys have brown eyes.

Select the correct answer using the code given below:

(a) II only
(b) I and III only
(c) II and III only
(d) None of the statements is a known fact

[ESE Prelims : 2022]

14.10 Children are in pursuit of a dog whose leash has broken. James is directly behind the dog. Ruby is behind James. Rachel is behind Ruby. Max is ahead of the dog walking down the street in the opposite direction. As the children and dog pass, Max turns around and joins the pursuit. He runs in behind Ruby. James runs faster and is alongside the dog on the left. Ruby runs faster and is alongside the dog on the right. Which child is directly behind the dog?

(a) James (b) Ruby
(c) Rachel (d) Max

[ESE Prelims : 2022]

14.11 At the baseball game, Henry was sitting in seat 253. Marla was sitting to the right of Henry in seat 254. In the seat to the left of Henry was George. Inez was sitting to the left of George. Which seat is Inez sitting in?

(a) 251 (b) 254
(c) 255 (d) 256

[ESE Prelims : 2022]

14.12 Most serious students are happy students, and most serious students go to graduate school. Furthermore, all students who go to graduate school are overworked.

Which one of the following can be properly inferred from the statements above?

(a) Most overworked students are happy students
(b) Some happy students are overworked
(c) All overworked students are serious students
(d) Some unhappy students go to graduate school

[ESE Prelims : 2022]

14.13 Of the five boys A, B, C, D and E two are good, one is poor and two are average in studies. Two of them study in post-graduate classes and three in under graduate classes. One comes from a rich family two from middle class families and two from poor families. One of them is interested in music, two in acting and one in sports. Of those studying in under graduate classes, two are average and one is poor in studies. Of the two boys interested in acting, one is a post-graduate student. The one who is interested in music comes from a middle class family. Both of the boys interested in acting are not industrious, good in studies come from middle class families, are average in studies and one of them is interested in acting. The boy interested in sports comes from a poor family, while the one interested in music in industrious. E is industrious, good in studies comes from a poor family and is not interested in acting, music or sports. Cis poor in studies in spite of being industrious. A comes from a rich family, is not industrious but good in studies. B is industrious and comes from a middle class family. Name the boy who is not industrious and is average in studies.

(a) A (b) B
(c) C (d) D

[ESE Prelims : 2023]

14.14 At an electric Data Processing Unit five out of the eight program sets, P, Q, R, S, T, U, V and W are to be operated daily. On any one day except for the first day of the month only three of the program sets must be the ones that were operated on the previous day. The program operating must also satisfy the following conditions:

1. If program P is to be operated on a day, V cannot be operated on that day.
2. If Q is to be operated on a day, T must be one of the programs to be operated after Q.
3. If R is to be operated on a day, V must be one of the programs to be operated after R.
4. The last program to be operated on any day must be either S or U.

If the program sets R and W are to be operated on the first day which of the following could be the other programs on that day?

(a) Q, V, S (b) Q, T, V
(c) T, S, U (d) T, S, V

[ESE Prelims : 2023]

14.15 Read the following information carefully and answer the question give below it:

1. Eight doctors *P, Q, R, S, T, U, V* and *W* visit charitable dispensary every day except on a holiday i.e., Monday.
2. Each doctor visits for one hour from Tuesday to Sunday except Saturday. The timings are 9 A.M. to 1 P.M. and 2 P.M. to 6 P.M., 1 P.M. to 2 P.M. is lunch break.
3. On Saturday it is opened only in the morning i.e. 9 A.M. to 1 P.M. and each doctor visits for only half an hour.
4. No other doctor visits the dispensary, before doctor *Q* and after *U*.
5. Doctor *W* comes immediately after the lunch break is followed by *R*.
6. *S* comes in the same order as *P* in the afternoon session.

If the lunch break and subsequent visiting hours are reduced by 15 minutes at what time doctor *U* is expected to attend the dispensary?

(a) 3.15 P.M. (b) 4 P.M.
(c) 4.15 P.M. (d) 4.45 P.M.

[ESE Prelims : 2023]

14.16 Study the following information carefully and answer the question given below it:

1. *P, Q, R, S, T* and *U* are six members in a family in which there are two married couples.
2. *T*, a teacher, is married to the doctor who is mother or *R* and *U*.
3. *Q*, the lawyer, is married to *P*.
4. *P* has one son and one grandson.
5. Of the two married ladies on is housewife.
6. There is one student and one male engineer in the family.

How is *R* related to *U*?

(a) Brother only (b) Sister only
(c) Brother or Sister (d) Mother

[ESE Prelims : 2023]

14.17 Read the following information carefully and answer the question that follow:

1. Madhu and Shobha are good in Dramatics and Computer Science.
2. Anjali and Madhu are good in Computer Science and Physics.
3. Anjali, Poonam and Nisha are good in physics and History.
4. Nisha and Anjali are good in Physics and Mathematics.
5. Poonam and Shobha are good in History and Dramatics.

Who is good in History, Physics, Computer Science and Mathematics?

(a) Poonam (b) Nisha
(c) Madhu (d) Anjali

[ESE Prelims : 2023]

14.18 The question given below the three statements followed by three conclusions numbered I, II and III. You have to take the given statements to be true even if they seem to be at variance from commonly known facts. Read all the conclusions and then decide which of the given conclusions logically follows from the given statements disregarding commonly known facts.

Statements:

All lions are tigers
All tigers are leopards
Some leopards are wolves

Conclusions:

I. No elephant is lion.
II. Some wolves are lions.
III. Some leopards are lions.

(a) Only I follows (b) Only II follows
(c) Only III follows (d) Only I and III follows

[ESE Prelims : 2023]

14.19 Cryptic language is popular since ages, mostly in the field of espionage and sending classified messages. If 'I LOVE YOU' is coded as 7, then how would you code 'GO TO HELL' in the same language?

(a) 1 (b) 4
(c) 3 (d) 5

[ESE Prelims : 2023]

14.20 What letter should replace the question mark?

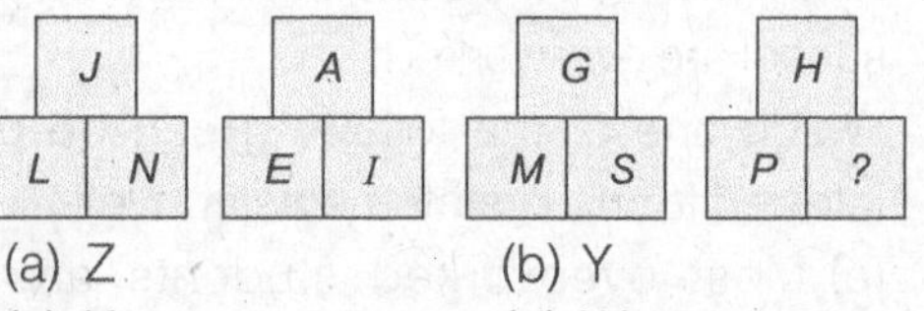

(a) Z (b) Y
(c) X (d) W

[ESE Prelims : 2023]

14.21 In the first two circles, the number inside the circle is written according to a particular relation. What is the number inside the third circle which follows the same relation as that of the first two circles?

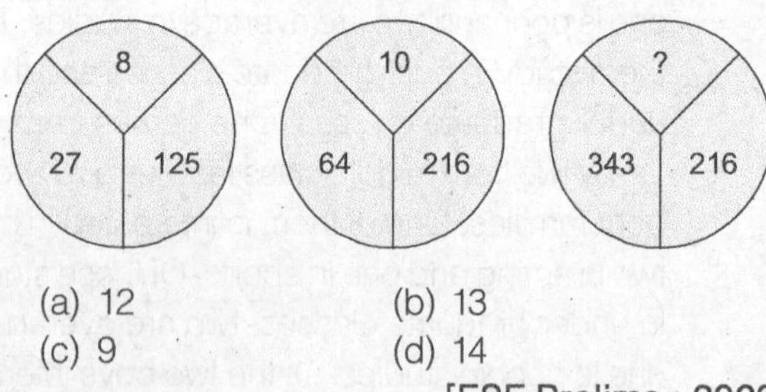

(a) 12 (b) 13
(c) 9 (d) 14

[ESE Prelims : 2023]

14.22 Deepthi is playing a treasure hunt game. At the first stage, Deepthi needs to choose a five-digit code to unlock the vault which contains the treasure. She gets the following codes to choose from

15342	26540	35415
23105	15320	13402
35047	71024	28305

The following clues are given to her to help her to find the code

P. The code number is not an even number.
Q. The product of the first two digits is odd.
R. The sum of the first four digits is 12.
S. The code number is not a multiple of 5.
If Deepthi had the option of selecting only one clue, which of the four clues will give her the best chance of finding out the five digit code?
(a) S (b) R
(c) Q (d) P
[ESE Prelims : 2023]

14.23 A number series is given with one term missing. Choose the correct alternative from options.
0.5, 0.55, 0.65, 0.8, ?
(a) 0.9 (b) 0.95
(c) 0.82 (d) 1
[ESE Prelims : 2023]

14.24 The soccer club is putting together a mural using lightly colored transparent paper. This paper then is cut into squares of different sizes that are placed next to each other to make the designs for the mural. Of course, the club wants to save money, so its members are trying to buy the minimum number of sheets of colored paper. Below is one of the designs they are going to use. What is the minimum number of squares they will need to make this design?

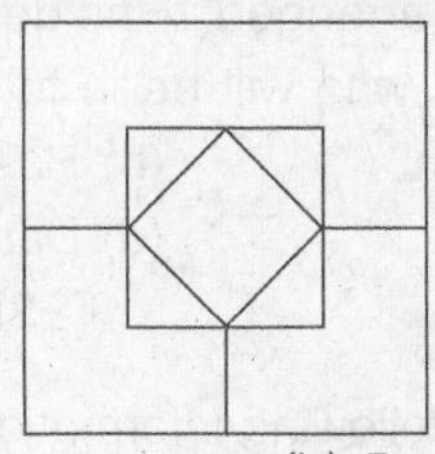

(a) 4 (b) 5
(c) 6 (d) 7
[ESE Prelims : 2023]

14.25 What is the correct alternative for the question mark?
2, 3, 8, 63, ?
(a) 1038 (b) 1998
(c) 3008 (d) 3968
[ESE Prelims : 2023]

14.26 Brianna, Ryan, Tyler, and Ashley were recently elected as the new class officers (president, vice president, secretary, treasurer) of the sophomore class at Summit College. From the following clues, determine which position each holds.
1. Ashley is younger than the president but older than the treasurer.
2. Brianna and the secretary are both the same age, and they are the youngest members of the group.
3. Tyler and the secretary are next door neighbors.
(a) Tyler is the president, Ashley is the vice president, Ryan is the secretary, and Brianna is the treasurer.
(b) Tyler is the president, Ashley is the vice president, Brianna is the secretary, and Ryan is the treasurer.
(c) Tyler is the president, Ryan is the vice president, Ashley is the secretary, and Brianna is the treasurer.
(d) Tyler is the president, Ryan is the vice president, Brianna is the secretary, and Ashley is the treasurer.
[ESE Prelims : 2024]

14.27 You need to buy groceries at the supermarket, deposit a cheque at credit union, and purchase a book at the bookstore. You can complete the errands in any order, however, you must start and end at your home. The driving time, in minutes, between each of these locations is given in the following figure.

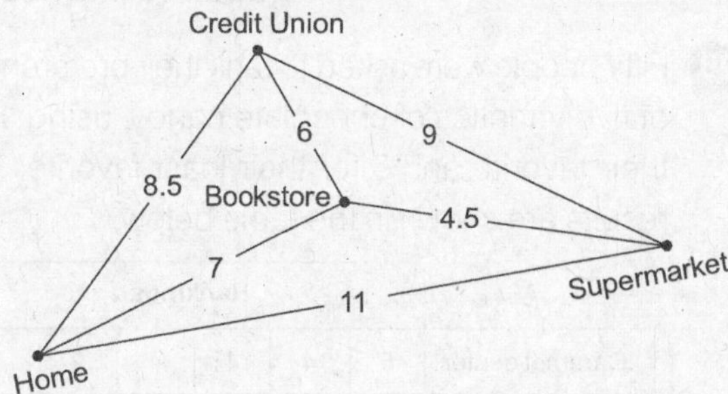

What is the route whose total driving time is less than 30 minutes?
(a) home, bookstore, credit union, supermarket, home
(b) home, supermarket, bookstore, credit union, home
(c) home, bookstore, supermarket, credit union, home
(d) home, supermarket, credit union, bookstore, home
[ESE Prelims : 2024]

14.28 Each of four siblings (Anita, Tony, Maria and Jose) is given ₹5000 to invest in the stock market. Each chooses a different stock. One chooses a utility stock, another an automotive stock, another a technology stock, and the other an oil stock.

1. Anita and the owner of the utility stock purchased their shares through an online brokerage, whereas Tony and the owner of the automotive stock did not.
2. The gain in value of Maria's stock is twice the gain in value of the automotive stock.
3. The technology stock is traded on NASDAQ, whereas the stock that Tony bought is traded on the New York Stock Exchange.

From the above clues, match the name of the sibling and stock bought.

(a) Maria: the utility stock; Jose: the automotive stock; Anita: the technology stock; Tony: the oil stock
(b) Maria: the utility stock; Anita: the automotive stock; Jose: the technology stock; Tony: the oil stock
(c) Maria: the utility stock; Tony: the automotive stock; Anita: the technology stock; Jose: the oil stock
(d) Jose: the utility stock; Maria: the automotive stock; Anita: the technology stock; Tony: the oil stock

[ESE Prelims : 2024]

14.29 Fifty people were asked to rank their preferences of five varieties of chocolate candy, using 1 for their favorite and 5 for their least favorite. The results are shown in the table below.

	Rankings					
Caramel center	5	4	4	4	2	4
Vanilla center	1	5	5	5	5	5
Almond center	2	3	2	1	3	3
Toffee center	4	1	1	3	4	2
Solid chocolate	3	2	3	2	1	1
Number of voters	17	11	9	8	3	2

According to the table (see the column in grey), three voters ranked solid chocolate first, caramel centers second, almond centers third, toffee centers fourth, and vanila centers fifth. According to this table, which variety of candy would win the taste test using the plurality voting system?

(a) Almond centers (b) Vanilla centers
(c) Toffee centers (d) Caramel centers

[ESE Prelims : 2024]

14.30 Study the given information carefully and answer the question :

There are seven books, one each of Psychology, Hindi, English, Sociology, Economics, Education and Accountancy lying on a table one above the other. Sociology is on the top of all books. Accountancy is immediately below Education which is immediate below Sociology. Economics is immediately above Psychology but not in the middle. Hindi is immediately below Psychology. Which three books are between Accountancy and Hindi?

(a) English, Economics and Psychology
(b) Economics, Psychology and Education
(c) Economics, Psychology and Hindi
(d) Cannot be determined

[ESE Prelims : 2024]

14.31 Read the information given below and answer the questions :

There is a group of five girls. Hasini is second in height but younger than Madhavi. Pooja is taller than Pranati but younger in age. Madhavi and Pranati are of the same age but Madhavi is tallest among them. Neelam is taller than Pooja and elder to Madhavi.

If they are arranged in the descending order of their ages who will be in fourth position?

(a) Neelam (b) Hasini
(c) Pranati (d) Data inadequate

[ESE Prelims : 2024]

14.32 Read the following information and answer the question :

Seven students P, Q, R, S, T, U and V take a series of tests. No two students get similar marks. V always scores more than P. P always scores more than Q. Each time either R scores the highest and T gets the least, or alternatively S scores the highest and U or Q scores the least.

If V is ranked fifth, which one of the following is correct?

(a) S scores the highest
(b) R is ranked second
(c) T is ranked third
(d) Q is ranked fourth

[ESE Prelims : 2024]

15. PRINCIPLES OF EQUATION

15.1 Three hundred passengers are travelling in white, silver and black cars; each of these cars is carrying 6, 5 and 3 passengers, respectively. If the number of white and silver cars is equal and three is only one black car, what is the total number of cars?
(a) 52 (b) 53
(c) 54 (d) 55
[ESE Prelims : 2017]

15.2 Given that 0.8 is one root of the equation, $x^3 - 0.6x^2 - 1.84x + 1.344 = 0$. The other roots of this equation will be
(a) 1.1 and –1.4 (b) –1.2 and 1.4
(c) 1.2 and –1.4 (d) –1.1 and 1.4
[ESE Prelims : 2018]

15.3 The equation, $x^3 - 8x^2 + 37x - 50 = 0$ is factored and it has $(3 + 4i)$ as one of its roots. What is the real root of this equation?
(a) 2 (b) 4
(c) 6.5 (d) 13
[ESE Prelims : 2018]

15.4 What is the form of the function $f(x)$ for the following data?

x	0	1	2	3
$f(x)$	3	6	11	18

(a) $x^2 + 2x + 3$ (b) $x^2 - 2x + 3$
(c) $x^2 + 2x - 3$ (d) $x^2 - 2x - 3$
[ESE Prelims : 2018]

15.5 What is the maximum value of z, if $z = 10x + 6y$ subject to the constraints
$3x + y \le 12,\ 2x + 5y \le 34,\ x \ge 0,\ y \ge 0$?
(a) 56 (b) 52
(c) 50 (d) 40
[ESE Prelims : 2018]

15.6 The objective function $z = 3x_1 + 5x_2$ is to be maximized subjected to constraints.
$x_1 + 2x_2 \le 200$
$x_1 + x_2 \le 150$
$x_1, x_2 \ge 0$
The values of x_1 and x_2 in this context are, respectively
(a) 100 and 75 (b) 125 and 75
(c) 100 and 50 (d) 125 and 50
[ESE Prelims : 2018]

15.7 If A X B means $(A^2 + B^2)$, then the value of 5 X (4 X 3) is
(a) 60 (b) 300
(c) 650 (d) 710
[ESE Prelims : 2021]

15.8 If $\frac{x+1}{x-1} = \frac{a}{b}$ and $\frac{1-y}{1+y} = \frac{b}{a}$, then the value of $\frac{x-y}{1+xy}$ is
(a) $\frac{2ab}{a^2 - b^2}$ (b) $\frac{a^2 - b^2}{2ab}$
(c) $\frac{a^2 + b^2}{2ab}$ (d) $\frac{a^2 - b^2}{ab}$
[ESE Prelims : 2021]

15.9 Suppose you enter an elevator at a certain floor. Then the elevator moves up 5 floors, down 3 floors, and up 2 floors. If you are then at the 8th floor, on what floor did you first enter the elevator?
(a) 8 (b) 7
(c) 6 (d) 9
[ESE Prelims : 2023]

16. AREA AND VOLUME

16.1 Consider a trapezoidal lamina *ABCD*, with *AB* parallel to *DC*, 6 cm apart; *AB* is 8 cm; *CD* is 12 cm; *CD* extends outwards by 1 cm from the foot of the perpendicular from *B* on *DC*. The centre of gravity of the lamina will be
(a) Along *AC* at a height of 3 cm from *DC*
(b) Along *BD* at a height of 3 cm from *DC*
(c) Along the line joining the mid-point of *AB* to the mid-point of *DC*; at a height of 2.8 cm from *DC*.
(d) At the intersection point of *AC* and *DB*
[ESE Prelims : 2018]

16.2 Circle *A* is 4 cm in diameter; circle *B* is 5 cm in diameter. Circle *C* has its circumference equal to the sum of the circumferences of both *A* and *B* together. What will be the ratio of the area of circle *C*, with respect to the area of circle *A* and circle *B* respectively?
(a) 5.0625 and 1.84 (b) 3.875 and 1.84
(c) 5.0625 and 3.24 (d) 3.875 and 3.24
[ESE Prelims : 2018]

16.3 A wall, rectangular in shape, has a perimeter of 72 m. If the length of its diagonal is 18 m, what is the area of the wall?

(a) 224 m^2 (b) 486 m^2
(c) 572 m^2 (d) 606 m^2

[ESE Prelims : 2018]

16.4 To isolate an enclosed area for conservation, an open traverse is run keeping close to (but outside of) the exterior boundary of the area through ground points $A \to B \to C \to D \to E \to F \to G \to$ towards H (to be eventually located). AB is 80° to the East of the North line at A. Deflection/Interior angles at B, C, D, E, F are indicated. What would be the magnitude of the deflection angle at G (as marked) so that GH may run parallel to BA? (Lengths are immaterial in this case.)

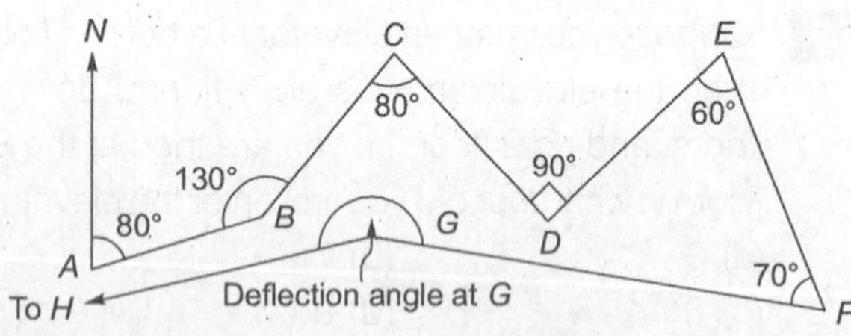

(a) 190° (b) 210°
(c) 200° (d) 230°

[ESE Prelims : 2018]

16.5 Consider the length of a room is 15 m and width is 10 m. If the sum of the areas of the floor and ceiling is equal to the sum of the areas of the four walls, then the volume of the room is

(a) 900 m^3 (b) 1000 m^3
(c) 1200 m^3 (d) 1500 m^3

[ESE Prelims : 2018]

16.6 A hemisphere depression is cutout from one face of the cubical wooden block such that the radius r of the hemisphere is equal to half of the edge of the cube. What will be the surface area of the remaining solid?

(a) $2r^2(\pi + 24)$ (b) $r^2(\pi + 24)$
(c) $2r^2(\pi + 36)$ (d) $r^2(\pi + 36)$

[ESE Prelims : 2019]

16.7 A rod of length l is to be divided into two parts, such that if 5 times the smaller portion is added to half of the larger portion, it will always be less than l. This can be achieved by taking length of the larger portion more than

(a) $\frac{9}{10}l$ (b) $\frac{7}{8}l$
(c) $\frac{6}{7}l$ (d) $\frac{5}{6}l$

[ESE Prelims : 2019]

16.8 In a lake, the tip of a bud of lotus is seen 10 cm above the surface of water. Forced by the wind, it gradually moved, and just submerged at a distance of 30 cm. The depth of water at the root of the lotus plant will be

(a) 40 cm (b) 50 cm
(c) 60 cm (d) 70 cm

[ESE Prelims : 2019]

16.9 In two concentric circles, a chord length 80 cm of larger circle becomes a tangent to the smaller circle whose radius is 9 cm. The radius of the larger circle will be

(a) 13 cm (b) 41 cm
(c) 52 cm (d) 75 cm

[ESE Prelims : 2019]

16.10 Consider following diagram: AC is a diameter of the large circle and $AB = BC$.
The ratio of areas of the large circle to the small circle of a square is

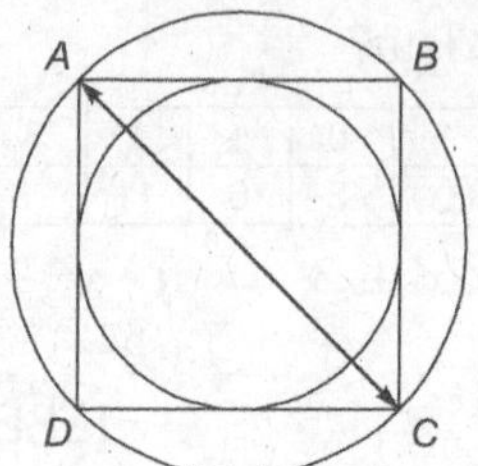

(a) 4 : 1 (b) 1 : 4
(c) 2 : 1 (d) 1 : 2

[ESE Prelims : 2019]

16.11 Consider the rectangle ABCD with $DE = \frac{1}{3}DC$ in the figure:

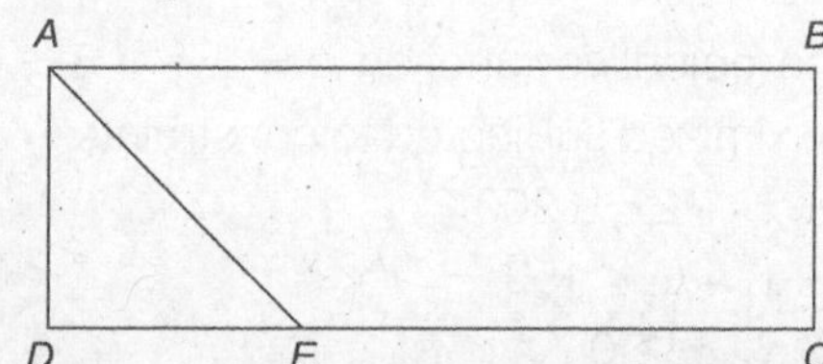

When the area of the triangle ADE is 20 cm^2, the area of the rectangle ABCD will be

(a) 60 cm^2 (b) 80 cm^2
(c) 100 cm^2 (d) 120 cm^2

[ESE Prelims : 2020]

16.12 Four metal rods of lengths 78 cm, 104 cm, 117 cm and 169 cm are to be cut into parts of equal length. Each part must be as long as possible. The maximum number of pieces that can be cut will be
(a) 27 (b) 36
(c) 43 (d) 52
[ESE Prelims : 2020]

16.13 The diagonal of a rectangle is $\sqrt{41}$ cm and its area is 20 sq.cm. The perimeter of the rectangle is
(a) 9 cm (b) 18 cm
(c) 20 cm (d) 41 cm
[ESE Prelims : 2021]

16.14 One side of a rhombus is 10 cm and one of its diagonals is 12 cm. The area of the rhombus is
(a) 24 sq. cm (b) 48 sq. cm
(c) 72 sq. cm (d) 96 sq. cm
[ESE Prelims : 2022]

■■■■

Reasoning & Aptitude : Answer Key

1.1	(d)	**1.2**	(a)	**1.3**	(b)	**1.4**	(b)	**1.5**	(d)	**1.6**	(a)	**1.7**	(b)	**1.8**	(a)
1.9	(d)	**1.10**	(a)	**1.11**	(a)	**1.12**	(b)	**2.1**	(c)	**2.2**	(b)	**2.3**	(b)	**3.1**	(c)
3.2	(c)	**4.1**	(b)	**4.2**	(b)	**5.1**	(a)	**5.2**	(c)	**6.1**	(b)	**6.2**	(d)	**7.1**	(d)
7.2	(d)	**7.3**	(a)	**7.4**	(c)	**7.5**	(c)	**7.6**	(a)	**8.1**	(b)	**8.2**	(c)	**8.3**	(c)
9.1	(a)	**10.1**	(d)	**10.2**	(c)	**10.3**	(c)	**11.1**	(c)	**11.2**	(d)	**12.1**	(b)	**13.1**	(a)
13.2	(c)	**13.3**	(b)	**13.4**	(b)	**14.1**	(b)	**14.2**	(d)	**14.3**	(c)	**14.4**	(a)	**14.5**	(c)
14.6	(a)	**14.7**	(c)	**14.8**	(d)	**14.9**	(a)	**14.10**	(d)	**14.11**	(a)	**14.12**	(b)	**14.13**	(d)
14.14	(d)	**14.15**	(b)	**14.16**	(c)	**14.17**	(d)	**14.18**	(c)	**14.19**	(b)	**14.20**	(c)	**14.21**	(b)
14.22	(b)	**14.23**	(d)	**14.24**	(b)	**14.25**	(d)	**14.26**	(a)	**14.27**	(c)	**14.28**	(a)	**14.29**	(c)
14.30	(a)	**14.31**	(d)	**14.32**	(a)	**15.1**	(d)	**15.2**	(c)	**15.3**	(a)	**15.4**	(a)	**15.5**	(a)
15.6	(c)	**15.7**	(c)	**15.8**	(a)	**15.9**	(*)	**16.1**	(c)	**16.2**	(c)	**16.3**	(b)	**16.4**	(a)
16.5	(a)	**16.6**	(b)	**16.7**	(a)	**16.8**	(a)	**16.9**	(b)	**16.10**	(c)	**16.11**	(d)	**16.12**	(b)
16.13	(b)	**16.14**	(d)												

EXPLANATIONS

1.1 (d)

Sum of square of first n natural number

$$\Sigma n^2 = \frac{n(n+1)(2n+1)}{6}$$

$$\Sigma_{n=1}^{n=16} n^2 - \Sigma_{n=1}^{n=7} n^2$$

$$= \frac{16(16+1)(32+1)}{6} - \frac{7(7+1)(14+1)}{6}$$

$$= 1496 - 140 = 1356$$

1.2 (a)

$$x \times y = 96 \quad \text{...(i)}$$

$$(x-4)(y+4) = 96 \quad \text{...(ii)}$$

Solving eq. (i) and (ii) we get,

$$x = 12,\ y = 8$$

$$\therefore \quad 12 - 4 = 8$$

Alternative Solution:

Using option (a)

$$8 \times 12 = 96$$

$$(12-4) \times (8+4) = 8 \times 12 = 96$$

1.3 (b)

$$\text{Sum} = 5 + 10 + 15 + \dots + 100$$

$$= 5(\Sigma 20) = 5 \times \frac{20 \times 21}{2} = 1050$$

1.4 (b)

$$2^2 + 4^2 + 6^2 + \dots + 20^2$$

$$\Rightarrow 2^2(1^2 + 2^2 + 3^2 + \dots + 10^2)$$

$$\because \Sigma n^2 = \frac{n(n+1)(2n+1)}{6}$$

$$\Rightarrow 2^2 \times \frac{10 \times 11 \times 21}{6}$$

$$\Rightarrow 4 \times \frac{10 \times 11 \times 21}{6} = 1540$$

1.5 (d)

Minimum number of row contain same number of tree of one variety

$$= \frac{\text{Total}}{\text{HCF}(21,42,56)} = \frac{21+42+56}{7}$$

$$= 3 + 6 + 8 = 17$$

1.6 (a)

$$(191)^2 + 81 = 36562$$

81 soldiers are left.

```
   1 | 3 65 62
  +1 | 1
  29 | 2 65
   9 | 2 61
 381 |    4 62
   1 |    3 81
     |      81
```

1.7 (b)

Total number of lines = 817

Both meet when :

Time when 1st boy complete x number of lines = Time when 2nd boy complete $(817 - x)$ number of lines

$$\frac{x}{200} = \frac{817 - x}{150}$$

$$150x = 200 \times 817 - 200x$$

$$x = \frac{200 \times 817}{350} = 466.857 \approx 467^{\text{th}} \text{ line}$$

1.8 (a)

Assuming anuhya picks the number 'n', we have

$$\frac{(2n)^2}{3} - 30 = 18$$

$$\Rightarrow \quad \frac{4n^2}{3} = 48$$

$$\Rightarrow \quad n^2 = 36$$

$$\Rightarrow \quad n = +6 \text{ or } -6$$

1.9 (d)

Let the age at which Diophantus died be 'x'. From the given information, the age of Diophantus at the time of marriage,

$$y = \frac{x}{6} + \frac{x}{12} + \frac{x}{7}$$

$$= \frac{(14+12+7)x}{84} = \frac{33x}{84} = \frac{11x}{28}$$

We can write,

$$(x-4) - \left(\frac{11x}{25} + 5\right) = \frac{x}{2}$$

$$\Rightarrow \quad \frac{x}{2} - \frac{11x}{28} = 9$$

$$\Rightarrow \quad 3x = 28 \times 9$$

$$\Rightarrow \quad x = 84 \text{ years}$$

1.10 **(a)**

Select a random number between 50 and 100, say 90. Adding 83 to this number, we get

$$83 + 90 = 173$$

Forming a new number by adding the digit in the hundreds place to the number formed by the other two digits, we get

$$1 + 73 = 74$$

Subtracting from the original number, the final result is obtained as:

$$90 - 74 = 16$$

1.11 **(a)**

Total points received by different members:

Avalon: $(5 \times 3) + 2 = 17$ points

Branson: $(2 \times 4) + 3 + (3 \times 1) = 14$ points

Columbus: $(2 \times 4) + (4 \times 2) = 16$ points

Dunkrirk: $(2 \times 4) + 2 + (3 \times 1) = 13$ points

Hence, Avalon will be the elected president.

1.12 **(b)**

The required number should be such that it leads a remainder of 1 when divided by 3, 4, 5 or 6 and no remainder when divided by 7.

LCM (3, 4, 5, 6) = 60

So, number in the form = $60x + 1$

The minimum value of x for which the number leaves no remainder when divided by 7 is 5. Hence,

Required number = $60 \times 5 + 1 = 301$

2.1 **(c)**

Given that Rajiv spends on

Food → 40%

House rent → 20%

Entertainment → 10%

Conveyance → 10%

$\therefore$ His savings = 100 – (40 + 20 + 10 + 10)
= 20%

The remaining percentage after expenditure 20% which equals Rs. 2000 leading to total monthly income = Rs. 10000.

2.2 **(b)**

Let the number of males = x

Then number of females = $(5500 - x)$

$$x(1 + 0.11) + (5500 - x) \times 1.2 = 6330$$

$$1.11x + 6600 - 1.2x = 6330$$

$$x = \frac{270}{9} \times 100$$

$$x = 3000$$

So, Number of females = 5500 – 3000 = 2500

2.3 **(b)**

Let Income = $100k$ per month

$$\text{Spend on food} = \frac{40}{100} \times 100k = 40k$$

Spend on cloth and conveyance

$$= \frac{60}{100} \times \frac{50}{100} \times 100k = 30k$$

$$\text{Saving} = \left(\frac{1}{3}\right) \times (100 - 40k - 30k)$$

$$= \frac{1}{3} \times 30k = 10k$$

$$\text{Monthly saving} = \frac{19200}{12} = ₹\ 1600 = 10k$$

$\Rightarrow$ $k = 160$

$\therefore$ Monthly income = $100k = 100 \times 160$
= ₹ 16,000

3.1 **(c)**

We can form 2 linear equations taking the number of Rs. 20 and Rs. 50 notes as T and F

$$T + F = 324 \quad \text{...(1)}$$

$$20T + 50F = 12450 \quad \text{...(2)}$$

Solving the 2 equations, we get $T = 125$

Alt 1: If all the notes are Rs. 50 notes, total amount = 324 × 50 = Rs. 16200 which is 16200 – 12450 = 3750

3750/(50 – 20) = 125 notes of Rs. 20.

Alt 2: This question can also be solved by putting options.

3.2 **(c)**

Using options

$3000 \times 1.25 + 3500 \times 1.1 = 7600$

$3000 \times 1.10 + 3500 \times 1.20 = 7500$

4.1 **(b)**

Let the number of year is n

Then,
$$A = P\left(1 + \frac{\frac{R}{2}}{100}\right)^{2n}$$

$$926.10 = 800\left(1+\frac{5}{100}\right)^{2n}$$

$$\frac{926.10}{800} = \left(\frac{21}{20}\right)^{2n}$$

$$\frac{92610}{80000} = \left(\frac{21}{20}\right)^{2n}$$

$$\left(\frac{21}{20}\right)^{3} = \left(\frac{21}{20}\right)^{2n}$$

$$2n = 3$$

$$n = \frac{3}{2}$$

4.2 (b)

For interest compounded annually

$$\left[P\left(1+\frac{R}{100}\right)^2 - P\right] - \frac{P \times R \times 2}{100} = 16$$

$\Rightarrow P(1.08)^2 - P - P(0.16) = 16$

$\Rightarrow P(1.1664) - P - P(0.16) = 16$

$\Rightarrow \quad P(0.0064) = 16$

$$P = \frac{16}{0.0064} = 2500$$

For compounded half yearly

$$= P\left(1+\frac{R/2}{100}\right)^{2\times 2} - \frac{P \times R \times 2}{100} - P$$

$= P(1.04)^4 - P(0.16) - P$

$=$ ₹24.6464

ADDITIONAL INFORMATION

Simple interest, S.I. $= \frac{P \times R \times T}{100}$

Compound interest, C.I. $= P\left(1+\frac{R}{100}\right)^T - P$

where, P: Principal ; R: Rate ; T: Time period

5.1 (a)

Let $3x$, $4x$, $5x$ be the present age of 3 brothers. After 10 years, sum of their age is $12x + 30 = 78$ which gives $x = 4$

Hence the present age of three brothers is 12, 16 and 20 years.

Alt: Can be directly solved using the given options since the sum of their present age = 78 − 30 = 48 which is satisfied by option 'A' alone.

5.2 (c)

Given, $\frac{2x+3y}{3x+5y} = \frac{18}{29}$

$$58x + 87y = 54x + 90y$$

$$4x = 3y$$

$$\Rightarrow \quad \frac{x}{y} = \frac{3}{4}$$

6.1 (b)

Let x be the number of books

Cost of x number of books = 1600

Each book cost, $c_1 = \frac{1600}{x}$

If cost of $(x + 8)$ books = 1600

Each book cost, $c_2 = \frac{1600}{x+8}$

According to question

$$c_1 - c_2 = 10$$

Using options

$$\frac{1600}{x} - \frac{1600}{x+8} = 10$$

$$x = 32$$

Alternatively

$$32 \times 50 = 40 \times 40$$

6.2 (d)

Let, F = Father's age

S = Son's age

$$\frac{F-10}{S-10} = 12 \quad \text{...(i)}$$

$$\frac{F+10}{S+10} = 2 \quad \text{...(ii)}$$

Solving equations (i) and (ii),

F = 34; S = 12

Alternative solution:

Using option (d)

Father's age (F) = 34 years

Son' age (S) = 12 years

$$\frac{F-10}{S-10} = \frac{34-10}{12-10} = \frac{24}{2} = 12$$

$$\frac{F+10}{S+10} = \frac{34+10}{12+10} = \frac{44}{22} = 2$$

7.1 (d)

5 men in 20 days = 100 man days

8 women in 25 days = 200 woman days

10 boys in 30 days = 300 boy days

which means 100 MD = 200 WD = 300 BD or 1 man day = 3 boy days; 1 woman day = 1.5 Boy days 2 men, 6 women and 5 boys = 2 × 3 + 6 × 1.5 + 5 = 20 boys deployed to complete the task which will get completed in 300/20 = 15 days

Alt: [2(1/100) + 6(1/200) + 5(1/300)]x = 1

Which gives x = 15

7.2 (d)

If '*W*' is the original number of workers estimated to finish the work, we can form an equation as: 10 W = (*W* – 5) × 12 which gives *W* = 30.

7.3 (a)

In the first case, total number of hours per week = 3.5 × 6 + 3 × 5 = 36 hours

In the second case, total number of hours = 7.5 ×5 = 37.5 hours

With increased wages, the effective hourly rate is $\frac{(36 \times 1.08)}{37.5} = 1.0368$ implying 3.68% as the increase in expected productivity.

7.4 (c)

Time		No. of strikes
1	→	1
2	→	2
⋮		⋮
12	→	12

Time		No. of strikes
13	→	1
14		2
⋮		⋮
24	→	12

$$2[1 + 2 + 3 \ldots +12] = 2 \times \frac{12 \times 13}{2} = 156$$

7.5 (c)

Let B finish the work in x days,

Then his one day work = $\frac{1}{x}$

∵A is twice as good as B.

Then A's one day work = $\frac{2}{x}$

So, $\left(\frac{2}{x} + \frac{1}{x}\right) \times 18 = 1$

$$\frac{3}{x} \times 18 = 1$$

$$x = 54$$

$$\text{A's one day work} = 2 \times \frac{1}{54} = \frac{1}{27}$$

So, A will complete the work alone in 27 days.

7.6 (a)

Total volume of the tank = 13.5 × 20 = 270 litres

Volume of the tank will remain same.

Let number of buckets = x

$$\therefore \quad 9x = 270$$

$$\therefore \quad x = \frac{270}{9} = 30 \text{ litres.}$$

8.1 (b)

Let the length of train = x

Length of the bridge = l

Speed of train = s

Train x — Bridge l

Speed (s)

$$\frac{l + x}{s} = t_1 \quad \ldots\text{(i)}$$

$$\frac{\frac{l}{2} + x}{s} = t_2 \quad \ldots\text{(ii)}$$

Option (1)

If $t_2 = \frac{t_1}{2}$

Replacing eq. (i) and (ii)

$$\frac{\frac{l}{2} + x}{s} = \frac{1}{2}\left(\frac{l + x}{s}\right)$$

$$2\left(\frac{l}{2} + x\right) = (l + x)$$

$$x = 0 \quad \text{(Not possible)}$$

Length of train must be greater than zero.

Therefore option (1) is false.

Option (2)

$$2(t_2) > t_1$$

$$\frac{2\left(\frac{l}{2} + x\right)}{s} > \frac{l + x}{s}$$

$$l + 2x > l + x$$

$$x > 0$$

Length of train has to be positive.

Therefore option (2) is true.

Option (3)

$$t_2 < \frac{t_1}{2}$$

$$\frac{\frac{l}{2}+x}{s} < \frac{1}{2}\left(\frac{l+x}{s}\right)$$

$$2\left(\frac{l}{2}+x\right) < l + x$$

$$x < 0$$

Length of train cannot be negative.
Therefore option (3) is false.

Option (4)

$$s = \frac{l}{10} \text{ if } t_1 - t_2 = 5$$

$$\frac{l+x}{s} = t_1 \quad \text{...(i)}$$

$$\frac{\frac{l}{2}+x}{s} = t_2 \quad \text{...(ii)}$$

Replacing $s = \frac{l}{10}$ in eq. (i) and (ii)

$$\frac{l+x}{\frac{l}{10}} = t_1 \quad \text{...(iii)}$$

$$\Rightarrow \quad x = \frac{lt_1 - 10l}{10} \quad \text{...(iv)}$$

$$\frac{\frac{l}{2}+x}{\frac{l}{10}} = t_2 \quad \text{...(v)}$$

$$\Rightarrow \quad x = \frac{lt_2 - 5l}{10} \quad \text{...(vi)}$$

Eq. (iv) = eq. (vi) solve $t_1 - t_2 = 5$
Option (2) and (4) are true.

8.2 (c)

Let total distance $= D$

$$\text{Average speed} = \frac{\text{Total distance}}{\text{Total time}}$$

$$= \left(\frac{D}{\frac{\frac{D}{2}}{60} + \frac{\frac{D}{4}}{30} + \frac{\frac{D}{4}}{10}}\right) = 24 \text{ kmph}$$

8.3 (c)

$$\left|\frac{11}{2}M - 30H\right| = \theta$$

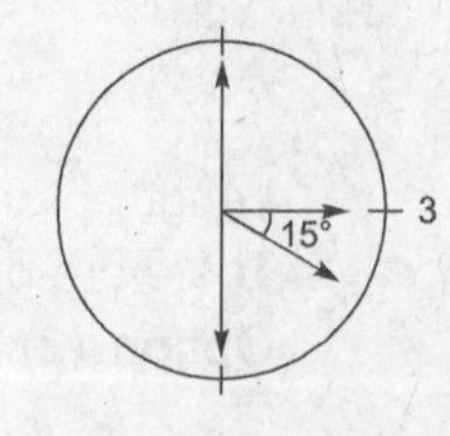

$$\theta = \left|\frac{11}{2}\times 30 - 30\times 3\right|$$

$$= |165 - 90| = 75^\circ$$

Alternate solution :

$$\left[5x \pm \frac{D}{6}\right]\times\frac{12}{11} = \text{M}$$

$$\left[5\times 3 \pm \frac{D}{6}\right]\times\frac{12}{11} = 30$$

$$\left[15 + \frac{D}{6}\right] = \frac{55}{2}$$

$$\frac{D}{6} = \frac{25}{2}$$

$$D = 75^\circ$$

9.1 (a)

Ignoring 1st term,

$$\frac{c}{s} = \frac{\Sigma n^3}{\Sigma n^2} = \frac{\left\{\frac{n(n+1)}{2}\right\}^2}{\frac{n(n+1)(2n+1)}{6}}$$

Putting, $n = 8$ we get $\frac{c}{s} = \frac{108}{17}$

10.1 (d)

The chance that exactly 2 of them are children

$$= \frac{{}^4C_2 \times {}^5C_2}{{}^9C_4} = \frac{10\times 6}{126} = \frac{60}{126} = \frac{10}{21}$$

10.2 (c)

(3) B_1 (3) B_2 (2) B_3 (2) B_4 (2) B_5

Arrange in shelf so that copies of the same book are not separates

$= 5! = 120$

Let book B_1 and B_2 have 3 copies each.
Book B_3, B_4 and B_5 have 2 copies each.

10.3 (c)

In the handshake between two people, the order doesn't matter. If 6 people shakes hands with each other,

Total number of handshakes $= {}^6C_2$

$$= \frac{6!}{2!4!} = \frac{5\times 6}{2} = 15$$

11.1 (c)

$$1000 = 750 + 400 - n(H \cap E)$$

$$\Rightarrow \quad n(H \cap E) = 150$$

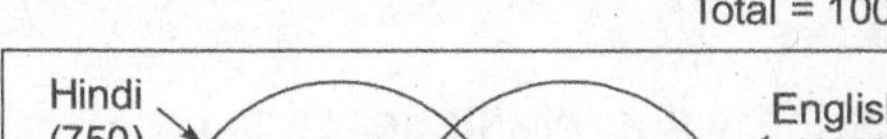

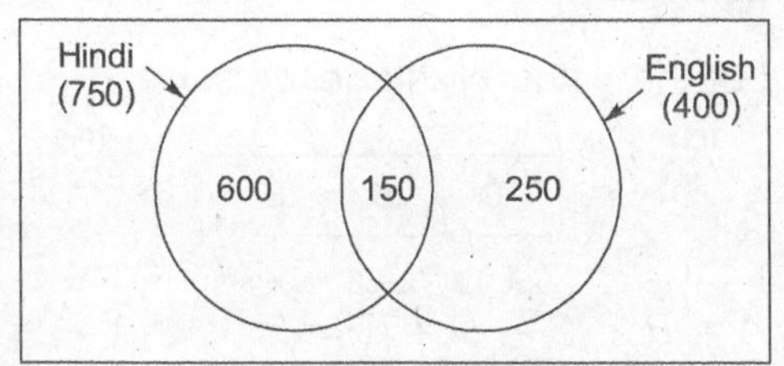

Only Hindi = 600

11.2 (d)

The given information can be represented in the Venn diagram as below:

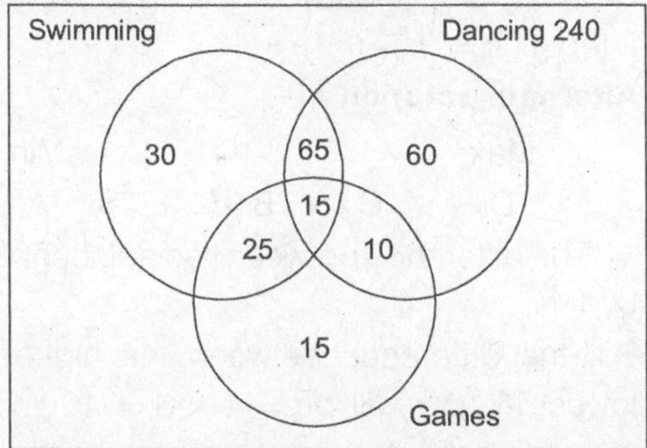

Number of passengers who like exactly two of the three types of activities.

$= 65 + 25 + 10$

$= 100$

12.1 (b)

The possible relationship diagram is

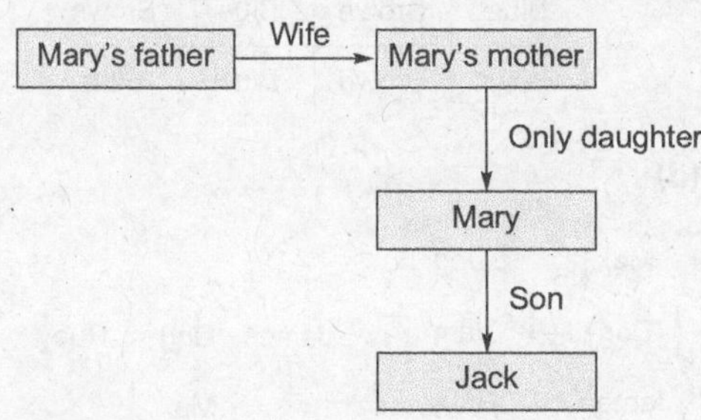

From the diagram it can be observed that Jack is son of Mary.

13.1 (a)

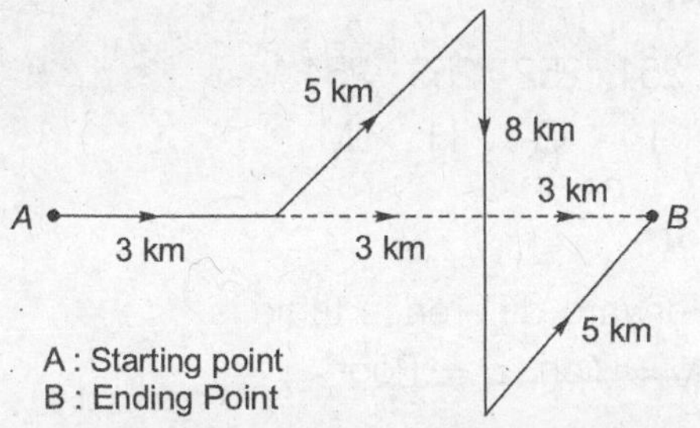

$AB = 9$ km

13.2 (c)

The possible direction diagram is

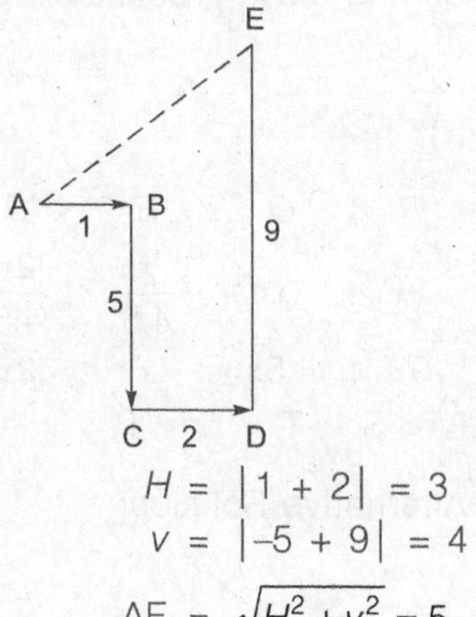

$H = |1 + 2| = 3$

$V = |-5 + 9| = 4$

Hence, $AE = \sqrt{H^2 + V^2} = 5$

13.3 (b)

$25 - 15 = 10$ km

13.4 (b)

Starting from North Bay and given that each bridge can be jogged exactly once, the path for jogging would be I → F → B → A → J → H → G → C → D → E.

Hence, the workout concludes over the 'E' bridge.

14.1 (b)

Care is to be taken to meet the given conditions i.e. least number of 9s and or √9 to be used and preference to be given to use of 9.

$$1=\frac{9}{9};\ 2=\frac{(9+9)}{9};\ 3=\sqrt{9};\ 4=\sqrt{9}+\frac{9}{9};$$

$$5=9-\sqrt{9}-\frac{9}{9};\ 6=9-\sqrt{9};\ 7=9-\frac{9+9}{9};\ 8=9-\frac{9}{9};$$

$$9=9;\ 10=9+\frac{9}{9};\ 11=\frac{99}{9};\ 12=9+\sqrt{9};$$

14.2 (d)

$A + B = 100$

$C + D = A$

$B = 4D$

$D = C - 10$ or $C = D + 10$

$$A + 4D = 100$$

and $\quad D + D + 10 = A$

or $\quad A = 2D + 10$

Solving these 2 equations given $D = 15$ and $C = 25$

14.3 (c)

$$P : Q : R : S : T$$

$$2x : x : \frac{x}{4.5} : \frac{2x}{4.5} : \frac{4x}{4.5}$$

$$9x : 4.5x : x : 2x : 4x$$

R < S < T < Q < P.

Alternative Solution:

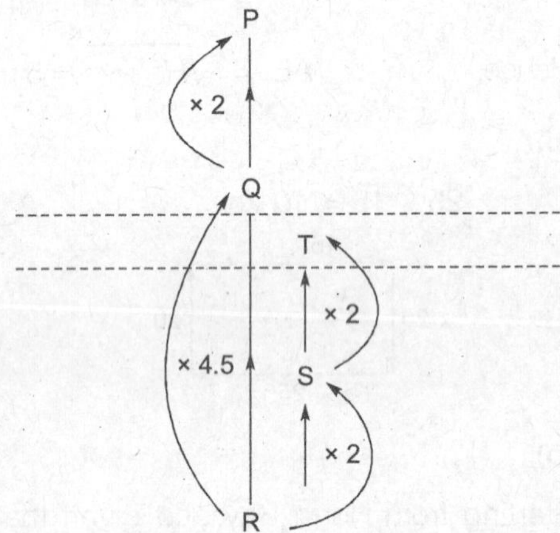

14.4 (a)

$$A > B > C = D > F > E$$

14.5 (c)

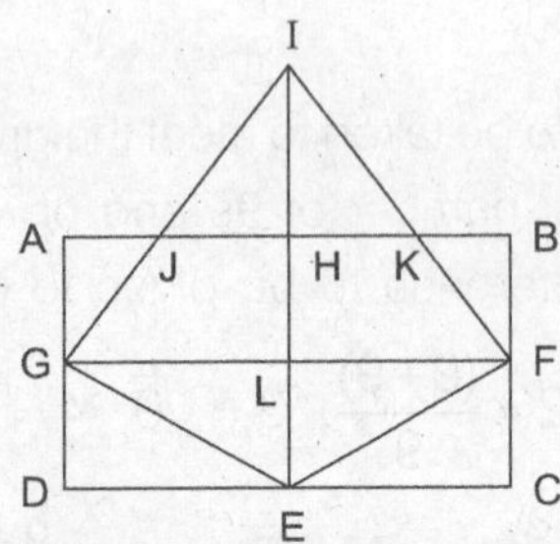

Possible triangles are: IJH, IHK, IJK, IGL, ILF, IGF, IGE, IFE, GEL, EFL, GEF, AJG, BKF, ECF, GDE.

So, number of triangles = 15.

14.6 (a)

A > B is true

B < C is false so B ≥ C

Hence, $\quad$ A > B ≥ C

So, statement 3 is true.

14.7 (c)

Let the value of integer A is x

max				min
D >	C >	B >	E	> A
$x + 16$	$x + 12$	$x + 8$	$x + 4$	x

$$A + B + C + D + E = E$$

$$A + B + C + D = 0$$

$$x + (x + 8) + (x + 12) + (x + 16) = 0$$

So, $\quad x = -9$

D C B E A

7 3 –1 –5 –9

So the difference between the highest and lowest integer is = 7 – (–9) = 16

Alternate solution :

Max $\qquad$ Min

D > C > B > E > A

∵ The difference between two adjacent integer is 4.

So, the difference between the highest and lowest integer will be = 4 × 4 = 16

14.9 (a)

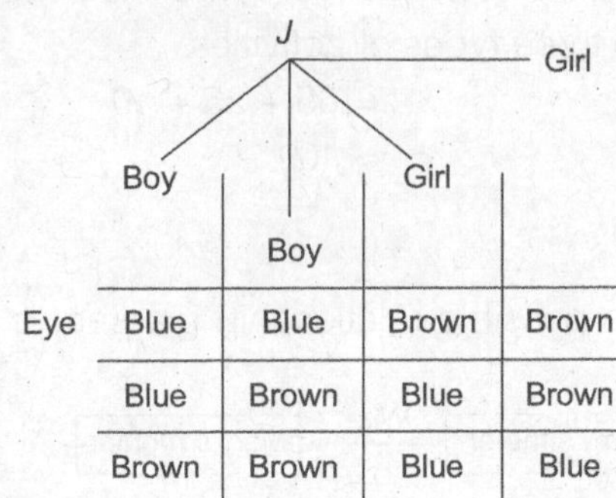

Eye	Blue	Blue	Brown	Brown
	Blue	Brown	Blue	Brown
	Brown	Brown	Blue	Blue

14.10 (d)

Max↓ $\quad$ ↑Dog

↑Dog → Jam $\qquad$ James Dog Ruby

James $\quad$ Ruby → $\quad$ Max

Ruby $\quad$ Max $\qquad$ Rachel

Rachel $\quad$ Rachel

14.11 (a)

251 252 253 254

I G H M

14.13 (d)

Forward difference table is

A = Rich, E = Poor

A, E = good in studies and C = poor in studies
B, D - Average in studies
B, D, C - middle class family
A, E in PG
A, D - Interested in acting
B, D - middle class family
D - Music
C - Sports
B, C, E - Industrious
A and D - Not industrious
So, D is average in studies and NOT industrious.

14.14 (d)

(PV) not together on any day.
$(\overrightarrow{R\ V})$ as V after R
Last S or U
$(\overrightarrow{Q\ T})$ Q followed by T
T, S, V only correct (d).

14.15 (b)

Q visit first and U visit last
After break there are four people W, R, P, U and total time reduced by 15 minute for break and 15 mins each for Dr.'s visiting after break total reduction of 75 min.
6 PM – 75 min = 4 : 45 PM
So, U came @ 4 : 00 - 4 : 45
As break is from 1 to 2 reduced by 15 mins.
So, break only from 1 to 1:45
Next 1:45 to 2:30
Next 2:30 to 3:15
Next 3:15 to 4:00
Next 4:00 to 4:45
U attended from 4 to 4:45
So, he began @ 4 PM = B

14.17 (d)

Following table can be plotted

	Dramatics	C.S.	Physics	History	Maths
Madhu	✓	✓	✓		
Shobha	✓	✓		✓	
Anjali		✓	✓	✓	✓
Poonam	✓		✓	✓	
Nisha			✓	✓	✓

Only Anjali good in History, Physics, C.S. and Maths.

14.18 (c)

(i) No mention of elephant
(ii) May or may not
(iii) Always follows

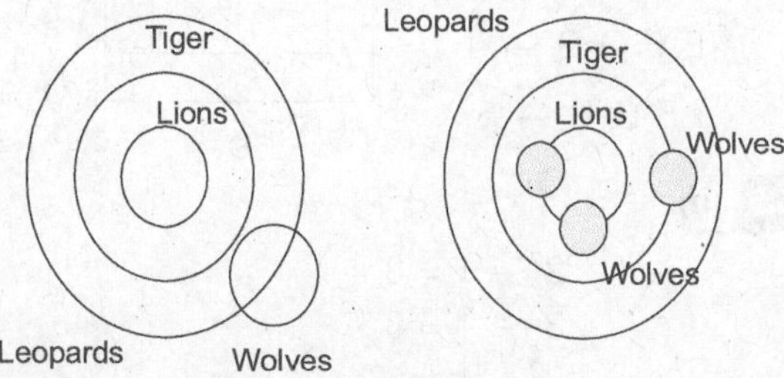

14.19 (b)

I LOVE YOU
9 12 15 22 5 25 15 21 = 124 = 1 + 2 + 4 = 7
7 15 20 15 8 5 12 12
GOTO HELL = 94 = 13 = 4

14.20 (c)

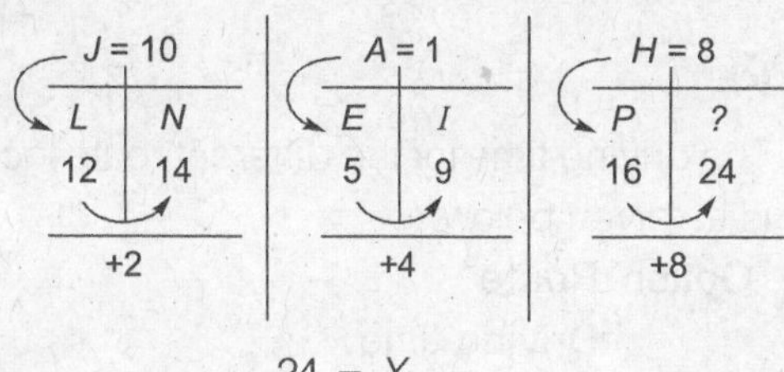

24 = X

14.21 (b)

8		10		13 = ?	
27	125	64	216	343	216
↓	↓	↓	↓	↓	↓
3^3	5^3	4^3	6^3	7^3	6^3
3 + 5 = 8		4 + 6 = 10		7 + 6 = 13	

14.22 (b)

As 3504
$R \rightarrow$ first four digits
3504 = 3 + 5 + 0 + 4 = 12
Only one number (code) 35047 satisfies it

14.23 (d)

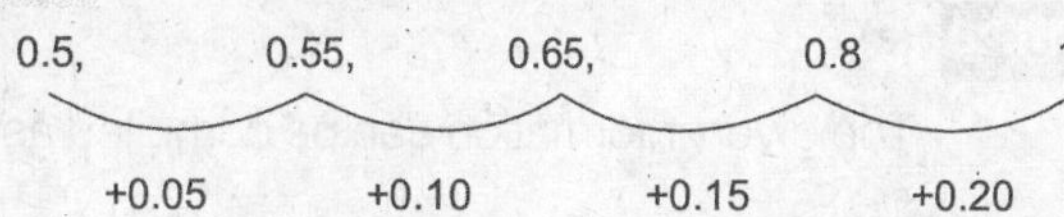

14.24 (b)

$PQRS \to 1$

$ABCD \to 2$

$\alpha\,\beta\,\gamma\,\delta \to 3$

$EFGS \to 4$

$i\,FGR \to 5$

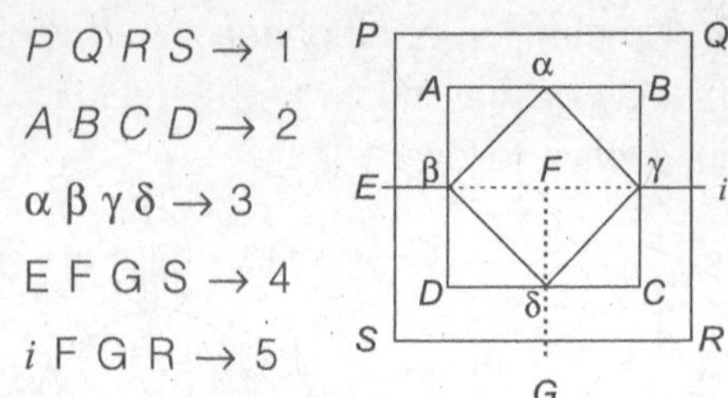

14.25 (d)

$$2^2 - 1 = 3$$
$$3^2 - 1 = 8$$
$$8^2 - 1 = 63$$
$$63^2 - 1 = 3968$$

Ans. = D = 3968

14.26 (a)

The given information can be compiled as below:

	President	Vice President	Secretary	Treasurer
Brianna	✗ (S2)	✗ (S2)	✗ (S2)	✓
Ryan	✗ (S3)	✗ (S2)	✓	✗ (S2)
Tayler	✓	✗ (S2)	✗ (S3)	✗ (S2)
Ashley	✗ (S1)	✓	✗ (S1)	✗ (S1)

14.27 (c)

The driving time for the different route locations is as given below:

Option Route

Driving time

(a) home, bookstore, credit union, supermarket, home
7+6+9+11=33 mins

(b) home, supermarket, bookstore, credit union, home
11+4.5+6+8.5=30 mins

(c) home, bookstore, supermarket, credit union, home
7+4.5+9+8.5=29 mins

(d) home, supermarket, credit union, bookstore, home
11+9+6+7=33 mins

Hence, the drive time of route (c) has driving time less than 30 minutes.

14.28 (a)

The given information can be compiled as below:

	Utility stock	Automotive Stock	Technology Stock	Oil stock
Anita	✗ (S1)	✗ (S1)	✓	✗
Tony	✗ (S1)	✗ (S1)	✗ (S3)	✓
Marita	✓	✗ (S2)	✗	✗
Jose	✗	✓	✗	✗

14.29 (c)

In plurality voting system, one who gets the most votes wins. Considering the total voters, the number of votes ranking '1' received by different varieties of chocolate candy are as detailed below:

Caramel center0

Vanilla center 17

Almond center8

Toffee center 20

Solid chocolate2

Hence, Toffee center candy would win the taste test using the plurality voting system.

14.30 (a)

Based on the given information, the order in which the books are arranged can be determined as below:

Sociology Education Accountancy English Economics Psychology Hindi

14.31 (d)

Based on the given information, the girls can be arranged in the descending order of ages as below:

Neelam, Madhavi = Pranati, Pooja, Hasini

or

Neelam, Madhavi = Pranati, Hasini, Pooja

Hence, data is inadequate to determine who is sitting in the fourth position.

14.32 (a)

In terms of scores, we have : V > P, P > Q i.e. V > P > Q.

If R scores the highest, we have R > .....> T,

If S scores the highest, we have S > .....> Q or S > .....> U.

If V ranks fifth, P and Q coming after it will occupy sixth and seventh places respectively i.e. Q ranks least. So, S will score the highest.

15.1 (d)

Since the number of passengers in Black car is 3, it leaves 300 – 3 = 297 passengers to be accommodated in White and Silver coloured cars which are equal in number ($W = S$).
We can form a linear equation:
$(6 + 5)W = 297$ giving $W = 27$ as also $S = 27$
Total number of cars being 27 + 27 + 1 = 55

15.2 (c)

Use the formulae for sum and product of roots
Given equation is $x^3 - 0.6x^2 - 1.84x + 1.344 = 0$
Let the roots of this equation be α, β, γ

$$\alpha + \beta + \gamma = -\frac{b}{a} = 0.6 \quad \text{....(i)}$$

$$\alpha\beta + \beta\gamma + \gamma\alpha = \frac{c}{a} = 1.84 \quad \text{....(ii)}$$

$$\alpha\beta\gamma = -\frac{d}{a} = 1.334 \quad \text{....(iii)}$$

If $\alpha = 0.8$ then $\beta + \gamma = 0.6 - 0.8 = -0.2$ which can be picked up as option since numerically greater value should have negative sign.

15.3 (a)

Use the formulae for sum and product of roots
Given equation is $x^3 - 8x^2 + 37x - 50 = 0$
Let the roots of this equation be α, β, γ
With one root α as $3 + 4i$; the other root β is its conjugate i.e. $3 - 4i$

$$\alpha + \beta + \gamma = -\frac{b}{a} = 8 \quad \text{.....(i)}$$

this gives the third root = 8 – 6 = 2

15.4 (a)

This question can be solved by just putting values given in various options.

15.5 (a)

This question can be solved by finding point of intersection of the 2 linear equations representing the constraints.

$$3x + y \le 12$$
and $$2x + 5y \le 34$$
which is $x = 2$ and $y = 6$
Putting this value in the given objective function i.e. $z = 10x + 6y$, we get maximum value of $z = 56$.

15.6 (c)

This question can be solved by finding point of intersection of the 2 linear equations representing the constraints

$$x_1 + 2x_2 \le 200$$
and $$x_1 + x_2 \le 150$$
which is $x_1 = 100$ and $x_2 = 50$

15.7 (c)

$$\because \quad A \times B = A^2 + B^2$$
$$\therefore \quad 5 \times (4 \times 3) = 5 \times (4^2 + 3^2)$$
$$= 5 \times (25) = 5^2 + 25^2$$
$$= 25 + 625 = 650$$

15.8 (a)

$$\because \quad \frac{x+1}{x-1} = \frac{a}{b} \text{ and } \frac{1 ?y}{1+y} = \frac{b}{a}$$

So, $$x = \frac{a+b}{a-b}$$

and $$y = \frac{a-b}{a+b}$$

and, $$\frac{x+1}{x-1} = \frac{1+y}{1-y}$$

$$(x + 1)(1 - y) = (x - 1)(1 + y)$$
$$x - xy + 1 - y = x + xy - 1 - y$$
$$2xy = 2$$
$$xy = 1$$

So, $$\frac{x-y}{1+xy} = \frac{x-y}{1+1} = \frac{1}{2}(x-y)$$

Putting value of x and y.

$$= \frac{1}{2}\left(\frac{a+b}{a-b} - \frac{a-b}{a+b}\right) = \frac{1}{2}\left(\frac{(a+b)^2 - (a-b)^2}{a^2 - b^2}\right)$$

$$= \frac{2ab}{a^2 - b^2}$$

15.9 (*)

$$x + 5 - 3 + 2 = 8$$
$$x = 4$$

* No option matched

16.1 (c)

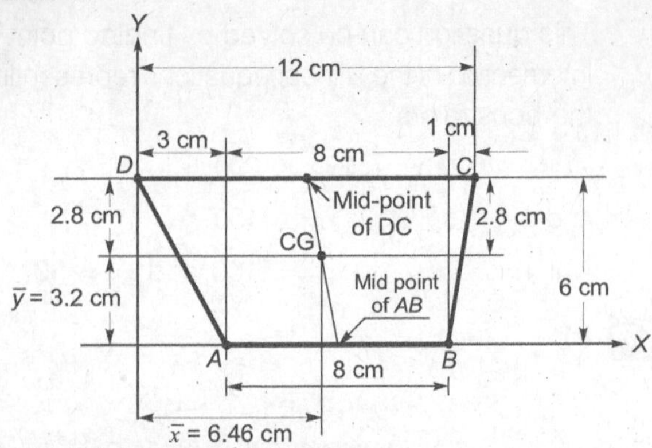

Total area of trapezium

$= \frac{1}{2} \times$ (Sum of parallel sides) $\times$ Distance between parallel sides

$$= \frac{1}{2} \times (8+12) \times 6 = 60 \text{ cm}^2$$

Equating moment about X axis of total area and component area gives

$$60 \times \bar{y} = \left(\frac{1}{2} \times 3 \times 6 \times 4\right) + (8 \times 6 \times 3) + \left(\frac{1}{2} \times 1 \times 6 \times 4\right)$$

$$\bar{y} = 3.2 \text{ cm}$$

Equating moment about *Y* axis of total area and component area gives

$$60 \times \bar{x} = \left(\frac{1}{2} \times 3 \times 6 \times 2\right) + \{8 \times 6 \times (3+4)\} + \left\{\frac{1}{2} \times 1 \times 6 \times \left(11 + \frac{1}{3}\right)\right\}$$

$$\bar{x} = 6.466 \text{ cm}$$

Hence centre of gravity of the lamina lies along line joining the mid-point of *AB* to the mid-point of *DC* at a perpendicular distance of 2.8 cm from DC.

16.2 (c)

Circumference of circle A = 4π as its diameter = 4 cm; Area of circle A = 4π or $\frac{16}{4}\pi$

Circumference of circle B = 5π as its diameter = 5 cm; Area of circle B = $\left(\frac{5}{2}\right)^2 \pi$ or $\frac{25}{4}\pi$

Circumference of circle C = Circumference of circle A + Circumference of circle *B* = $4\pi + 5\pi = 9\pi$

Diameter of circle C = 9 cm

Area of circle C = $\left(\frac{9}{2}\right)^2 \pi = \frac{81}{4}\pi$

Ratio of are of circle C w.r.t. area of circle A and that of circle B will be $\frac{81}{16}, \frac{81}{25}$ or 5.0625 and 3.24.

16.3 (b)

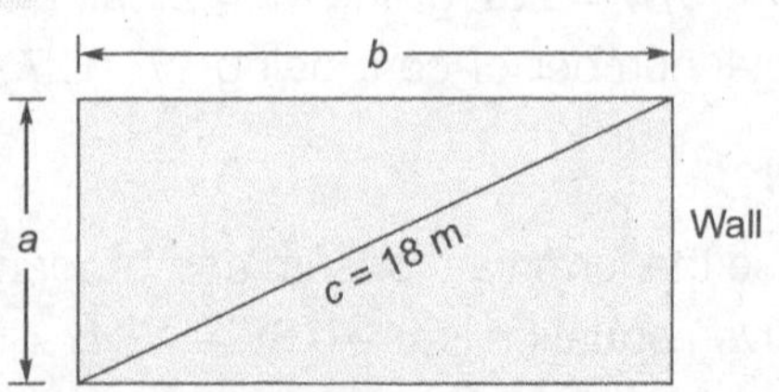

$$\text{Perimeter} = 72 \text{ m}$$

$$\Rightarrow \quad (a+b) \times 2 = 72 \text{ m} \quad \text{... (i)}$$

$$a + b = 36 \text{ m}$$

$$a^2 + b^2 = c^2$$

$$\Rightarrow \quad a^2 + b^2 = 18^2 = 324 \quad \text{... (ii)}$$

$$\text{Area} = a \times b \quad \text{... (iii)}$$

we know, $(a+b)^2 = a^2 + b^2 + 2ab$

$$\Rightarrow \quad 36^2 = 324 + 2 \times (\text{Area})$$

$$\Rightarrow \quad \text{Area} = 486 \text{ m}^2$$

16.4 (a)

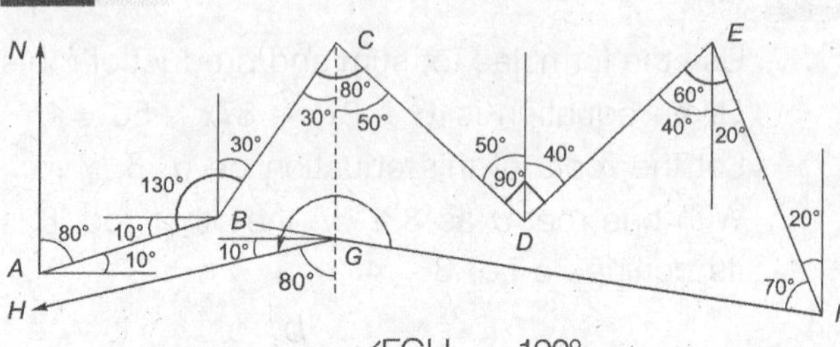

$$\angle FGH = 190°$$

Note: Line FG will be horizontal and not inclined as given in question.

Alternate Solution :

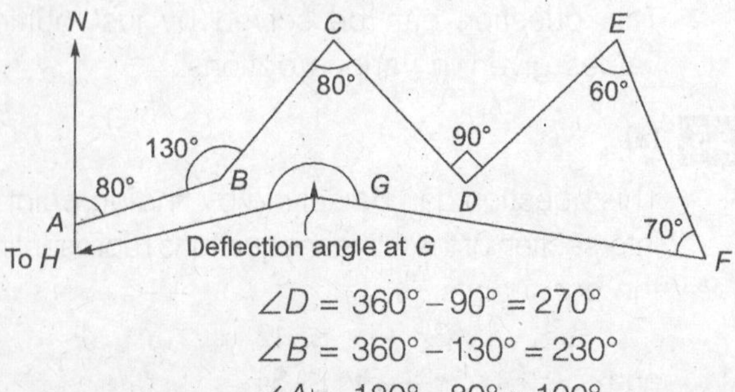

$$\angle D = 360° - 90° = 270°$$

$$\angle B = 360° - 130° = 230°$$

$$\angle A = 180° - 80° = 100°$$

$$\angle H = 180° - 100° = 80°$$

Sum of internal angles of polygon,

$$= (n-2) \times 180°$$

$$= (8-2) \times 180° = 1080°$$

$\angle A + \angle B + \angle C + \angle D + \angle E + \angle F + \angle G + \angle H = 1080°$

$100° + 230° + 80° + 270° + 60° + 70° + \angle G + 80° = 1080°$

$\therefore \quad \angle G = 190°$

16.5 **(a)**

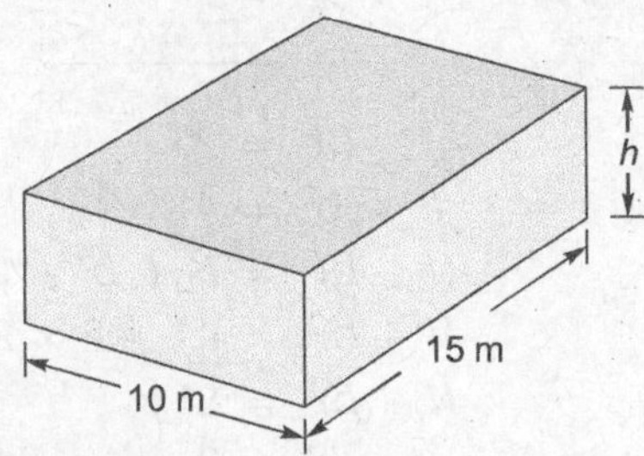

Given information is

Area of floor + Area of ceiling = Sum of area of four walls

$\Rightarrow (10 \times 15) + (10 \times 15) = (15 \times h \times 2) + (10 \times h \times 2)$

$\Rightarrow 300 = 50h$

$\Rightarrow h = 6 \text{ m}$

Volume of room = 10 m × 15 m × 6 m = 900 m^3

16.6 **(b)**

Required area = Trough area of hemisphere [A]

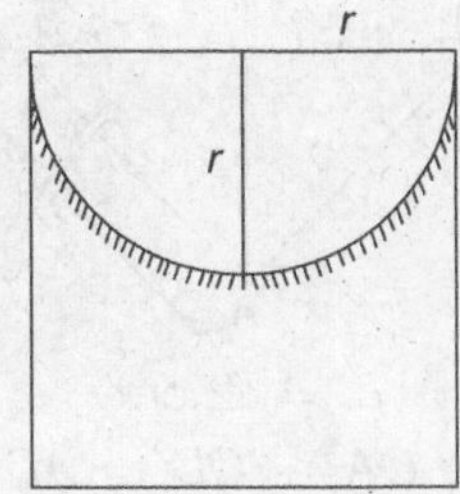

+ Area of five remaining faces of cube [B]

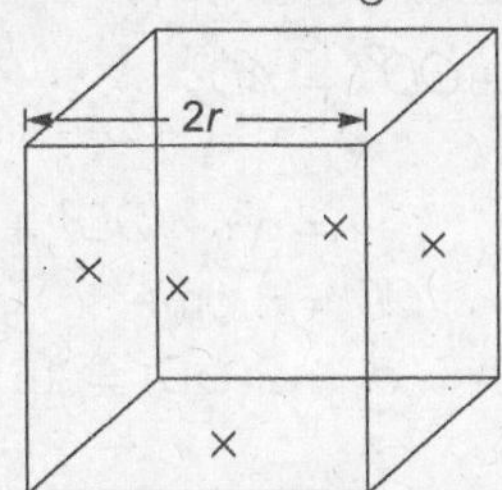

+ Left out area on the face of cube after taking out hemisphere [C]

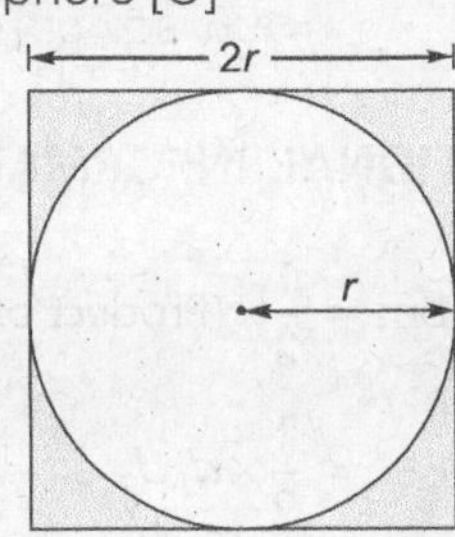

$A = 2\pi r^2$

$B = 5 \times (2r)^2 = 20r^2$

$C = 4r^2 - \pi r^2$

Required area $= A + B + C$

$= \pi r^2 + 24r^2$

$= r^2(\pi + 24)$

16.7 **(a)**

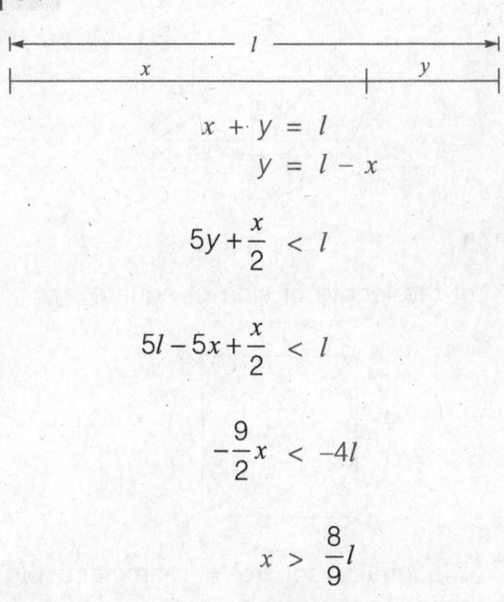

$x + y = l$

$y = l - x$

$5y + \frac{x}{2} < l$

$5l - 5x + \frac{x}{2} < l$

$-\frac{9}{2}x < -4l$

$x > \frac{8}{9}l$

Only option greater than $\frac{8}{9}l$ is option (a)

Alternatively

In option (a), let length of rod = 10

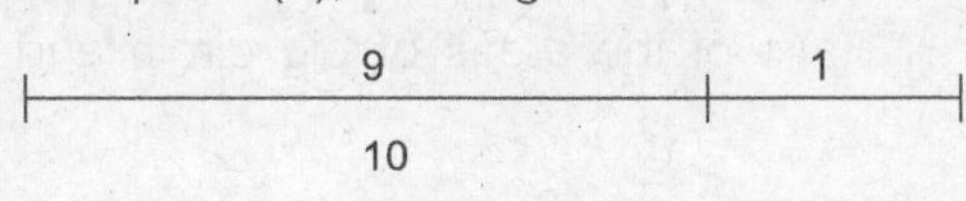

$5 + 4.5 = 9.5 < 10$

In option (b) let length of rod = 8 and likewise. Only option (a) satisfies.

16.8 **(a)**

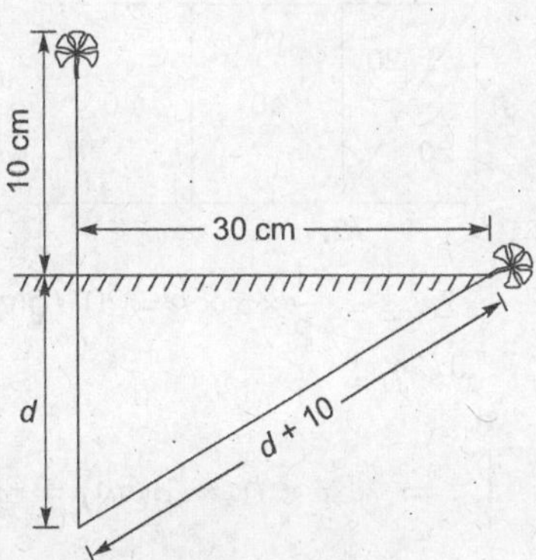

Applying Pythagoras Theorem

$d^2 + 30^2 = (d + 10)^2$

Using options, $d = 40$ (Pythagorean Triplet)

16.9 **(b)**

$$R = \sqrt{40^2 + 9^2} = 41$$

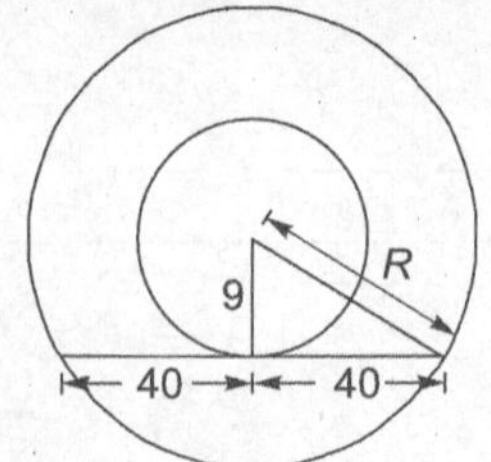

16.10 **(c)**

Let the length of side of square = x

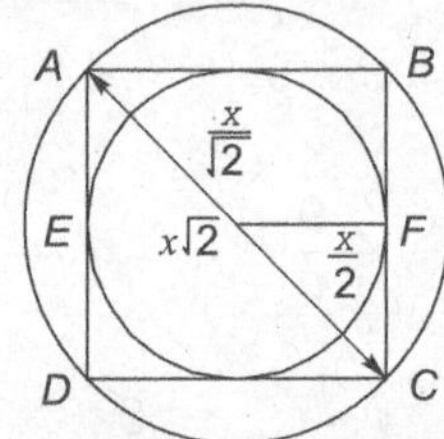

Diagonal of square = Diameter of big circle

$$= x\sqrt{2}$$

$$\text{Radius of big circle} = \frac{x\sqrt{2}}{2} = \frac{x}{\sqrt{2}}$$

$$\text{Radius of small circle} = \frac{1}{2}(\text{side of square}) = \frac{x}{2}$$

$\therefore$ Ratio of the areas of big circle and small

$$\text{circle} = \frac{\pi \frac{x^2}{2}}{\pi \frac{x^2}{4}} = \frac{2}{1} = 2:1$$

16.11 **(d)**

$$\Delta_{ADE} = \frac{1}{2} \times x \times h = 20 \text{ (given)}$$

$$x \times h = 40$$

$$\Rightarrow 3x \times h = 3(xh) = 3(40) = 120$$

16.12 **(b)**

HCF of 78, 104, 117, 169 is 13 (i.e. maximum length).

$$\text{Total pieces} = \frac{78}{13} + \frac{104}{13} + \frac{117}{13} + \frac{169}{13}$$

$$= 6 + 8 + 9 + 13 = 36$$

16.13 **(b)**

Let the side of rectangle is l, b

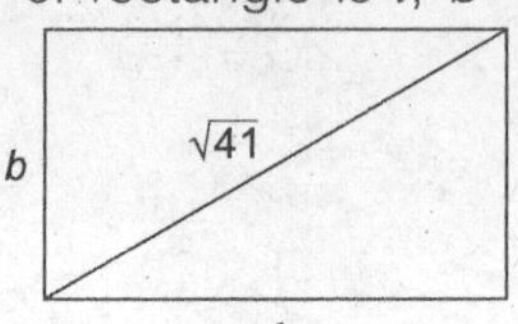

Then, $lb = 20$

$$l^2 + b^2 = 41$$

So, $(l + b)^2 = l^2 + b^2 + 2lb$

$$(l + b)^2 = 41 + 2 \times 20$$

$$(l + b)^2 = 81$$

$$(l + b) = 9$$

$$2(l + b) = 18$$

16.14 **(d)**

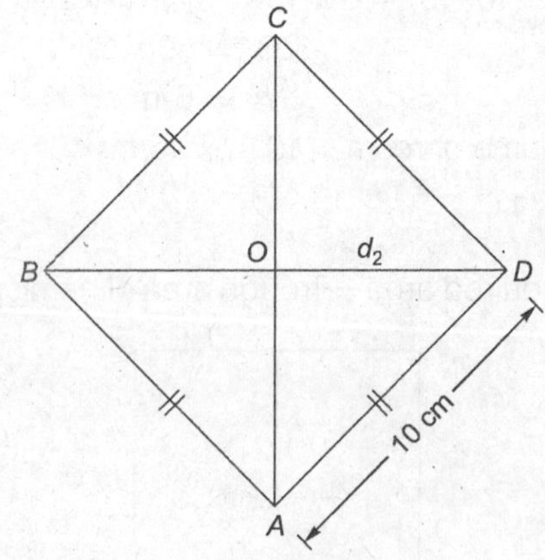

$$d_1 = 12 \text{ cm}$$

For ΔODA, $OA^2 + OD^2 = AD^2$

$$\left(\frac{d_1}{2}\right)^2 + OD^2 = AD^2$$

$$= 6^2 + OD^2 + 10^2$$

$$\Rightarrow OD = 8 \text{ cm}$$

$$\Rightarrow d_2 = 2(OD) = 16 \text{ cm}$$

$$\therefore \text{Area of } ABCD = \frac{1}{2} d_1 \times d_2 = \frac{1}{2} \times 12 \times 18$$

$$= 96 \text{ sq. cm.}$$

ADDITIONAL INFORMATION

$$\text{Area of rhombus} = \frac{1}{2} \times (\text{Product of its diagonals})$$

$$= \frac{1}{2} \times d_1 d_2$$

Engineering Mathematics and Numerical Analysis

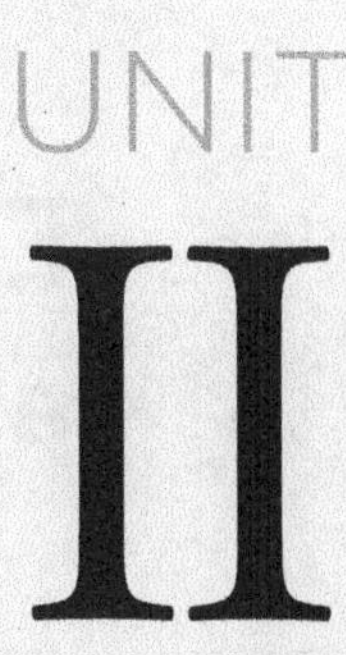

CONTENTS

Engineering Mathematics and Numerical Analysis

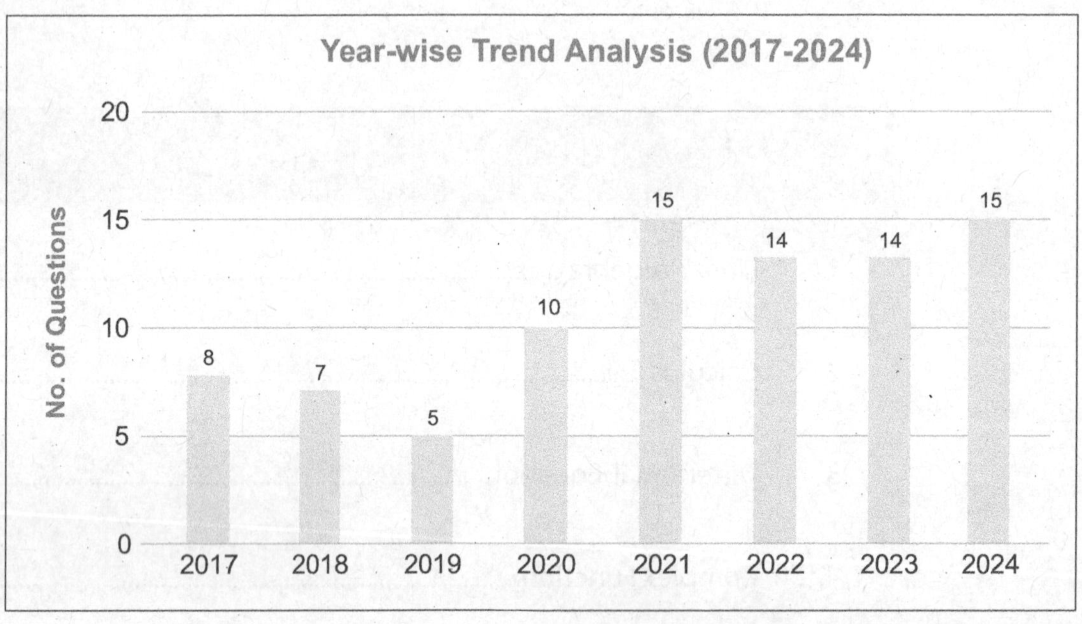

TOPIC-WISE TREND ANALYSIS (2017-2024)		
Sl. No.	**Topics**	**Total Number of Questions**
1.	Linear Algebra	12
2.	Calculus	28
3.	Differential Equation	6
4.	Complex Function	9
5.	Probability and Statics	10
6.	Numerical Methods	10
7.	Fourier Series	4
8.	Partial Differential Equations	3
9.	Laplace Transform	6

UNIT

II

Engineering Mathematics and Numerical Analysis

1. LINEAR ALGEBRA

1.1 The solution of the system of equations $x + y + z = 4$, $x - y + z = 0$, $2x + y + z = 5$ is

(a) $x = 2, y = 2, z = 0$
(b) $x = 1, y = 4, z = 1$
(c) $x = 2, y = 4, z = 3$
(d) $x = 1, y = 2, z = 1$

[ESE Prelims : 2017]

1.2 Let the Eigen vector of the matrix $\begin{bmatrix} 1 & 2 \\ 0 & 2 \end{bmatrix}$ be written in the form $\begin{bmatrix} 1 \\ a \end{bmatrix}$ and $\begin{bmatrix} 1 \\ b \end{bmatrix}$. What is the value of $(a + b)$?

(a) 0 (b) $\frac{1}{2}$
(c) 1 (d) 2

[ESE Prelims : 2018]

1.3 The lowest Eigen value of the 2×2 matrix $\begin{bmatrix} 4 & 2 \\ 1 & 3 \end{bmatrix}$ is

(a) 1 (b) 2
(c) 3 (d) 5

[ESE Prelims : 2019]

1.4 For the matrix $A = \begin{bmatrix} 1 & 4 \\ 2 & 3 \end{bmatrix}$ the expression

$$A^5 - 4A^4 - 7A^3 + 11A^2 - A - 10I$$

is equivalent to

(a) $A^2 + A + 5I$ (b) $A + 5I$
(c) $A^2 + 5I$ (d) $A^2 + 2A + 6I$

[ESE Prelims : 2020]

1.5 The highest Eigen value of the 2×2 matrix $\begin{bmatrix} 1 & 2 \\ 4 & 3 \end{bmatrix}$ is

(a) –1 (b) –5
(c) 5 (d) 1

[ESE Prelims : 2021]

1.6 If $\Delta = \begin{vmatrix} p & p^2 & (p^3 - 1) \\ q & q^2 & (q^3 - 1) \\ r & r^2 & (r^3 - 1) \end{vmatrix} = 0$, in which p, q, r are different. The value of pqr is

(a) 3 (b) 1
(c) 2.5 (d) 3.5

[ESE Prelims : 2021]

1.7 If $A = \begin{bmatrix} -1 & 2 & 3 & -2 \\ 2 & -5 & 1 & 2 \\ 3 & -8 & 5 & 2 \\ 5 & -12 & -1 & 6 \end{bmatrix}$, then the rank of the matrix A is

(a) 2 (b) 5
(c) 4 (d) 3

[ESE Prelims : 2021]

1.8 If $A = \begin{bmatrix} 1 & 3 & 2 \\ 2 & 0 & -1 \\ 1 & 2 & 3 \end{bmatrix}$, then which one of the following is correct?

(a) $A^3 - 3A^2 - 4A + 11I = 0$
(b) $A^3 - 4A^2 - 3A + 11I = 0$
(c) $A^3 + 4A^2 - 3A + 11I = 0$
(d) $A^3 - 3A^2 + 4A + 11I = 0$

[ESE Prelims : 2021]

1.9 What is the largest eigen value in modulus of the matrix $A = \begin{bmatrix} 2 & 3 & 2 \\ 4 & 3 & 5 \\ 3 & 2 & 9 \end{bmatrix}$ with an initial vector $(1, 1, 1)^T$ by power method?

(a) 11.84 (b) 12.84
(c) 13.84 (d) 14.84

[ESE Prelims : 2022]

1.10 Reduce the matrix $A = \begin{bmatrix} 1 & 3 & 4 \\ 3 & 2 & -1 \\ 4 & -1 & 1 \end{bmatrix}$ to the tridiagonal form.

(a) $\begin{bmatrix} 1 & -5 & 0 \\ -5 & \frac{2}{5} & \frac{1}{5} \\ 0 & \frac{1}{5} & \frac{13}{5} \end{bmatrix}$ (b) $\begin{bmatrix} 1 & 0 & -5 \\ -5 & \frac{2}{5} & -\frac{1}{5} \\ 0 & -\frac{13}{5} & \frac{1}{5} \end{bmatrix}$

(c) $\begin{bmatrix} 1 & -5 & 0 \\ -5 & -\frac{2}{5} & -\frac{13}{5} \\ 0 & \frac{1}{5} & \frac{1}{5} \end{bmatrix}$ (d) $\begin{bmatrix} 1 & -5 & 0 \\ -5 & -\frac{2}{5} & \frac{1}{5} \\ 0 & \frac{13}{5} & \frac{1}{5} \end{bmatrix}$

[ESE Prelims : 2022]

1.11 If $X_1 = \begin{bmatrix} 1 \\ 1 \\ 1 \end{bmatrix}, X_2 = \begin{bmatrix} 1 \\ 0 \\ 1 \end{bmatrix}$ and $X_3 = \begin{bmatrix} 0 \\ 1 \\ 1 \end{bmatrix}$ are the eigenvectors of the matrix $A = \begin{bmatrix} 2 & 1 & -1 \\ 3 & 2 & -3 \\ 3 & 1 & -2 \end{bmatrix}$, then $A^5 =$

(a) $\begin{bmatrix} 32 & 31 & -31 \\ 33 & 32 & -33 \\ 33 & 31 & -32 \end{bmatrix}$ (b) $\begin{bmatrix} 32 & 31 & -33 \\ 33 & 32 & -31 \\ 33 & 31 & -32 \end{bmatrix}$

(c) $\begin{bmatrix} 32 & 31 & -32 \\ 33 & 32 & -33 \\ 33 & 31 & -31 \end{bmatrix}$ (d) $\begin{bmatrix} 32 & 33 & -31 \\ 33 & 32 & -33 \\ 33 & 31 & 32 \end{bmatrix}$

[ESE Prelims : 2023]

1.12 The equation for the ellipsoid of inertia of a solid body is

$$P(x) \equiv 4x_1^2 + 4x_2^2 + x_3^2 - 2x_1x_2.$$

What is the standard from in terms of a new orthogonal set of axes $O\{y_1, y_2, y_3\}$?

(a) $y_1^2 - 3y_2^2 + 3y_3^3$
(b) $y_1^2 + 5y_2^2 + 3y_3^2$
(c) $y_1^2 - 5y_2^2 + 3y_3^2$
(d) $y_1^2 - 5y_2^2 - 3y_3^2$

[ESE Prelims : 2023]

2. CALCULUS

2.1 The minimum value of the function $f(x) = \left(\frac{x^3}{3}\right) - x$ occurs at

(a) $x = 1$ (b) $x = -1$
(b) $x = 0$ (d) $x = \frac{1}{\sqrt{3}}$

[ESE Prelims : 2017]

2.2 The value of the integral $\int_0^{2\pi} \left(\frac{3}{9 + \sin^2 \theta}\right) d\theta$ is

(a) $\frac{2\pi}{\sqrt{10}}$ (b) $2\sqrt{10}\,\pi$
(c) $\sqrt{10}\,\pi$ (d) 2π

[ESE Prelims : 2017]

2.3 If $u = \log\left(\frac{x^2 + y^2}{x + y}\right)$, what is the value of $x\frac{\partial u}{\partial x} + y\frac{\partial u}{\partial y}$?

(a) 0 (b) 1
(c) u (d) eu

[ESE Prelims : 2018]

2.4 Find the absolute maximum and minimum values of

$$f(x, y) = 2 + 2x + 2y - x^2 - y^2$$

on triangular plate in the first quadrant, bounded by the lines $x = 0$, $y = 0$ and $y = 9 - x$.

(a) –4 (b) –2
(c) 4 (d) 2

[ESE Prelims : 2020]

2.5 The value of

$$\Delta^{10}\left[(1-ax)(1-bx^2)(1-cx^3)(1-dx^4)\right]$$

is

(a) $abcd(10!)$ (b) $abcd(9!)$
(c) $abcd(8!)$ (d) $abcd(7!)$

[ESE Prelims : 2020]

2.6 If $u = \log_e\left(\frac{x^4+y^4}{x+y}\right)$, the value of $x\frac{\partial u}{dx}+y\frac{\partial u}{dy}$ is

(a) 6 (b) 5
(c) 4 (d) 3

[ESE Prelims : 2020]

2.7 The value of $\int_0^1\int_0^x (x^2+y^2)dA$, where dA indicates small area in xy-plane, is

(a) $\frac{1}{2}$ sq. units (b) $\frac{1}{3}$ sq. units
(c) $-\frac{1}{2}$ sq. units (d) $-\frac{1}{3}$ sq. units

[ESE Prelims : 2021]

2.8 If $x = uv$, $y = \frac{u+v}{u-v}$, then $\frac{\partial(u,v)}{\partial(x,y)}$ is

(a) $\frac{(u-v)^2}{4uv}$ (b) $\frac{(u+v)^2}{4uv}$
(c) $\frac{(u-v)}{4uv}$ (d) $\frac{(u+v)}{4uv}$

[ESE Prelims : 2021]

2.9 If $u = x^3 + y^3$ where $x = a\cos t$, $y = b\sin t$, then $\frac{du}{dt} =$

(a) $-3a^3\cos^2 t\sin t + 3b^3\sin^2 t\cos t$
(b) $3a^3\sin^2 t\cos t + 3b^3\cos^2 t\sin t$
(c) $3b\sin^2 t\cos t + 3a^3\sin^2 t\cos t$
(d) $-3a^3\sin t + 3b^3\cos^2 t\sin t$

[ESE Prelims : 2021]

2.10 The Maclaurin's series expansion of $e^{\sin x}$ is

(a) $1+x-\frac{x^2}{2}+\frac{x^4}{12}-.........$
(b) $1-x+\frac{x^2}{2}-\frac{x^4}{8}+.........$
(c) $1+x+\frac{x^2}{2}-\frac{x^4}{8}+.........$
(d) $1+x+\frac{x^2}{2}-\frac{x^4}{12}+.........$

[ESE Prelims : 2021]

2.11 The value of $\sum_1^n \frac{1}{(x+3)(x+4)}$ is

(a) $\frac{n}{n+2}$ (b) $\frac{2n}{n+1}$
(c) $\frac{n}{4(n+4)}$ (d) $\frac{n}{2(n+2)}$

[ESE Prelims : 2021]

2.12 The surface area of that portion of the surface $z=\sqrt{4-x^2}$ that lies above the rectangle R in the xy-plane whose coordinates satisfy $0 \le x \le 1$ and $0 \le y \le 4$ is equal to

(a) $4-\pi$ (b) $\frac{3}{4}\pi^2$
(c) $\frac{\sqrt{3}}{5}\pi$ (d) $\frac{4}{3}\pi$

[ESE Prelims : 2021]

2.13 By integration the area bounded by the three straight lines $y = 4 - x$, $y = 3x$ and $3y = x$ is

(a) 2 square units (b) 3 square units
(c) 4 square units (d) 5 square units

[ESE Prelims : 2022]

2.14 The power series for $\ln\left(\frac{1+x}{1-x}\right)$ is

(a) $\left(x+\frac{x^3}{3}+\frac{x^5}{5}+...\right)$
(b) $2\left(x-\frac{x^3}{3}+\frac{x^5}{5}-...\right)$
(c) $2\left(x+\frac{x^3}{3}+\frac{x^5}{5}+...\right)$
(d) $\left(x-\frac{x^3}{3}+\frac{x^5}{5}-...\right)$

[ESE Prelims : 2022]

2.15 The mean value of $y = 3x^2 + 4x + 1$ between $x = -1$ and $x = 2$ is

(a) 2 (b) 4
(c) 6 (d) 8

[ESE Prelims : 2022]

2.16 From the Taylor series for $y(x)$, what is the value of $y(0.1)$ correct to four decimal places if $y(x)$ satisfies $y' = x - y^2$ and $y(0) = 1$?

(a) 0.9138 (b) 0.7254
(c) 0.5286 (d) 0.3524

[ESE Prelims : 2022]

2.17 What is the shape of the curve represented by

$$\frac{x}{5} = \sqrt{\left[1 + \left(\frac{y}{2}\right)^2\right]}?$$

(a) Hyperbola (b) Rectangular hyperbola
(c) Parabola (d) Ellipse

[ESE Prelims : 2022]

2.18 The value of $\int_0^4 \sqrt{(16 - x^2)}\,dx$ is

(a) π (b) 2π
(c) 3π (d) 4π

[ESE Prelims : 2022]

2.19 The value of $\int_0^2 \frac{3x}{\sqrt{(2x^2 + 1)}}dx$ is (take positive values of square roots only)

(a) 1 (b) 2
(c) 3 (d) 4

[ESE Prelims : 2022]

2.20 What is the length of the curve $x = 2\cos^3\theta$, $y = 2\sin^3\theta$ between the points corresponding to $\theta = 0$ and $= \frac{\pi}{2}$?

(a) 2 units (b) 3 units
(c) 4 units (d) 5 units

[ESE Prelims : 2022]

2.21 For what values of a and b is the vector field $F = (x + z)i + a(y + z)j + b(x + y)k$ a conservative field?

(a) $a = b = 1$ (b) $a = b = -1$
(c) $a = 1, b = -1$ (d) $a = -1, b = 1$

[ESE Prelims : 2023]

2.22 Let S be the surface of the paraboloid of revolution $z = 1 - x^2 - y^2$ with the domain of definition $x^2 + y^2 \le 1$, and let Γ be the boundary of the paraboloid.

Given: $F = x^3 i + (x + y - z)\,j + yzk$.

What is the value of $\iint_S \text{curl}\, F \cdot dS$?

(a) 2π (b) π
(c) $\frac{\pi}{2}$ (d) π^2

[ESE Prelims : 2023]

2.23 The plane $x = 1$ intersects the paraboloid $z = x^2 + y^2$ in parabola. The slope of the tangent line to the parabola at (1, 2, 5) is

(a) 2 (b) 6
(c) 4 (d) 5

[ESE Prelims : 2024]

2.24 What is the absolute minimum value of $f(x, y) = 2 + 2x + 4y - x^2 - y^2$ on the triangular region in the first quadrant bounded by the lines $x = 0$, $y = 0$, and $y = 9 - x$?

(a) –11 (b) –43
(c) –61 (d) –41

[ESE Prelims : 2024]

2.25 What is the centroid ($\delta = 1$) of the solid enclosed by the cylinder $x^2 + y^2 = 4$, bounded above by the paraboloid $z = x^2 + y^2$, and bounded below by the xy-plane?

(a) $\left(0, 0, \frac{3}{4}\right)$ (b) $\left(0, 0, \frac{4}{3}\right)$
(c) $\left(0, 0, \frac{5}{4}\right)$ (d) $\left(0, 0, \frac{4}{5}\right)$

[ESE Prelims : 2024]

2.26 What is the integral

$$\int_1^2 \int_{1/y}^{y} \sqrt{\frac{y}{x}} e^{\sqrt{xy}}\, dx dy?$$

(a) $2e(e + 2)$ (b) $2e(1 - e)$
(c) $2e(e - 2)$ (d) $2e(1 + e)$

[ESE Prelims : 2024]

2.27 Suppose we do not know the path of a hang glider, but only its acceleration vector $a(t) = -(3\cos t)i - (3\sin t)j + 2k$. We also know that initially (at time $t = 0$) the glider departed from the point (4, 0, 0) with velocity $v(0) = 3j$. What is the glider's position as a function of t?

(a) $r(t) = (1 + 3\cos t)i - 3\sin tj + t^2k$

(b) $r(t) = (-1 + 3\cos t)i + 3\sin tj + t^2k$

(c) $r(t) = (1 - 3\cos t)i + 3\sin tj + t^2k$

(d) $r(t) = (1 + 3\cos t)i + 3\sin tj + t^2k$

[ESE Prelims : 2024]

2.28 Fourier transform of

$$f(t) = \begin{cases} \sin at, |t| \le \pi/a \\ 0, |t| > \pi/a \end{cases} \text{ is}$$

(a) $\dfrac{2j\sin\pi\omega/a}{-a^2-\omega^2}$ (b) $\dfrac{j\sin\pi\omega/a}{-a^2-\omega^2}$

(c) $\dfrac{j\sin\pi\omega/a}{a^2-\omega^2}$ (d) $\dfrac{2aj\sin\pi\omega/a}{\omega^2-a^2}$

[ESE Prelims : 2024]

3. DIFFERENTIAL EQUATION

3.1 The solution of the differential equation

$$(1+y^2)dx = (\tan^{-1}y - x)dy$$

is

(a) $x = \tan^{-1}y + 1 + ce^{-\tan^{-1}y}$

(b) $x = \tan^{-1}y - 1 + ce^{-\tan^{-1}y}$

(c) $x = \frac{1}{2}\tan^{-1}y - 1 + ce^{-\tan^{-1}y}$

(d) $x = \frac{1}{2}\tan^{-1}y + 1 + ce^{-\tan^{-1}y}$

[ESE Prelims : 2020]

3.2 The surface which intersects the surfaces of the system $z(x + y) = c(3z + 1)$ orthogonally and which passes through the circle $x^2 + y^2 = 1$, $z = 1$, is given by

(a) $x^2 + y^2 = 2z^3 + z^2 - 2$

(b) $x^2 - y^2 = z^3 + z + 1$

(c) $x^2 - y^2 = z^2 + 4$

(d) $x^2 + y^2 = z^3 + z^2 + 4$

[ESE Prelims : 2021]

3.3 What is the particular solution of the differential equation $5\dfrac{dy}{dx} + 2x = 3$ if the boundary conditions are $y = \dfrac{7}{5}$ and $x = 2$?

(a) $y = \dfrac{3x}{5} - \dfrac{x^2}{5} + 1$ (b) $y = \dfrac{3x}{5} + \dfrac{x^2}{5} - 2$

(c) $y = \dfrac{5x}{3} - \dfrac{x^2}{3} + 1$ (d) $y = \dfrac{5x}{3} + \dfrac{x^2}{3} - 2$

[ESE Prelims : 2022]

3.4 What is the general solution of a homogeneous differential equation with the characteristic equation?

$$\lambda^3(\lambda + 4)^2 (\lambda^2 + 2\lambda + 5)^2 = 0$$

(a) $y(x) = c_1 + c_2x + c_3x^2 + c_4e^{-4x} + c_5xe^{4x} + e^x\{c_6 \cos 2x + c_7 \sin 2x + c_8 x\cos 2x + c_9 x \sin 2x\}$

(b) $y(x) = c_1 + c_2x + c_3x^2 + c_4e^{-4x} + c_5xe^{-4x} + e^{-x}\{c_6 \cos 2x + c_7 \sin 2x\} + e^x\{c_8 x\cos 2x + c_9 x \sin 2x\}$

(c) $y(x) = c_1 + c_2x + c_3x^2 + c_4e^{-4x} + c_5xe^{4x} + e^x\{c_6 \cos 2x + c_7 \sin 2x\} + e^{-x}\{c_8 x\cos 2x + c_9 x \sin 2x\}$

(d) $y(x) = c_1 + c_2x + c_3x^2 + c_4e^{-4x} + c_5xe^{-4x} + e^{-x}\{c_6 \cos 2x + c_7 \sin 2x + c_8 x\cos 2x + c_9 x \sin 2x\}$

[ESE Prelims : 2023]

3.5 What is the initial value if $\dfrac{d^2y}{dx^2} + 4\dfrac{dy}{dx} + 3y = e^{-x}$, with $y(0) = 2$, $\left(\dfrac{dy}{dx}\right)_{x=0} = 1$?

(a) $y(x) = \left(\dfrac{13}{4} + \dfrac{1}{2}x\right)e^{-x} - \dfrac{5}{4}e^{-3x}$

(b) $y(x) = \left(\dfrac{13}{4} + \dfrac{1}{2}x\right)e^{-3x} - \dfrac{5}{4}e^{-x}$

(c) $y(x) = \left(\dfrac{13}{4} + \dfrac{1}{2}x\right)e^{-x} + \dfrac{5}{4}e^{-3x}$

(d) $y(x) = \left(\dfrac{13}{4} - \dfrac{1}{2}x\right)e^{-x} - \dfrac{5}{4}e^{-3x}$

[ESE Prelims : 2023]

3.6 What is the solution of the equation $\dfrac{d^2y}{dt^2} + y(t) = \int_0^t \sin\tau\, y(t-\pi)d\tau$, subject to the initial conditions $y(0) = 1$ and $\left(\dfrac{dy}{dt}\right)_{t=0} = 0$?

(a) $y(t) = \dfrac{1}{2}(1 - \cos\sqrt{2}t)$, for $t > 0$

(b) $y(t) = \frac{1}{2}\left(1+\cos\sqrt{2}t\right)$, for $t > 0$

(c) $y(t) = \frac{1}{2}\left(-1-\cos\sqrt{2}t\right)$, for $t > 0$

(d) $y(t) = -\frac{1}{2}\left(1-\cos\sqrt{2}t\right)$, for $t > 0$

[ESE Prelims : 2023]

4. COMPLEX FUNCTION

4.1 If $W = \phi + i\psi$ represents the complex potential for an electric field.

Given $\psi = x^2 - y^2 + \frac{x}{x^2+y^2}$, then the function ϕ is

(a) $-2xy + \frac{y}{x^2+y^2} + C$

(b) $2xy + \frac{y}{x^2+y^2} + C$

(c) $-2xy + \frac{x}{x^2+y^2} + C$

(d) $2xy + \frac{x}{x^2+y^2} + C$

[ESE Prelims : 2017]

4.2 The residue of $f(z) = \frac{z^3}{(z-1)^4(z-2)(z-3)}$ at $z = 3$ is

(a) -8 (b) $\frac{101}{16}$

(c) 0 (d) $\frac{27}{16}$

[ESE Prelims : 2017]

4.3 In the Laurent expansion of $f(z) = \frac{1}{(z-1)(z-2)}$ valid in the region $1 < |z| < 2$, the coefficient of $\frac{1}{z^2}$ is

(a) 0 (b) $\frac{1}{2}$

(c) 1 (d) -1

[ESE Prelims : 2018]

4.4 What is the residue of the function $\frac{1-e^{2z}}{z^4}$ at its pole?

(a) $\frac{4}{3}$ (b) $-\frac{4}{3}$

(c) $-\frac{2}{3}$ (d) $\frac{2}{3}$

[ESE Prelims : 2018]

4.5 Polar form of the Cauchy-Riemann equations is

(a) $\frac{\partial u}{\partial r} = r\frac{\partial v}{\partial \theta}$ and $\frac{\partial v}{\partial r} = -r\frac{\partial u}{\partial \theta}$

(b) $\frac{\partial u}{\partial r} = \frac{1}{r}\frac{\partial v}{\partial \theta}$ and $\frac{\partial v}{\partial r} = -\frac{1}{r}\frac{\partial u}{\partial \theta}$

(c) $\frac{\partial u}{\partial r} = \frac{1}{r}\frac{\partial v}{\partial \theta}$ and $\frac{\partial v}{\partial r} = -r\frac{\partial u}{\partial \theta}$

(d) $\frac{\partial u}{\partial r} = r\frac{\partial v}{\partial \theta}$ and $\frac{\partial v}{\partial r} = -\frac{1}{r}\frac{\partial u}{\partial \theta}$

[ESE Prelims : 2019]

4.6 If $f(z)$ has a pole of order n at $z = a$, then Residue of function $f(z)$ at a is

(a) $\text{Res } f(a) = \frac{1}{(n)!}\left\{\frac{d^{n-1}}{dz^{n-1}}\left((z-a)^{n-1}f(z)\right)\right\}_{z=a}$

(b) $\text{Res } f(a) = \frac{1}{(n-1)!}\left\{\frac{d^{n-1}}{dz^{n-1}}\left((z-a)^{n-1}f(z)\right)\right\}_{z=a}$

(c) $\text{Res } f(a) = \frac{1}{(n)!}\left\{\frac{d^{n-1}}{dz^{n-1}}\left((z-a)^{n}f(z)\right)\right\}_{z=a}$

(d) $\text{Res } f(a) = \frac{1}{(n-1)!}\left\{\frac{d^{n-1}}{dz^{n-1}}\left((z-a)^{n}f(z)\right)\right\}_{z=a}$

[ESE Prelims : 2019]

4.7 The general value of $\log(1+i) + \log(1-i)$ is

(a) $\log 2 - 4n\pi i$ (b) $\log 2 + 4n\pi i$

(c) $\log 2 + 2n\pi i$ (d) $\log 2 - 2n\pi i$

[ESE Prelims : 2020]

4.8 What is the contour integral $\int_C z^2 dz$ along the path C from $-1 + j$ to $5 + 3j$ and composed of two straight line segments, the first from $-1 + j$ to $5 + j$ and the second from $5 + j$ to $5 + 3j$?

(a) $-4 + \frac{196}{3}j$ (b) $-4 - \frac{196}{3}j$

(c) $4 - \frac{196}{3}j$ (d) $4 + \frac{196}{3}j$

[ESE Prelims : 2024]

4.9 The image in the w plane of the circle $\left|z+\frac{3}{4}+j\right|=\frac{7}{4}$ under the inversion mapping $w = 1/z$ is

(a) a circle centre (1/2, 2/3) and radius 7/6
(b) a circle centre (1/2, –2/3) and radius 7/6
(c) a circle centre (–1/2, 2/3) and radius 7/6
(d) a circle centre (–1/2, –2/3) and radius 7/6

[ESE Prelims : 2024]

5. PROBABILITY AND STATICS

5.1 A committee of 4 is to be formed from among 4 girls and 5 boys. What is the probability that the committee will have number of boys less than number of girls?

(a) 2/9 (b) 4/9
(c) 4/5 (d) 1/6

[ESE Prelims : 2019]

5.2 A bag contains 4 white and 2 black balls and another bag contains 3 of each colour. A bag is selected at random and a ball is drawn at random from the bag chosen. The probability of the white ball drawn is

(a) $\frac{1}{3}$ (b) $\frac{1}{4}$
(c) $\frac{5}{12}$ (d) $\frac{7}{12}$

[ESE Prelims : 2020]

5.3 X is a continuous random variable with probability density function given by

$$f(x) = kx \quad (0 \le x < 2)$$
$$= 2k \quad (2 \le x < 4)$$
$$= -kx + 6k \quad (4 \le x < 6)$$

The value of k will be

(a) $\frac{2}{3}$ (b) $\frac{1}{8}$
(c) 1 (d) 8

[ESE Prelims : 2020]

5.4 The first moment about origin of binomial distribution is

(a) np (b) npq
(c) $n(1-p)$ (d) $n(1-p)q$

[ESE Prelims : 2020]

5.5 For the regression equations

$$y = 0.516x + 33.73$$

and $$x = 0.512y + 32.52$$

the means of x and y are nearly

(a) 67.6 and 68.6 (b) 68.6 and 68.6
(c) 67.6 and 58.6 (d) 68.6 and 58.6

[ESE Prelims : 2020]

5.6 Suppose that a book of 600 pages contains 40 printing mistakes. Assume that these errors are randomly distributed throughout the book and x, the number of errors per page has a Poisson distribution. What is the probability that 10 pages selected at random will be free of errors?

(a) $\frac{1}{3}e^{-1}$ (b) $2e^{-\frac{1}{3}}$
(c) $e^{-\frac{2}{3}}$ (d) $\frac{1}{3}e^{-2}$

[ESE Prelims : 2021]

5.7 A batch of 100 capacitors contains 73 which are within the required tolerance values, 17 which are below the required tolerance values, and the remaining are above the required tolerance values. What is the probability that when randomly selecting a capacitor and then a second capacitor, if both are within the required tolerance values when selecting with replacement?

(a) 0.3319 (b) 0.5329
(c) 0.7239 (d) 0.9249

[ESE Prelims : 2022]

5.8 The standard deviation of the exponential distribution of

$$f_X(x) = \begin{cases} \lambda e^{-\lambda x}, & x \ge 0 \\ 0, & x < 0 \end{cases} \text{ is}$$

(a) $\frac{1}{\lambda}$ (b) $\frac{2}{\lambda^2}$
(c) $\frac{3}{\lambda^3}$ (d) $\frac{2}{\lambda^3}$

[ESE Prelims : 2024]

5.9 Suppose that 0.1% of the people in a certain area have a disease D and that a mass screening test is used to detect cases. The test gives either a positive or a negative result for each person. Ideally, the test would always give a positive result for a person who has D, and would never do so for a person who has not. In practice the test gives a positive result with probability 99.9% for a person who has *D*, and with probability 0.2% for a person who has not. What is the probability that a person for whom the test is positive actually has the disease?

(a) $\frac{1}{3}$ (b) $\frac{5}{3}$

(c) $\frac{4}{3}$ (d) $\frac{2}{3}$

5.10 Let the random variables *X* and *Y* have joint density function given by

$$f_{X,Y}(x, y) = \begin{cases} c(1-y), & 0 \le x \le y \le 1 \\ 0, & \text{otherwise} \end{cases}$$

Then the marginal density function for *X* is

(a) $f_X(x) = 6\left(\frac{1}{2} - x - \frac{x^2}{2}\right)$ for $0 \le x \le 1$

(b) $f_X(x) = 6\left(\frac{1}{2} + x + \frac{x^2}{2}\right)$ for $0 \le x \le 1$

(c) $f_X(x) = 6\left(\frac{1}{2} + x - \frac{x^2}{2}\right)$ for $0 \le x \le 1$

(d) $f_X(x) = 6\left(\frac{1}{2} - x + \frac{x^2}{2}\right)$ for $0 \le x \le 1$

[ESE Prelims : 2024]

6. NUMERICAL METHODS

6.1 What is the cube root of 1468 to 3 decimal places?

(a) 11.340 (b) 11.353
(c) 11.365 (d) 11.382

[ESE Prelims : 2018]

6.2 What is the value of $(1525)^{0.2}$ to 2 decimal places?

(a) 4.33 (b) 4.36
(c) 4.38 (d) 4.30

[ESE Prelims : 2018]

6.3 The real root of $x^3 + x^2 + 3x + 4 = 0$ correct to four decimal places, obtained using Newton Raphson method is

(a) –1.3334 (b) 1.3221
(c) –1.2229 (d) 1.2929

[ESE Prelims : 2021]

6.4 The value of $\int_0^6 \frac{dx}{1+x^2}$ by Simpson's $\frac{1}{3}$ rule is

(a) 1.3111 (b) 1.3941
(c) 1.3735 (d) 1.3662

[ESE Prelims : 2021]

6.5 The value of *y* at *x* = 0.1 to five places of decimals, by Taylor's series method, given that

$\frac{dy}{dx} = x^2y - 1$, $y(0) = 1$, is

(a) 0.68281 (b) 0.81122
(c) 0.90033 (d) 0.70127

[ESE Prelims : 2021]

6.6 The fixed point iterative scheme for determining $\sqrt{2}$ is

(a) $x_{n+1} = \frac{1}{2}\left(x_n - \frac{2}{x_n}\right)$

(b) $x_{n+1} = \frac{1}{2}\left(-x_n + \frac{2}{x_n}\right)$

(c) $x_{n+1} = -\frac{1}{2}\left(x_n + \frac{2}{x_n}\right)$

(d) $x_{n+1} = \frac{1}{2}\left(x_n + \frac{2}{x_n}\right)$

[ESE Prelims : 2023]

6.7 The Gauss-Seidel iterative method for the system of equations :

$$-\frac{1}{4}x_2 - \frac{1}{4}x_3 + x_4 = \frac{1}{4}$$

$$-\frac{1}{4}x_1 + x_3 - \frac{1}{4}x_4 = \frac{1}{4}$$

$$x_1 - \frac{1}{4}x_2 - \frac{1}{4}x_3 = \frac{1}{2}$$

$$-\frac{1}{4}x_1 + x_2 - \frac{1}{4}x_4 = \frac{1}{2}$$ is

(a) $x_1^{(n+1)} = 0.5 - 0.25x_2^{(n)} + 0.25x_3^{(n)}$,
$x_1^{(n+1)} = 0.5 + 0.25x_1^{(n+1)} + 0.25x_4^{(n)}$,
$x_3^{(n+1)} = 0.25 + 0.25x_1^{(n+1)} + 0.25x_4^{(n)}$,
$x_4^{(n+1)} = 0.25 - 0.25x_2^{(n+1)} + 0.25x_3^{(n+1)}$

(b) $x_1^{(n+1)} = 0.5 + 0.25x_2^{(n)} + 0.25x_3^{(n)}$,
$x_2^{(n+1)} = 0.5 + 0.25x_1^{(n+1)} + 0.25x_4^{(n)}$,
$x_3^{(n+1)} = 0.25 + 0.25x_1^{(n+1)} + 0.25x_4^{(n)}$,
$x_4^{(n+1)} = 0.25 + 0.25x_2^{(n+1)} + 0.25x_3^{(n+1)}$

(c) $x_1^{(n+1)} = 0.5 + 0.25x_2^{(n)} + 0.25x_3^{(n)}$,
$x_2^{(n+1)} = 0.5 + 0.25x_1^{(n+1)} - 0.25x_4^{(n)}$,
$x_3^{(n+1)} = 0.25 + 0.25x_1^{(n+1)} - 0.25x_4^{(n)}$,
$x_4^{(n+1)} = 0.25 + 0.25x_2^{(n+1)} + 0.25x_3^{(n+1)}$

(d) $x_1^{(n+1)} = 0.5 + 0.25x_2^{(n)} + 0.25x_3^{(n)}$,
$x_2^{(n+1)} = 0.5 - 0.25x_1^{(n+1)} + 0.25x_4^{(n)}$,
$x_3^{(n+1)} = 0.25 + 0.25x_1^{(n+1)} + 0.25x_4^{(n)}$,
$x_4^{(n+1)} = 0.25 + 0.25x_2^{(n+1)} + 0.25x_3^{(n+1)}$

[ESE Prelims : 2023]

6.8 What is the missing figure in the following table?

x	1	2	3	4	5
$y = f(x)$	2	5	7	–	32

(a) 10 (b) 13
(c) 14 (d) 17

[ESE Prelims : 2023]

6.9 What is $f'(0.2)$ from the following tabular data?

x	0.0	0.2	0.4	0.6	0.8	1.0
$y(x)$	1.00	1.16	3.56	13.96	41.96	101.00

(a) 4.2 (b) 2.2
(c) 5.2 (d) 3.2

[ESE Prelims : 2023]

6.10 The nth coefficient of a series is given by $a_n = \dfrac{1.5.9.13...(4n+1)}{2^n}$. What is the expression a_n in terms of the gamma function?

(a) $a_n = 2^{n+2}\dfrac{\Gamma\left(n+\frac{5}{4}\right)}{\Gamma\left(\frac{1}{4}\right)}$

(b) $a_n = 2^{n+1}\dfrac{\Gamma\left(n+\frac{5}{4}\right)}{\Gamma\left(\frac{1}{4}\right)}$

(c) $a_n = 2^{n}\dfrac{\Gamma\left(n+\frac{5}{4}\right)}{\Gamma\left(\frac{1}{4}\right)}$

(d) $a_n = 2^{n+3}\dfrac{\Gamma\left(n+\frac{5}{4}\right)}{\Gamma\left(\frac{1}{4}\right)}$

[ESE Prelims : 2023]

7. FOURIER SERIES

7.1 For the function

$$f(x) = \begin{cases} -2, & -\pi < x < 0 \\ 2, & 0 < x < \pi \end{cases}$$

The value of a_n in the Fourier series expansion of $f(x)$ is
(a) 2 (b) 4
(c) 0 (d) –2

[ESE Prelims : 2017]

7.2 Let $f(x) = \begin{cases} -\pi, \text{ if } & -\pi < x \le 0 \\ \pi, \text{ if } & 0 < x \le \pi \end{cases}$ be a periodic function of period 2π. The coefficient of $\sin 5x$ in the Fourier series expansion of $f(x)$ in the interval $[-\pi, \pi]$ is

(a) $\dfrac{4}{5}$ (b) $\dfrac{5}{4}$

(c) $\dfrac{4}{3}$ (d) $\dfrac{3}{4}$

[ESE Prelims : 2018]

7.3 The temperature distribution $T(x)$ at a distance x, measured from one end, along a bar of length L is given by $T(x) = Kx\,(L - x)$ $(0 \le x \le L)$, K = constant. A Fourier series expansion consisting of sine terms only for $T(x)$ is

(a) $\dfrac{8KL^2}{\pi^3}\sum_{n=1}^{\infty}\dfrac{1}{(2n-1)^3}\sin\dfrac{(2n-1)\pi x}{L}$

(b) $\dfrac{8KL^2}{\pi^3}\sum_{n=1}^{\infty}\dfrac{1}{(2n-1)^2}\sin\dfrac{(2n-1)\pi x}{L}$

(c) $\dfrac{8KL^3}{\pi^3}\sum_{n=1}^{\infty}\dfrac{1}{(2n-1)^3}\sin\dfrac{(2n-1)\pi x}{L}$

(d) $\dfrac{8KL^3}{\pi^3}\sum_{n=1}^{\infty}\dfrac{1}{(2n-1)^2}\sin\dfrac{(2n-1)\pi x}{L}$

[ESE Prelims : 2024]

7.4 Passing a sinusoidal voltage $A \sin \omega t$ through a half-wave rectifier produces the clipped sine wave shown in the following figure.

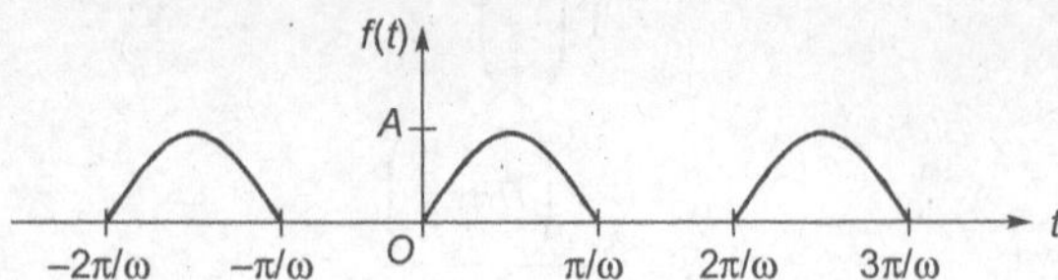

A Fourier series expansion of the rectified wave is

(a) $f(t) = \frac{A}{\pi}\left[1 + \frac{\pi}{2}\sin\omega t + 2\sum_{n=1}^{\infty}\frac{\cos 2n\omega t}{4n^2 - 1}\right]$

(b) $f(t) = \frac{A}{\pi}\left[1 + \frac{\pi}{2}\sin\omega t - 2\sum_{n=1}^{\infty}\frac{\cos 2n\omega t}{4n^2 - 1}\right]$

(c) $f(t) = \frac{A}{\pi}\left[1 - \frac{\pi}{2}\sin\omega t - 2\sum_{n=1}^{\infty}\frac{\cos 2n\omega t}{4n^2 - 1}\right]$

(d) $f(t) = \frac{A}{\pi}\left[1 - \frac{\pi}{2}\sin\omega t + 2\sum_{n=1}^{\infty}\frac{\cos 2n\omega t}{4n^2 - 1}\right]$

[ESE Prelims : 2024]

8. PARTIAL DIFFERENTIAL EQUATIONS

8.1 The complete integral of $(z - px - qy)^3 = pq + 2(p^2 + q)^2$ is

(a) $z = ax + by + \sqrt[3]{pq + 2(p^2 + q)^2}$

(b) $z = ax + by + \sqrt[3]{ab + 2(a^2 + b)^2}$

(c) $z = ax + by + \sqrt[3]{ab} + \sqrt[3]{2(a^2 + b)^2}$

(d) $z = ax + by + c$

[ESE Prelims : 2017]

8.2 The solution of the following partial differential equation $\frac{\partial^2 u}{\partial x^2} = 9\frac{\partial^2 u}{\partial y^2}$ is:

(a) $\sin(3x - y)$ (b) $3x^2 + y^2$
(c) $\sin(3x - 3y)$ (d) $(3y^2 - x^2)$

[ESE Prelims : 2017]

8.3 The solution of initial value problem; $\frac{\partial u}{\partial x} = 2\frac{\partial u}{\partial t} + u$, where $u(x, 0) = 6e^{-3x}$ is

(a) $u = 6e^{-3x + t}$ (b) $u = 6e^{-(2x + 2t)}$
(c) $u = 6e^{-(3x + 2t)}$ (d) $u = 6e^{-(3x + 2t)}$

[ESE Prelims : 2019]

9. LAPLACE TRANSFORM

9.1 The value of $L^{-1}\left\{\frac{5s^2 + 8s - 1}{(s+3)(s^2+1)}\right\}$ is

(a) $2e^{-3t} + 3\cos t - \sin t$
(b) $2e^{-3t} - 3\cos t + \sin t$
(c) $3e^{-3t} + 2\cos t - \sin t$
(d) $3e^{-3t} - 2\cos t + \sin t$

[ESE Prelims : 2022]

9.2 What is the Laplace transform of $2e^{3t}(4\cos 2t - 5\sin 2t)$?

(a) $\frac{8s + 44}{s^2 + 6s - 13}$ (b) $\frac{4s - 44}{s^2 - 6s + 13}$

(c) $\frac{4s + 44}{s^2 + 6s - 13}$ (d) $\frac{8s - 44}{s^2 - 6s + 13}$

[ESE Prelims : 2022]

9.3 Fourier series representation of $f(x) = x + 1$ for $-1 \le x \le 1$ is

(a) $f(x) = 1 - \frac{2}{\pi}\sum_{n=1}^{\infty}\frac{(-1)^{n+1}}{n}\sin n\pi x$

(b) $f(x) = -1 - \frac{2}{\pi}\sum_{n=1}^{\infty}\frac{(-1)^{n+1}}{n}\sin n\pi x$

(c) $f(x) = 1 + \frac{2}{\pi}\sum_{n=1}^{\infty}\frac{(-1)^{n+1}}{n}\sin n\pi x$

(d) $f(x) = -1 + \frac{2}{\pi}\sum_{n=1}^{\infty}\frac{(-1)^{n+1}}{n}\sin n\pi x$

[ESE Prelims : 2023]

9.4 Let $f(x) = \begin{cases} 1, & |x| < a \\ 0, & |x| > a, \end{cases}$ and

$g(x) = \begin{cases} 1, & 0 < x < a \\ 0, & \text{otherwise,} \end{cases}$ then the Fourier transform of $3f(x) - 2g(x)$ is

(a) $\sqrt{\frac{2}{\pi}}\left\{\frac{3\sin\omega a}{\omega} + \left(\frac{1 - e^{i\omega a}}{i\omega}\right)\right\}$

(b) $\sqrt{\frac{2}{\pi}}\left\{\frac{3\sin\omega a}{\omega} - \left(\frac{1 + e^{-i\omega a}}{i\omega}\right)\right\}$

(c) $\sqrt{\frac{2}{\pi}}\left\{\frac{-3\sin\omega a}{\omega}-\left(\frac{1-e^{-i\omega a}}{i\omega}\right)\right\}$

(d) $\sqrt{\frac{2}{\pi}}\left\{\frac{3\sin\omega a}{\omega}-\left(\frac{1-e^{-i\omega a}}{i\omega}\right)\right\}$

[ESE Prelims : 2023]

9.5 The continuous-time signal $f(t) = e^{-2\omega t}$, where ω is a real constant, is sampled when $t \geq 0$ at intervals T. What is the z transform of the resulting sequence of samples?

(a) $\frac{z}{z-e^{-2\omega T}}$ (b) $\frac{z}{1-e^{-2\omega T}}$

(c) $\frac{z}{z-e^{-\omega T}}$ (d) $\frac{z}{z-e^{2\omega T}}$

[ESE Prelims : 2024]

9.6 If $(z) = \frac{z}{z^2 - z + 1}$, then the inverse z transform of $Y(z)$ is

(a) $\sqrt{\frac{1}{3}}\sin\frac{1}{3}k\pi$ (b) $2\sqrt{\frac{1}{3}}\sin\frac{1}{3}k\pi$

(c) $2\sqrt{\frac{1}{3}}\cos\frac{1}{3}k\pi$ (d) $2\sqrt{\frac{1}{3}}\sin k\pi$

[ESE Prelims : 2024]

■■■■

Engineering Mathematics and Numerical Analysis: Answer Key

1.1 (d)	1.2 (b)	1.3 (b)	1.4 (b)	1.5 (c)	1.6 (b)	1.7 (a)	1.8 (b)
1.9 (a)	1.10 (a)	1.11 (a)	1.12 (b)	2.1 (a)	2.2 (a)	2.3 (b)	2.4 (c)
2.5 (a)	2.6 (d)	2.7 (b)	2.8 (a)	2.9 (a)	2.10 (c)	2.11 (c)	2.12 (d)
2.13 (c)	2.14 (c)	2.15 (c)	2.16 (a)	2.17 (a)	2.18 (d)	2.19 (c)	2.20 (b)
2.21 (a)	2.22 (b)	2.23 (c)	2.24 (c)	2.25 (b)	2.26 (c)	2.27 (d)	2.28 (d)
3.1 (b)	3.2 (a)	3.3 (a)	3.4 (d)	3.5 (a)	3.6 (b)	4.1 (a)	4.2 (d)
4.3 (d)	4.4 (b)	4.5 (b)	4.6 (d)	4.7 (c)	4.8 (a)	4.9 (c)	5.1 (d)
5.2 (d)	5.3 (b)	5.4 (a)	5.5 (a)	5.6 (c)	5.7 (b)	5.8 (a)	5.9 (a)
5.10 (d)	6.1 (c)	6.2 (a)	6.3 (c)	6.4 (d)	6.5 (c)	6.6 (d)	6.7 (b)
6.8 (c)	6.9 (d)	6.10 (a)	7.1 (c)	7.2 (a)	7.3 (a)	7.4 (b)	8.1 (b)
8.2 (a)	8.3 (c)	9.1 (a)	9.2 (d)	9.3 (c)	9.4 (d)	9.5 (a)	9.6 (b)

EXPLANATIONS

1.1 (d)

$$x + y + z = 4 \quad ...(1)$$
$$x - y + z = 0 \quad ...(2)$$
$$2x + y + z = 5 \quad ...(3)$$

Adding (1) and (2) & (2) and (3) gives $2x + 2z = 4$ and $3x + 2z = 5$ which gives $x = 1$, $z = 1$ and $y = 2$

Alt: Option (b) can be eliminated since they do not satisfy 1st condition. Only (d) satisfies 3rd equation.

Given equations,

$$x + y + z = 4 \quad ...(1)$$
$$x - y + z = 0 \quad ...(2)$$
$$2x + y + z = 5 \quad ...(3)$$

Alternate :

$$\begin{bmatrix} 1 & 1 & 1 \\ 1 & -1 & 1 \\ 2 & 1 & 1 \end{bmatrix}\begin{bmatrix} x \\ y \\ z \end{bmatrix} = \begin{bmatrix} 4 \\ 0 \\ 5 \end{bmatrix}$$

Matrix equation, $AX = B$

$$\text{Augmented matrix} = [A \;\; B] = \begin{bmatrix} 1 & 1 & 1 & 4 \\ 1 & -1 & 1 & 0 \\ 2 & 1 & 1 & 5 \end{bmatrix}$$

Apply, $R_2 \rightarrow R_2 - R_1$

$R_3 \rightarrow R_3 - 2R_1$

$$[A \;\; B] = \begin{bmatrix} 1 & 1 & 1 & -4 \\ 0 & -2 & 0 & -4 \\ 0 & -1 & -1 & -3 \end{bmatrix}$$

Apply, $R_3 \rightarrow R_3 - \left(\frac{1}{2}\right)R_2$

$$[A \;\; B] = \begin{bmatrix} 1 & 1 & 1 & 4 \\ 0 & -2 & 0 & -4 \\ 0 & 0 & -1 & -1 \end{bmatrix}$$

Matrix equation, $AX = B$

$$\begin{bmatrix} 1 & 1 & 1 \\ 0 & -2 & 0 \\ 0 & 0 & -1 \end{bmatrix}\begin{bmatrix} x \\ y \\ z \end{bmatrix} = \begin{bmatrix} 4 \\ -4 \\ -1 \end{bmatrix}$$

Modified equations,

$$x + y + z = 4 \quad ...(4)$$
$$-2y = -4 \quad ...(5)$$
$$-z = -1$$

From equation (5), $y = 2$

From equation (y), $z = 1$

Therefore, $x + 2 + 1 = 4$

$$x = 1$$

1.2 (b)

Given, $A = \begin{bmatrix} 1 & 2 \\ 0 & 2 \end{bmatrix}$

Eigen values of A are 1, 2.

Let $\begin{bmatrix} 1 \\ a \end{bmatrix}$ be eigen vector for $\lambda_1 = 1$ eigen value

and $\begin{bmatrix} 1 \\ b \end{bmatrix}$ be eigen vector for $\lambda_2 = 2$ eigen value

$$Ax = \lambda x$$

$$\begin{bmatrix} 1 & 2 \\ 0 & 2 \end{bmatrix}\begin{bmatrix} 1 \\ a \end{bmatrix} = 1\begin{bmatrix} 1 \\ a \end{bmatrix}$$

$$1 + 2a = 1$$
$$2a = 0$$
$$a = 0$$

$$\begin{bmatrix} 1 & 2 \\ 0 & 2 \end{bmatrix}\begin{bmatrix} 1 \\ b \end{bmatrix} = 2\begin{bmatrix} 1 \\ b \end{bmatrix}$$

$$1 + 2b = 2$$
$$2b = 2b$$
$$b = \frac{1}{2}$$

$$\therefore \quad a + b = \frac{1}{2}$$

ADDITIONAL INFORMATION

In a triangular and diagonal matrix, eigen values are diagonal elements themselves.

1.3 (b)

Let λ_1 and λ_2 be the eigen values of matrix $\begin{bmatrix} 1 & 2 \\ 0 & 2 \end{bmatrix}$

$$\lambda_1 + \lambda_2 = \text{Trace}(A)$$
$$\lambda_1 + \lambda_2 = 7 \quad ...(i)$$

and $\lambda_1 \lambda_2 = |A| = ad - bc$

$$\lambda_1 \lambda_2 = 10 \quad ...(ii)$$

From equations (i) and (ii),

$$\lambda_1 = 5, \quad \lambda_2 = 2$$

Hence, $\lambda_{min} = 2$

Alternative : $A = \begin{bmatrix} 4 & 2 \\ 1 & 3 \end{bmatrix}$

$$[A - \lambda I] = 0$$

$$\begin{bmatrix} 4-\lambda & 2 \\ 1 & 3-\lambda \end{bmatrix} = 0$$

$$\Rightarrow \lambda^2 - 7\lambda + 10 = 0$$

$$\lambda = 2, 5$$

$$\lambda_{min} = 2$$

ADDITIONAL INFORMATION

- Sum of eigen values = Trace of A = Sum of diagonal elements.
- Product of eigen values = $|A|$.

1.4 (b)

$$A = \begin{bmatrix} 1 & 4 \\ 2 & 3 \end{bmatrix}$$

Characteristic equation of A is $|A - \lambda I| = 0$

$$\begin{vmatrix} (1-\lambda) & 4 \\ 2 & (3-\lambda) \end{vmatrix} = 0$$

$$(1-\lambda)(3-\lambda) - 8 = 0$$

$$3 - \lambda - 3\lambda + \lambda^2 - 8 = 0$$

$$\lambda^2 - 4\lambda - 5 = 0$$

By Cayley-Hamilton theorem we can replace λ by A

Hence, we have

$$A^2 - 4A - 5I = 0 \quad \text{...(i)}$$

Now given expression is

$A^5 - 4A^4 - 7A^3 + 11A^2 - A - 10I$

$= A^3(A^2 - 4A - 5I) - 2A^3 + 11A^2 - A - 10I$

$= A^3(0) - 2A(A^2 - 4A - 5I) + 3A^2 - 11A - 10I$

$= A^3(0) - 2A(0) + 3(A^2 - 4A - 5I) + A + 5I$

$= A^3(0) - 2A(0) + 3(0) + A + 5I$

$= A + 5I$

ADDITIONAL INFORMATION

Cayley-Hamilton theorem states that every square matrix satisfies its own characteristics equation.

1.5 (c)

Given matrix, $A = \begin{bmatrix} 1 & 2 \\ 4 & 3 \end{bmatrix}$

Characteristic equation $|A - \lambda I|$

$$= \begin{vmatrix} 1-\lambda & 2 \\ 4 & 3-\lambda \end{vmatrix} = 0$$

$$\Rightarrow \lambda^2 - 4\lambda - 5 = 0$$

$$\Rightarrow \lambda = 5, -1$$

$$\therefore \lambda_{max} = 5$$

1.6 (b)

$$\Delta = \begin{vmatrix} p & p^2 & p^3 \\ q & q^2 & q^3 \\ r & r^2 & r^3 \end{vmatrix} - \begin{vmatrix} p & p^2 & 1 \\ q & q^2 & 1 \\ r & r^2 & 1 \end{vmatrix} = 0$$

$$= (pqr)\begin{vmatrix} 1 & p & p^2 \\ 1 & q & q^2 \\ 1 & r & r^2 \end{vmatrix} - \begin{vmatrix} 1 & p & p^2 \\ 1 & q & q^2 \\ 1 & r & r^2 \end{vmatrix} = 0$$

$$= (pqr - 1)\begin{vmatrix} 1 & p & p^2 \\ 1 & q & q^2 \\ 1 & r & r^2 \end{vmatrix} = 0$$

$$\Rightarrow (pqr - 1) = 0 \text{ or } (p - q)(q - r)(r - p) = 0$$

$$\therefore pqr = 1$$

1.7 (a)

Given, $A = \begin{bmatrix} -1 & 2 & 3 & -2 \\ 2 & -5 & 1 & 2 \\ 3 & -8 & 5 & 2 \\ 5 & -12 & -1 & 6 \end{bmatrix}$

Apply $R_1 \to -R_1$

$$A = \begin{bmatrix} 1 & -2 & -3 & 2 \\ 2 & -5 & 1 & 2 \\ 3 & -8 & 5 & 2 \\ 5 & -12 & -1 & 6 \end{bmatrix}$$

Apply, $R_2 \to R_2 - 2R_1$; $R_3 \to R_3 - 3R_1$; $R_4 \to R_4 - 5R_1$

$$A = \begin{bmatrix} 1 & -2 & -3 & 2 \\ 0 & -1 & 7 & -2 \\ 0 & -2 & 14 & -4 \\ 0 & -2 & 14 & -4 \end{bmatrix}$$

Apply, $R_4 \to R_4 - 2R_2$; $R_3 \to R_3 - 2R_2$

$$A = \begin{bmatrix} 1 & -2 & -3 & 2 \\ 0 & -1 & 7 & -2 \\ 0 & 0 & 0 & 0 \\ 0 & 0 & 0 & 0 \end{bmatrix}$$

So, rank $[A] = 2$

1.8 (b)

Given, $A = \begin{bmatrix} 1 & 3 & 2 \\ 2 & 0 & -1 \\ 1 & 2 & 3 \end{bmatrix}$

Characteristic equation, $|A - \lambda I| = 0$

$$\begin{vmatrix} 1-\lambda & 3 & 2 \\ 2 & -\lambda & -1 \\ 1 & 2 & 3-\lambda \end{vmatrix} = 0$$

$(1-\lambda)(-3\lambda + \lambda^2 + 2) - 3(6 - 2\lambda + 1) + 2(4 + \lambda) = 0$

$(1-\lambda)(\lambda^2 - 3\lambda + 2) - 3(7 - 2\lambda) + 2(4 + \lambda) = 0$

$\lambda^3 - 4\lambda^2 - 3\lambda + 11 = 0$

By C-H theorem replace λ by A,

$$A^3 - 4A^2 - 3A + 11I = 0$$

1.9 (a)

By power method

$$AX = \begin{bmatrix} 2 & 3 & 2 \\ 4 & 3 & 5 \\ 3 & 2 & 9 \end{bmatrix}\begin{bmatrix} 1 \\ 1 \\ 1 \end{bmatrix} = \begin{bmatrix} 7 \\ 12 \\ 14 \end{bmatrix} = 14\begin{bmatrix} 0.5 \\ 0.857 \\ 1 \end{bmatrix}$$

Now,

$$\begin{bmatrix} 2 & 3 & 2 \\ 4 & 3 & 5 \\ 3 & 2 & 9 \end{bmatrix}\begin{bmatrix} 0.5 \\ 0.857 \\ 1 \end{bmatrix} = \begin{bmatrix} 5.571 \\ 9.571 \\ 12.214 \end{bmatrix} = 12.214\begin{bmatrix} 0.456 \\ 0.786 \\ 1 \end{bmatrix}$$

Now,

$$\begin{bmatrix} 2 & 3 & 2 \\ 4 & 3 & 5 \\ 3 & 2 & 9 \end{bmatrix}\begin{bmatrix} 0.456 \\ 0.786 \\ 1 \end{bmatrix} = \begin{bmatrix} 5.270 \\ 9.182 \\ 11.940 \end{bmatrix} = 11.94\begin{bmatrix} 0.441 \\ 0.769 \\ 1 \end{bmatrix}$$

Again,

$$\begin{bmatrix} 2 & 3 & 2 \\ 4 & 3 & 5 \\ 3 & 2 & 9 \end{bmatrix}\begin{bmatrix} 0.441 \\ 0.769 \\ 1 \end{bmatrix} = \begin{bmatrix} 5.189 \\ 9.071 \\ 11.861 \end{bmatrix} = 11.86\begin{bmatrix} 0.437 \\ 0.764 \\ 1 \end{bmatrix}$$

Again,

$$\begin{bmatrix} 2 & 3 & 2 \\ 4 & 3 & 5 \\ 3 & 2 & 9 \end{bmatrix}\begin{bmatrix} 0.437 \\ 0.764 \\ 1 \end{bmatrix} = \begin{bmatrix} 5.166 \\ 9.040 \\ 11.839 \end{bmatrix} = 11.839\begin{bmatrix} 0.436 \\ 0.7635 \\ 1 \end{bmatrix}$$

Hence, Largest eigen value = 11.839 ≈ 11.84

ADDITIONAL INFORMATION

Power method (also know as the power iteration) is an eigen value algorithm; given a diagonalizable matrix A, the algorithm will produce a number λ, which is the greatest (in absolute value) eigen value of A, and a non vector of λ, that is $4V = \lambda V$.

1.10 (a)

$$A = \begin{bmatrix} 1 & 3 & 4 \\ 3 & 2 & -1 \\ 4 & -1 & 1 \end{bmatrix}$$

Here, $a_{12} = 3,\ a_{13} = 4$

so, $\tan\alpha = \dfrac{a_{13}}{a_{12}} = \dfrac{4}{3}$

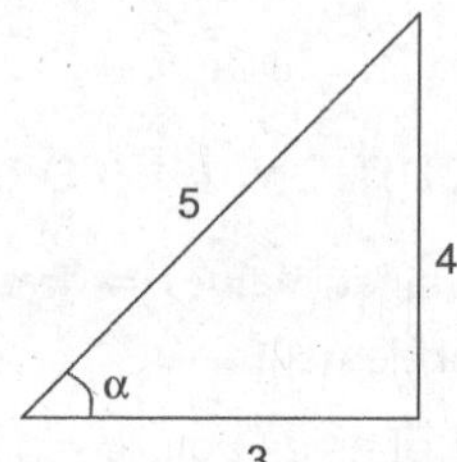

$\Rightarrow \quad \sin\alpha = \dfrac{4}{5}, \quad \cos\alpha = \dfrac{3}{5},$

So, Orthogonal matrix is

$$P = \begin{bmatrix} 1 & 0 & 0 \\ 0 & -\cos\alpha & -\sin\alpha \\ 0 & -\sin\alpha & \cos\alpha \end{bmatrix}$$

i.e. $P = \begin{bmatrix} 1 & 0 & 0 \\ 0 & -3/5 & -4/5 \\ 0 & -4/5 & 3/5 \end{bmatrix}$

and $P^{-1} = P^T = \begin{bmatrix} 1 & 0 & 0 \\ 0 & -3/5 & -4/5 \\ 0 & -4/5 & 3/5 \end{bmatrix}$

Now, $P^{-1}AP$

$$= \begin{bmatrix} 1 & 0 & 0 \\ 0 & -3/5 & -4/5 \\ 0 & -4/5 & 3/5 \end{bmatrix}\begin{bmatrix} 1 & 3 & 4 \\ 3 & 2 & -1 \\ 4 & -1 & 1 \end{bmatrix}\begin{bmatrix} 1 & 0 & 0 \\ 0 & -3/5 & -4/5 \\ 0 & -4/5 & 3/5 \end{bmatrix}$$

$$= \frac{1}{25}\begin{bmatrix} 5 & 0 & 0 \\ 0 & -3 & -4 \\ 0 & -4 & 3 \end{bmatrix}\begin{bmatrix} 1 & 3 & 4 \\ 3 & 2 & -1 \\ 4 & -1 & +1 \end{bmatrix}\begin{bmatrix} 5 & 0 & 0 \\ 0 & -3 & -4 \\ 0 & -4 & 3 \end{bmatrix}$$

$$= \frac{1}{25}\begin{bmatrix} 5 & 15 & 20 \\ -25 & -2 & -1 \\ 0 & -11 & 7 \end{bmatrix}\begin{bmatrix} 5 & 0 & 0 \\ 0 & -3 & -4 \\ 0 & -4 & 3 \end{bmatrix}$$

$$= \frac{1}{25}\begin{bmatrix} 25 & -125 & 0 \\ -125 & 10 & 5 \\ 0 & 5 & 65 \end{bmatrix} = \begin{bmatrix} 1 & -5 & 0 \\ -5 & 2/5 & 1/5 \\ 0 & 1/5 & 13/5 \end{bmatrix}$$

Which is in tridiagonal form, hence option (a) will be the answer.

ADDITIONAL INFORMATION

A tridiagonal matrix is a band matrix that has non zero elements only on the main diagonal, the subdiagonal/lower diagonal (the first diagonal below this), and the superdiagonal/upper diagonal (the first diagonalabove the main diagonal).

1.11 (a)

Using $AX_1 = \lambda_1 X_1 \Rightarrow \lambda_1 = 2$

and $AX_2 = \lambda_2 X_2 \Rightarrow \lambda_2 = 1$

and $AX_3 = \lambda_3 X_3 \Rightarrow \lambda_3 = -1$

∴ By diagonalization

$$D = P^{-1}AP$$

$$D = \begin{bmatrix} 1 & 1 & -1 \\ 0 & -1 & 1 \\ -1 & 0 & 1 \end{bmatrix}\begin{bmatrix} 2 & 1 & -1 \\ 3 & 2 & -3 \\ 3 & 1 & -2 \end{bmatrix}\begin{bmatrix} 1 & 1 & 0 \\ 1 & 0 & 1 \\ 1 & 1 & 1 \end{bmatrix}$$

$$D = \begin{bmatrix} 2 & 0 & 0 \\ 0 & 1 & 0 \\ 0 & 0 & -1 \end{bmatrix}$$

By diagonalization

$$D = P^{-1}AP$$

$$A^5 = PD^5P^{-1}$$

$$= \begin{bmatrix} 1 & 1 & 0 \\ 1 & 0 & 1 \\ 1 & 1 & 1 \end{bmatrix}\begin{bmatrix} 2^5 & 0 & 0 \\ 0 & 1^5 & 0 \\ 0 & 0 & (-1) \end{bmatrix}\begin{bmatrix} 1 & 1 & -1 \\ 0 & -1 & 1 \\ -1 & 0 & 1 \end{bmatrix}$$

$$A^5 = \begin{bmatrix} 32 & 1 & 0 \\ 32 & 0 & -1 \\ 32 & 1 & -1 \end{bmatrix}\begin{bmatrix} 1 & 1 & -1 \\ 0 & -1 & 1 \\ -1 & 0 & 1 \end{bmatrix} = \begin{bmatrix} 32 & 31 & -31 \\ 33 & 32 & -33 \\ 33 & 31 & -32 \end{bmatrix}$$

1.12 (b)

Given quadratic form,

$$P(x) = 4x_1^2 + 4x_2^2 + x_3^2 - 2x_1x_2$$

Symmetric matrix,

$$A = \begin{bmatrix} 4 & -1 & 0 \\ -1 & 4 & 0 \\ 0 & 0 & 1 \end{bmatrix}$$

Eigen values of A are

$$\lambda = 1, 3, 5$$

Orthogonal form of given quadratic form is given by y^TDy

$$y^TDy = [y_1\ y_2\ y_3]\begin{bmatrix} 1 & 0 & 0 \\ 0 & 5 & 0 \\ 0 & 0 & 3 \end{bmatrix}\begin{bmatrix} y_1 \\ y_2 \\ y_3 \end{bmatrix}$$

$$= y_1^2 + 5y_2^2 + 3y_3^2$$

2.1 (a)

Given, $$f(x) = \left(\frac{x^3}{3}\right) - x \quad \text{...(i)}$$

Differentiate equation (i),

$$f'(x) = x^2 - 1 \quad \text{...(ii)}$$

Critical value,

$$f'(x) = 0$$

$$x^2 - 1 = 0$$

$$x = +1, \ -1$$

Differentiate equation (ii),

$$f''(x) = 2x$$

$$f''(x)\big|_{x=1} = 2 \times 1 = 2 > 0$$

$$f''(x)\big|_{x=-1} = 2 \times (-1) = -2 < 0$$

$f''(x)\big|_{x=1} > 0$ which means minimum value of the function exits for $x = 1$.

Alternate : This question can be directly solved by putting given values of x.

2.2 (a)

Given, $$I = \int_0^{2\pi}\left(\frac{3}{9+\sin^2\theta}\right)d\theta$$

Divide numerator and denominator by $\cos^2\theta$

$$I = \int_0^{2\pi}\frac{3\sec^2\theta}{9\sec^2\theta+\tan^2\theta}d\theta = 4\int_0^{\pi/2}\frac{3\sec^2\theta}{9\sec^2\theta+\tan^2\theta}d\theta$$

Using property :

$$\int_0^{2a} f(x)dx = \begin{cases} 2\int_0^a f(x)dx & ; \text{ if } f(2a-x) = f(x) \\ 0 & ; \text{ if } f(2a-x) = -f(x) \end{cases}$$

$$I = 4\int_0^{\pi/2}\frac{3\sec^2\theta}{9+10\tan^2\theta}d\theta = \frac{12}{10}\int_0^{\pi/2}\frac{\sec^2\theta}{\frac{9}{10}+\tan^2\theta}d\theta$$

Let, $\tan\theta = t$, $\sec^2\theta\, d\theta = dt$

Limits: $\theta = 0 \quad t = 0$; $\theta = \frac{\pi}{2} \quad t \to \infty$

$$I = \frac{12}{10}\times\int_0^{\pi/2}\frac{dt}{\left(\sqrt{\frac{9}{10}}\right)^2+t^2}$$

$$I = \frac{12}{10}\cdot\left[\frac{1}{\frac{3}{\sqrt{10}}}\tan^{-1}\frac{t}{\frac{3}{\sqrt{10}}}\right]_0^{\infty} = \frac{4}{\sqrt{10}}\left[\frac{\pi}{2}-0\right] = \frac{2\pi}{\sqrt{10}}$$

2.3 (b)

Given, $u = \log\left(\frac{x^2+y^2}{x+y}\right)$ is a non-homogenous function.

$f(u) = e^4 = \frac{x^2+y^2}{x+y}$ is a homogenous function of degree $n = 1$.

By Euler theorem,

$$xu_x + y\,u_y = n\frac{f(u)}{f'(u)} = 1\cdot\frac{e^u}{e^u} = 1$$

2.4 (c)

Given, $f(x, y) = 2 + 2x + 2y - x^2 - y^2$

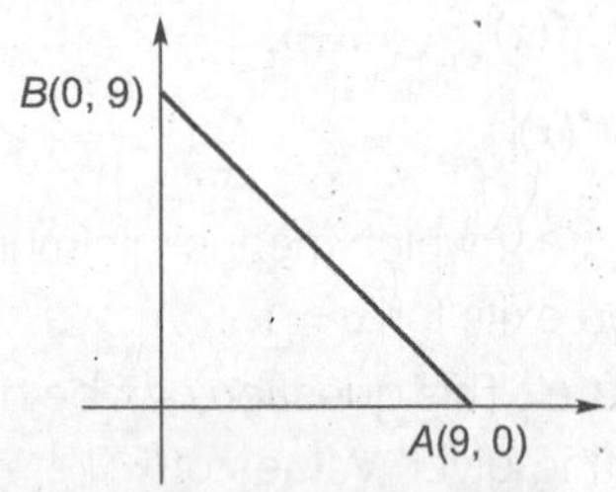

can have maximum (or) minimum values at critical values inside ΔOAB (or) on its boundary.

$$f(x, y) = 2 + 2x + 2y - x^2 - y^2$$
$$f_x = 2 - 2x$$
$$f_y = 2 - 2y$$
$$f_x = 0 \text{ and } f_y = 0$$

gives critical point (1, 1)

At (1, 1) $r = f_{xx} = -2$

$s = f_{xy} = 0$

$t = f_{yy} = -2$

$rt - s^2 = 4 > 0$

and $r = -2 < 0$ max at (1, 1)

∴ Maximum value $f(1, 1) = 4$ at critical point.

On OA, $y = 0$

$$g(x) = f(x, 0) = 2 + 2x - x^2 \quad \{0 \le x \le 9\}$$

Absolute maximum,

$g'(x) = 2 - 2x$

$g'(x) = 0$

$g(0) = 2$

$g(9) = -61$

$g(1) = 3$

∴ Absolute maximum = 3, absolute minimum = –61.

On OB, $x = 0$

$$f(0, y) = 2 + 2y - y^2 \quad \{0 \le y \le 9\}$$

Absolute maximum = 3,

Absolute minimum = –61

Along AB, $y = 9 - x$

$f(x, y) = f(x, 9 - x)$

$= -61 + 18x - 2x^2$

$= h(x) \quad \{0 \le x \le 9\}$

$h'(x) = 18 - 4x = 0$

$x = \frac{9}{2}$

$h(0) = -61$

$h(9) = -1$

$h\left(\frac{9}{2}\right) = -\frac{41}{2}$

∴ Absolute maximum = 4

Absolute minimum = –61

2.5 (a)

Given, $\Delta^{10}\left[(1-ax)(1-bx^2)(1-cx^3)(1-dx^4)\right]$

In case of continuous function, forward difference operator 'Δ' and differentiation operator $\frac{d}{dx}$ are similar operations and expression given in the bracket is a polynomial of degree 10 with coefficient (abcd) as the coefficient of x^{10}.

So we can concluded that

$$\Delta^{10}[(1 - ax)(1 - bx)(1 - cx)(1 - dx)]$$
$$= \frac{d^{10}}{dx^{10}}\left[(abcd)x^{10} + ...\right]$$
$$= abcd(10!) \quad \text{(other terms = 0)}$$

2.6 (d)

Given, $u = \log_e\frac{x^4+y^4}{x+y}$ is non-homogenous function

Can be written as $e^u = \frac{x^4+y^4}{x+y}$

is a homogeneous function of degree

$n = 4 - 1 = 3$

and $\phi(u) = e^u$

By Euler theorem

$$x\frac{\partial u}{dx} + y\frac{\partial u}{dy} = n\frac{\phi(u)}{\phi'(u)} = 3\frac{e^u}{e^u} = 3$$

2.7 (b)

Given, $I = \int_0^1\int_0^x (x^2 + y^2)\,dydx$

$$= \int_0^1 \left(x^2 y + \frac{y^3}{3} \right)_0^x dx = \int_0^1 \frac{4x^3}{3} dx = \left[\frac{x^4}{3} \right]_0^1 = \frac{1}{3}$$

2.8 (a)

Given, $x = uv,\ y = \dfrac{u+v}{u-v}$

$$\frac{\partial(x,y)}{\partial(u,v)} = \begin{vmatrix} x_u & x_v \\ y_u & y_v \end{vmatrix} = \begin{vmatrix} v & u \\ \dfrac{-2v}{(u-v)^2} & \dfrac{2u}{(u-v)^2} \end{vmatrix}$$

$$= \frac{4uv}{(u-v)^2}$$

$$\therefore \quad \frac{\partial(u,v)}{\partial(x,y)} = \frac{1}{\dfrac{\partial(x,y)}{\partial(u,v)}} = \frac{(u-v)^2}{4uv}$$

2.9 (a)

Given, $u = x^3 + y^3$ ...(i)

where, $x = a\cos t$ ...(ii)

$y = b\sin t$ ...(iii)

$$\frac{dy}{dt} = \frac{du}{dx} \times \frac{dx}{dt} + \frac{du}{dy} \times \frac{dy}{dt}$$

$= 3x^2(-a\sin t) + 3y^2(b\cos t)$

$= -3a^3 \cos^2 t \sin t + 3b^3 \sin^2 t \cos t$

2.10 (c)

Given, $f(x) = e^{\sin x}$

$\Rightarrow f(0) = e^0 = 1$

$f'(x) = e^{\sin x} (\cos x)$

$\Rightarrow f'(0) = 1$

$f''(x) = e^{\sin x} \cos^2 x + e^{\sin x} (-\sin x)$

$f''(0) = 1$

$= e^{\sin x} [\cos^2 x - \sin x]$

$f'''(x) = e^{\sin x} [-\sin 2x - \cos x]$

$+ e^{\sin x} [\cos^3 x - \sin x \cos x]$

$f'''(x) = -1 + 1 = 0$

$f^{iv}(x) = e^{\sin x} [-2\cos 2x + \sin x]$

$+ e^{\sin x} \cos x [-\sin 2x - \cos x]$

$$+e^{\sin x}\left[3\cos^2 x(-\sin x) - \frac{\cos 2x}{2} \times 2\right]$$

$+ e^{\sin x} [\cos^4 x - \sin x \cos^2 x]$

$f^{iv}(0) = -2 - 1 - 1 + 1 = -3$

$\vdots$

Substitute in Maclaurin Series

$$e^{\sin x} = 1 + x + \frac{x^2}{2} + \frac{x^3}{3!}(0) + \frac{x^4}{4!} \times (-3) + \ldots\ldots\ldots\ldots$$

$$= 1 + x + \frac{x^2}{2} - \frac{x^4}{8} + \ldots\ldots\ldots\ldots\ldots$$

2.11 (c)

Given, $f(x) = \displaystyle\sum_1^n \frac{1}{(x+3)(x+4)}$

$$\text{expand } f(x) = \left[\frac{1}{4.5} + \frac{1}{5.6} + \frac{1}{6.7} + \ldots + \frac{1}{(n+3)(n+4)}\right]$$

$$= \left[\left(\frac{1}{4} - \frac{1}{5}\right) + \left(\frac{1}{5} - \frac{1}{6}\right) + \left(\frac{1}{6} - \frac{1}{7}\right) + \ldots + \left(\frac{1}{n+3} - \frac{1}{n-4,}\right)\right]$$

$$= \frac{1}{4} - \frac{1}{n+4} = \frac{n}{4(n+4)}$$

2.12 (d)

$$z = \sqrt{4 - x^2}$$

$$\frac{\partial z}{\partial x} = \frac{-x}{\sqrt{4-x^2}}, \quad \frac{\partial z}{\partial y} = 0$$

Required surface area

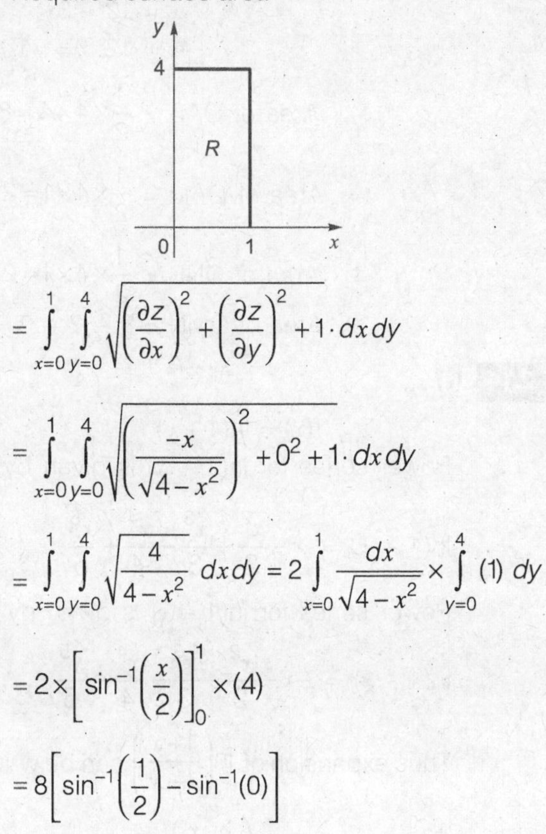

$$= \int_{x=0}^{1} \int_{y=0}^{4} \sqrt{\left(\frac{\partial z}{\partial x}\right)^2 + \left(\frac{\partial z}{\partial y}\right)^2 + 1} \cdot dx\,dy$$

$$= \int_{x=0}^{1} \int_{y=0}^{4} \sqrt{\left(\frac{-x}{\sqrt{4-x^2}}\right)^2 + 0^2 + 1} \cdot dx\,dy$$

$$= \int_{x=0}^{1} \int_{y=0}^{4} \sqrt{\frac{4}{4-x^2}}\, dx\,dy = 2\int_{x=0}^{1} \frac{dx}{\sqrt{4-x^2}} \times \int_{y=0}^{4} (1)\, dy$$

$$= 2 \times \left[\sin^{-1}\left(\frac{x}{2}\right)\right]_0^1 \times (4)$$

$$= 8\left[\sin^{-1}\left(\frac{1}{2}\right) - \sin^{-1}(0)\right]$$

$$= 8 \cdot \left(\frac{\pi}{6} - 0\right) = \frac{4}{3} \cdot \pi = \frac{4\pi}{3}$$

2.13 (c)

Given equation,

$$y = 4 - x$$
$$y = 3x$$
$$3y = x$$

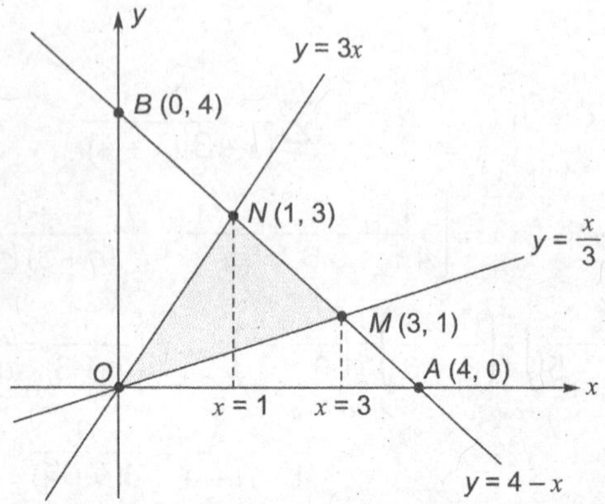

Points of intersection:

N:– $3x = 4 - x$

$4x = 4 \Rightarrow x = 1$

M:– $\frac{x}{3} = 4 - x$

$x = 12 - 3x$

$4x = 12 \Rightarrow x = 3$

$$\text{Area of OAB} = \frac{1}{2}\times 4\times 4 = 8$$

$$\text{Area of OAM} = \frac{1}{2}\times 4\times 1 = 2$$

$$\text{Area of ONB} = \frac{1}{2}\times 4\times 1 = 2$$

$$\text{Area of OMN} = 8 - 2 - 2 = 4$$

2.14 (c)

$$f(x) = \ln(1 + x)$$

Power series for $\ln(1 + x)$ is given by

$$\ln(1 + x) = x - \frac{x^2}{2} + \frac{x^3}{3} - \frac{x^4}{4} + \frac{x^5}{5} ....$$

Power series for $\ln(1 - x)$ is given by

$$\ln(1 - x) = -x - \frac{x^2}{2} - \frac{x^3}{3} - \frac{x^4}{4} - \frac{x^5}{5} ....$$

Thus expansion of $\ln\left(\frac{1+x}{1-x}\right)$ can be written as

$$\ln\left(\frac{1+x}{1-x}\right) = \ln(1 + x) - \ln(1 - x)$$

$$= \left(x - \frac{x^2}{2} + \frac{x^3}{3} - \frac{x^4}{4} + \frac{x^5}{5} ....\right) - \left(-x - \frac{x^2}{2} - \frac{x^3}{3} - \frac{x^4}{4} - \frac{x^5}{5} ....\right)$$

$$= 2\left(x + \frac{x^3}{3} + \frac{x^5}{5} ....\right)$$

By Maclaurin series

$$f(x) = f(0) + x \cdot f'(0) + \frac{x^2}{2} \cdot f''(0) + ...$$

2.15 (c)

Given function, $y = 3x^2 + 4x + 1$

By mean value theorem

$$f(c) = \frac{1}{b-a}\int_a^b y dx = \frac{1}{2-(-1)}\int_{-1}^{2} (3x^2 + 4x + 1) dx$$

$$= \frac{1}{3}\left[x^3 + 2x^2 + x\right]_{-1}^{2}$$

$$= \frac{1}{3}\left(2^3 + 1 + 2\left(2^2 - 1\right) + 3\right) = 6$$

2.16 (a)

Given: $x = 0$, $y(0) = 1$

$$y' = x - y^2$$

$$\Rightarrow \quad y'(0) = 0 - [y(0)]^2 = -1$$

$$y'' = 1 - 2yy'$$

$$\Rightarrow \quad y''(0) = 1 - 2\, y(0)$$

$$y'(0) = 1 - 2(1)(-1) = 3$$

$$y''' = -2\,[yy'' + (y')^2]$$

$$\Rightarrow \quad y'''(0) = -2\,[y(0)y''(0) + [y'(0)]^2]$$

$$= -2\,[1 \times 3 + (-1)^2] = -8$$

Taylor series expansion about $x = 0$ is given as,

$$y(x) = y(0) + xy'(0) + \frac{x^2}{2!}y''(0) + \frac{x^3}{3!}y'''(0) + ...$$

$$= 1 + x(-1) + \frac{x^2}{2!}(3) + \frac{x^3}{3!}(-8) + ...$$

$$y(x) = 1 - x + \frac{3}{2}x^2 - \frac{4}{3}x^3 + ...$$

Hence, $y(0.1) = 1 - 0.1 + \frac{3}{2}(0.1)^2 - \frac{4}{3}(0.1)^3 + ...$

[Neglecting higher power]

$= 1 - 0.1 + 1.5 \times 0.01 - 1.3333 \times 0.001$

$= 1 - 0.1 + 0.015 - 0.0013$

$= 1.015 - 0.1013 = 0.9137$

2.17 (a)

Given: $\frac{x}{5} = \sqrt{1 + \left(\frac{y}{2}\right)^2}$

$$\left(\frac{x}{5}\right)^2 = 1+\left(\frac{y}{2}\right)^2$$

$$\frac{x^2}{25} = 1+\frac{y^2}{4}$$

$$\Rightarrow \quad \frac{x^2}{25}-\frac{y^2}{4} = 1$$

$$\frac{x^2}{a^2}-\frac{x^2}{b^2} = 1$$

[General equation of hyperbola]

2.18 (d)

Comparing with $\int_a^b y dx$, for given integral,

$y=\sqrt{16-x^2}$

It can also be written as

$$x^2 + y^2 = 16$$

This is equation of a circle of radius 4 centred at origin as shown below.

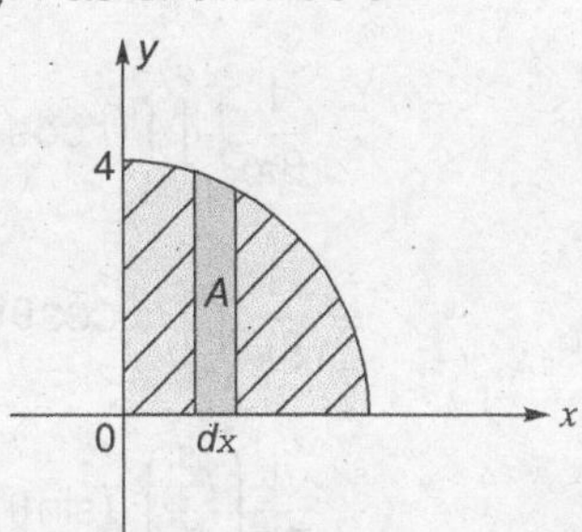

Thus, $\int_0^4 \sqrt{16-x^2}dx$ = Area of circle in I quadrant

$$= \frac{1}{4}[\pi(16)] = 4\pi$$

2.19 (c)

$$I = \int_0^2 \frac{3x}{\sqrt{2x^2+1}}dx$$

Let, $2x^2 + 1 = t$

$$\Rightarrow \quad 4xdx = dt$$

$$xdx = \frac{dt}{4}$$

$$\therefore \int_0^2 \frac{3x}{\sqrt{2x^2+1}}dx = \int_1^9 \frac{3}{\sqrt{t}}\left(\frac{dt}{4}\right) = \frac{3}{4}\int_1^9 \frac{1}{\sqrt{t}}dt$$

$$= \frac{3}{4}\left[\frac{\sqrt{t}}{1/2}\right]_1^9 = \frac{3}{2}(\sqrt{9}-\sqrt{1}) = 3$$

2.20 (b)

Given curve, $x = 2\cos^3\theta$

$y = 2\sin^3\theta$

Here, $\frac{dx}{d\theta} = 6\cos^2\theta\,(-\sin\theta)$ and $\frac{dy}{d\theta} = 6\sin^2\theta(\cos\theta)$

Length of curve within $\theta \in \left(0, \frac{\pi}{2}\right)$

$$L = \int_0^{\pi/2} \sqrt{\left(\frac{dx}{d\theta}\right)^2 + \left(\frac{dy}{d\theta}\right)^2}\, d\theta$$

$$= \int_0^{\pi/2} \sqrt{36\cos^4\theta\sin^2\theta + 36\sin^4\theta\cos^2\theta}\, d\theta$$

$$= \int_0^{\pi/2} \sqrt{36\cos^4\theta\sin^2\theta(\cos^2\theta+\sin^2\theta)}\, d\theta$$

$$= 6\int_0^{\pi/2} \cos\theta\cdot\sin\theta\, d\theta$$

$$= 3\int_0^{\pi/2} \sin 2\theta\, d\theta = -\frac{3}{2}[\cos 2\theta]_0^{\pi/2}$$

$$= \frac{-3}{2}(\cos\pi - \cos 0) = 3$$

2.21 (a)

$$\vec{F} = (x+z)\hat{i} + a(y+z)\hat{j} + b(x+y)\hat{k}$$

$$\text{Curl}\,\vec{F} = \vec{\nabla}\times\vec{F} = \begin{bmatrix} \hat{i} & \hat{j} & \hat{k} \\ \frac{\partial}{\partial x} & \frac{\partial}{\partial y} & \frac{\partial}{\partial z} \\ x+z & ay+az & bx+by \end{bmatrix}$$

$$= \hat{i}(b-a) - \hat{j}(b-1) + \hat{k}(0-0)$$

$$= (b-a)\hat{i} + (1-b)\hat{j} + 0\hat{k}$$

For conservative field Curl $\vec{F} = 0$

$$(b-a)\hat{i} + (1-b)\hat{j} + 0\hat{k} = 0\hat{i} + 0\hat{j} + 0\hat{k}$$

$\Rightarrow b = a$ and $b = 1$. So, $a = b = 1$

Hence, Answer is (a).

2.22 (b)

When $x^2 + y^2 = 1$, $z = 1 - x^2 - y^2 = 1 - 1 = 0$

i.e., Boundary curve is $x^2 + y^2 = 1$ and $z = 0$.

Let $x = \cos t$, $y = \sin t$, $0 \le t \le 2\pi$ then

$$r(t) = (\cos t)\hat{i} + (\sin t)\hat{j}$$

$$\Rightarrow \quad r'(t) = -(\sin t)\hat{i} + (\cos t)\hat{j}$$

Again $\vec{F} = x^3\hat{i} + (x+y-z)\hat{j} + (yz)\hat{k}$

$$= (\cos^3 t)\hat{i} + (\cos t + \sin t - 0)\hat{j} + (0)\hat{k}$$

So, $\vec{F}(r(t))\cdot r'(t) = -\sin t\cos^3 t + \cos^2 t + \cos t\sin t + 0$

So, by Stoke's theorem

$$\iint_S (\text{Curl } \vec{f}\cdot d\vec{s}) = \int_C (\vec{F}(r(t))\cdot r'(t))dt \approx \int_C \vec{F}\cdot d\vec{r}$$

$$= \int_{t=0}^{2\pi} (-\sin t\cos^3 t + \cos^2 t + \cos t\sin t)dt$$

$$= \left(\frac{\cos^4 t}{4}\right)_0^{2\pi} + \left(\frac{t}{2} + \frac{\sin 2t}{4}\right)_0^{2\pi} + \left(\frac{\sin^2 t}{2}\right)_0^{2\pi}$$

$$= 0 + \left(\frac{2\pi - 0}{2}\right) + 0 + 0 = \pi$$

Hence, correct option is (b).

2.23 (c)

At $x = 1, \quad z = x^2 + y^2$

$\Rightarrow$ $z = 1 + y^2$

Then $\frac{dz}{dy} = 2y$

So $\left(\frac{dz}{dy}\right)_{y=2} = 2(2) = 4$

2.24 (c)

$$f(x, y) = 2 + 2x + 4y - x^2 - y^2 \quad ...(1)$$

$$f_x = 2 - 2x, \quad f_y = 4 - 2y$$

$$f_{xx} = -2 = r, f_{yy} = -2 = t, f_{xy} = 0 = s$$

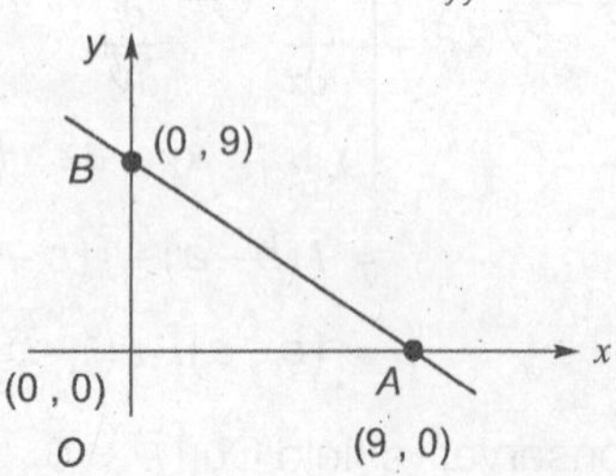

Turning points are

$$\frac{\partial f}{\partial x} = 0, \quad \frac{\partial f}{\partial y} = 0$$

$$2 - 2x = 0, \quad 4 - 2y = 0$$

$$x = 1, \quad y = 2 \text{ i.e. } P(1, 2)$$

Now at $P(1, 2)$; $rt - s^2 > 0$ and $r < 0$ so $P(1, 2)$ is point of maxima.

Hence, Minima will occur at corner points

$$f(0, 0) = 2 + 0 + 0 - 0 - 0 = 2$$

$$f(9, 0) = 2 + 18 + 0 - 81 - 0$$

$$= -61 \Rightarrow \textbf{Answer}$$

$$f(0, 9) = 2 + 0 + 36 - 0 - 81 = -43$$

2.25 (b)

Given solid bounded by $x^2 + y^2 = 4$ above $z = x^2 + y^2$ below xy-plane, $z = 0$ and density ($\delta = 1$).

So, the cylindrical coordinates will be $x = r\cos\theta$, $y = r\sin\theta$, $z = z$, $x^2 + y^2 = r^2$

$R = \{(x, y, z) \mid 0 \le z \le r^2, 0 \le r \le 2, 0 \le \theta \le 2\pi\}$

We know that,

$$\text{mass} = \iiint_R \delta dv = \iiint_0 1\, r\, dr\, d\theta\, dz$$

$$= \int_{\theta=0}^{2\pi}\int_{r=0}^{2}\int_0^{r^2} r\, dz\, dr\, d\theta$$

$$= \int_0^{2\pi}\int_0^2 r(r^2)\, dr\, d\theta$$

$$= \left(\frac{r^4}{4}\right)_0^2 (\theta)_0^{2\pi} = 8\pi$$

Centroid is given by center of mass $(\bar{x}, \bar{y}, \bar{z})$

Center of mass $\bar{x} = \frac{1}{m}\iiint_R x\delta dv$

$$= \frac{1}{8\pi}\int_0^{2\pi}\int_0^2\int_0^{r^2} (r\cos\theta) r\, dz\, d\theta\, dr$$

$$= \frac{1}{8\pi}\int_0^{2\pi}\int_0^2 r^4\cos\theta\, dr\, d\theta$$

$$= \frac{1}{8\pi}\left(\frac{r^5}{5}\right)_0^2 (\sin\theta)_0^{2\pi} = 0$$

and $\bar{y} = \frac{1}{m}\iiint_R y\delta dv$

$$= \frac{1}{8\pi}\int_0^{2\pi}\int_0^2\int_0^{r^2} r\sin\theta\, r\, dz\, dr\, d\theta$$

$$= \frac{1}{8\pi}\int_0^{2\pi}\int_0^2 r^4\sin\theta\, dr\, d\theta = 0$$

and $\bar{z} = \frac{1}{m}\iiint_R z\delta dv$

$$= \frac{1}{8\pi}\int_0^{2\pi}\int_0^2\int_0^{r^2} z\, r\, dz\, dr\, d\theta$$

$$= \frac{1}{8\pi}\int_0^{2\pi}\int_0^2 \left.\frac{z^2}{2}\right]_0^{r^2} dr\, d\theta$$

$$= \frac{1}{8\pi}\int_0^{2\pi}\left(\frac{r^5}{10}\right)_0^2 d\theta = \frac{4}{3}$$

$\therefore$ Centroid $(\bar{x}, \bar{y}, \bar{z}) = \left(0, 0, \frac{4}{3}\right)$

2.26 (c)

$$I = \int_1^2\int_{\frac{1}{y}}^{y} \sqrt{\frac{y}{x}} \cdot e^{\sqrt{xy}} \cdot dx\,dy$$

Put, $\sqrt{xy} = t$ — At $x = \frac{1}{y}$, $t = 1$; At $x = y$, $t = y$

$$\frac{1}{\sqrt{x}}dx = \frac{2dt}{\sqrt{y}} = \int_1^2\left[\int_{x=\frac{1}{y}}^{y} \frac{e^{\sqrt{xy}}}{\sqrt{x}}dx\right]\sqrt{y}\,dy$$

So, $$I = \int_1^2\left(e^t \cdot \frac{2dt}{\sqrt{y}}\right)_1^y \cdot \sqrt{y}\,dy$$

$$= 2\int_{y=1}^{2}\left(e^y - e\right)dy$$

$$= 2\left(e^y - y \cdot e\right)_1^2 = 2\left[\left(e^2 - 2e\right) - \left(e^1 - e\right)\right] = 2e(e-2)$$

2.27 (d)

$$a(t) = -(3\cos t)\hat{i} - (3\sin t)\hat{j} + 2\hat{k};$$

$v(0) = 3\hat{j}$, $r(0) = 4\hat{i}$

Integrating,

$v(t) = -(3\sin t)\hat{i} + (3\cos t)\hat{j} + (2t)\hat{k} + C_1$

Using $v(0) = 3\hat{j} \Rightarrow C_1 = 0$

So, $v(t) = -(3\sin t)\hat{i} + (3\cos t)\hat{j} + (2t)\hat{k}$

Again integrating,

$r(t) = (3\cos t)\hat{i} + (3\sin t)\hat{j} + (t^2)\hat{k} + C_2;$

Using $r(0) = 4\hat{i} \Rightarrow C_2 = \hat{i}$

i.e. $r(t) = (3\cos t + 1)\hat{i} + (3\sin t)\hat{j} + (t^2)\hat{k}$

2.28 (d)

$$f(t) = \begin{cases} \sin at, |t| \le \frac{\pi}{a} \\ 0, |t| > \frac{\pi}{a} \end{cases} = \begin{cases} \sin at, & -\frac{\pi}{a} \le t \le \frac{\pi}{a} \\ 0, & \text{otherwise} \end{cases}$$

$$F\{f(t)\} = \int_{-\infty}^{\infty} e^{-i\omega t} \cdot f(t)dt = \int_{-\frac{\pi}{a}}^{\frac{\pi}{a}} e^{-i\omega t} \cdot \sin at\,dt$$

$$\left\{\text{Using } \int e^{ax}\sin bx\,dx = \frac{e^{ax}}{a^2+b^2}(a\sin bx - b\cos bx)\right\}$$

$$= \left[\frac{e^{-i\omega t}}{-\omega^2 + a^2}(-i\omega\sin at - a\cos at)\right]_{-\frac{\pi}{a}}^{\frac{\pi}{a}}$$

$$= \frac{1}{a^2-\omega^2}\left[\left\{e^{-i\omega\frac{\pi}{a}}(0 - a\cos\pi)\right\} - \left\{e^{i\omega\frac{\pi}{a}}(0 - a\cos(-\pi))\right\}\right]$$

$$= \frac{1}{a^2-\omega^2}\left[ae^{-i\omega\frac{\pi}{a}} - ae^{i\omega\frac{\pi}{a}}\right] = \frac{1}{\omega^2 - a^2}\left[e^{i\omega\frac{\pi}{a}} - e^{-i\omega\frac{\pi}{a}}\right]$$

$$= \frac{1}{\omega^2 - a^2}\left(2a\sin\left(\frac{\pi\omega}{a}\right)\right) = \frac{2ai}{\omega^2 - a^2}\sin\left(\frac{\pi\omega}{a}\right)$$

3.1 (b)

Given : $(1 + y^2)\,dx = (\tan^{-1}y - x)\,dy$

$$\frac{dx}{dy} = \frac{\tan^{-1}y - x}{1+y^2}$$

$$\frac{dx}{dy} + \frac{x}{1+y^2} = \frac{\tan^{-1}y}{1+y^2} \quad \text{....(i)}$$

Which is L.D.E. in x and y so on comparison with

$$\frac{dx}{dy} + Px = Q$$

We have $P = \frac{1}{1+y^2}$

$$Q = \frac{\tan^{-1}y}{1+y^2}$$

$$\text{I.F.} = e^{\int P dy}$$

$$\text{I.F.} = e^{\int P dy} = e^{\int \frac{1}{1+y^2}} = e^{\tan^{-1}y}$$

$$\text{I.F.} = e^{\int P dy} = e^{\int \frac{dy}{1+y^2}} = e^{\tan^{-1}y}$$

Hence solution of eq. (i) is

$$x\,(\text{IF}) = \int Q(IF)dy + c$$

$$xe^{\tan^{-1}y} = \int\left(\frac{\tan^{-1}y}{1+y^2}\right)e^{\tan^{-1}y}dy + c$$

Put $\tan^{-1}y = t$

$$\Rightarrow \frac{dy}{1+y^2} = dt$$

So, $$xe^{\tan^{-1}y} = \int t \cdot e^t\,dt + c$$

$$= te^t - e^t + c$$

$$xe^{\tan^{-1}y} = e^{\tan^{-1}y}\left(\tan^{-1}y - 1\right) + c$$

$$x = \tan^{-1}y - 1 + ce^{-\tan^{-1}y}$$

3.2 (a)

Given surface is $f = \dfrac{z(x+y)}{c(3z+1)}$

Lagrange's subsidiary equation are

$$\frac{dx}{\frac{\partial f}{\partial x}} = \frac{dy}{\frac{\partial f}{\partial y}} = \frac{dz}{\frac{\partial f}{\partial z}}$$

$$\Rightarrow \frac{dx}{\frac{z}{c(3z+1)}} = \frac{dy}{\frac{z}{c(3z+1)}} = \frac{dz}{\frac{x+y}{c(3z+1)^2}}$$

$$\Rightarrow \frac{dx}{z} = \frac{dy}{z} = \left(\frac{3z+1}{x+y}\right)dz \quad \text{...(i)}$$

Taking 1st and 2nd : $\dfrac{dx}{z} = \dfrac{dy}{z}$

$$\Rightarrow \int dx = \int dy + c_1$$

$$\Rightarrow x - y = c_1 \quad \text{...(ii)}$$

Again using (i), $\dfrac{dx+dy}{2z} = \left(\dfrac{3z+1}{x+y}\right)dz$

$$\int (x+y)(dx+dy) = \int (6z^2+2z)dz + c_2$$

$$\Rightarrow \frac{(x+y)^2}{2} = 2z^3 + z^2 + c_2$$

or $(x+y)^2 = 4z^3 + 2z^2 + c_2$

$$x^2 + y^2 + 2xy = 4z^3 + 2z^2 + c_2$$

$$1 + 2xy = 4(1)^3 + 2(1)^2 + c_2$$

$$\Rightarrow 2xy = 5 + c_2 \quad \text{...(iii)}$$

Putting the value of $2xy$ from equation (ii) into equation (iii),

$$\{\because (x-y)^2 = c_1^2 \Rightarrow 2xy = 1 - c_1^2\}$$

By equation (iii),

$$1 - c_1^2 = 5 + c_2$$

$$\Rightarrow c_1^2 + c_2 + 4 = 0$$

$$(x-y)^2 + (x+y)^2 - 4z^3 - 2z^2 + 4 = 0$$

$$2(x^2+y^2) = 2(2z^3 - z^2 - 2)$$

or $x^2 + y^2 = 2z^3 - z^2 - 2$

3.3 (a)

Given, $5\dfrac{dy}{dx} + 2x = 3$

$$\Rightarrow 5\frac{dy}{dx} = 3 - 2x$$

$$\int 5dy = \int (3-2x)dx$$

$$5y = 3x - \frac{2x^2}{2} + C$$

$$y = \frac{3x}{5} - \frac{x^2}{5} + C$$

Boundary condition,

At $x = 2$, $y = \dfrac{7}{5}$

$$\frac{7}{5} = \frac{3\times 2}{5} - \frac{2^2}{5} + C = \frac{6}{5} - \frac{4}{5} + C$$

$$\frac{7}{5} = \frac{2}{5} + C$$

$$C = \frac{5}{5} = 1$$

$$y = \frac{3x}{5} - \frac{x^2}{5} + 1$$

3.4 (d)

Roots of characteristics equations are

$\lambda = 0$ (3 times)

$\lambda = -4$ (2 times)

$\lambda = -1 \pm i2$ (2 times)

$\therefore$ General solution of D.E is $y = y_{CF}$

$$y_{CF} = c_1 + c_2x + c_3x^2 + c_4e^{-4x} + c_5xe^{-4x} + e^{-x}(c_6\cos 2x + c_7\sin 2x) + xe^{-x}(c_8\cos 2x + c_9\sin 2x)$$

3.5 (a)

$$y'' + 4y' + 3y = e^{-x} \text{ with } y(0) = 2,\ y'(0) = 1$$

It is non-homogeneous LDE with constant coefficients.

$\therefore$ General solution is $y = y_{CF} + y_I$

Obtain y_{CF}: Given by $f(m) = m^2 + 4m + 3 = 0$

$$\Rightarrow m = -1, -3$$

$$\therefore y_{CF} = C_1e^{-x} + C_2e^{-3x}$$

Obtain y_{PI}: $y_{PI} = \dfrac{Q(x)}{f(D)} = \dfrac{e^{-x}}{D^2+4D+3}$

$$= x\frac{e^{-x}}{2D+4} = \frac{x}{2}e^{-x}$$

$$\therefore y = y_{CF} + y_{PI}$$

$$= C_1e^{-x} + C_2e^{-3x} + \frac{x}{2}e^{-x}$$

$$y' = -C_1e^{-x} - 3C_2e^{-3x} + \frac{e^{-x}}{2} - \frac{x}{2}e^{-x}$$

Put, $x = 0$, $y = 2$

$$2 = C_1 + C_2$$

Put, $x = 0$, $y' = 1$

$$1 = -C_1 - 3C_2 + \frac{1}{2}$$

Solving, $C_1 + C_2 = 2 \Rightarrow C_2 = 2 - C_1$

$$C_1 + 3(2 - C_1) = -\frac{1}{2}$$

$$\Rightarrow \quad 2C_1 = \frac{13}{2} \Rightarrow C_1 = \frac{13}{4}$$

$$C_2 = 2 - \frac{13}{4} = \frac{-5}{4}$$

$$\therefore \quad y = \frac{13}{4}e^{-x} - \frac{5}{4}e^{-3x} + \frac{x}{2}e^{-x}$$

3.6 (b)

By verifying boundary conditions and D.E with given options

We get $\quad y(t) = \frac{1}{2}(1 + \cos\sqrt{2}t), t > 0$

4.1 (a)

$$W = \phi + i\psi$$

$$\psi = x^2 - y^2 + \frac{x}{x^2 + y^2}$$

$$\psi_x = 2x + \frac{(x^2 + y^2) - x(2x)}{(x^2 + y^2)^2}$$

$$\psi_x = 2x + \frac{y^2 - x^2}{(x^2 + y^2)^2} = -\phi_y$$

$$\psi_y = -2y - \frac{x}{(x^2 + y^2)^2} \times (2y)$$

$$= -2y - \frac{2xy}{(x^2 + y^2)^2} = \phi_x$$

$$\phi = -2xy + \frac{y}{x^2 + y^2}$$

$$\phi_x = -2y - \frac{y}{(x^2 + y^2)^2}(2x) = -2y - \frac{2xy}{(x^2 + y^2)^2} = \psi_y$$

$$\phi_y = -2x + \frac{(x^2 + y^2) - y(2y)}{(x^2 + y^2)^2}$$

$$= -2x + \frac{x^2 - y^2}{(x^2 + y^2)^2} = -\psi_x$$

Satisfy CR equation,

$$\psi_x = -\phi_y$$

$$\psi_y = \phi_x$$

$$\therefore \quad \phi = -2xy + \frac{y}{x^2 + y^2} + C$$

4.2 (d)

Residue at $z = 3$ is

$$= \lim_{z \to 3}\left[(z-3)\frac{z^3}{(z-1)^4(z-2)(z-3)}\right]$$

$$= \frac{3^3}{(3-1)^4(3-2)} = \frac{27}{16}$$

4.3 (d)

$$f(z) = \frac{1}{(z-1)(z-2)} = \frac{-1}{z-1} + \frac{1}{z-2}$$

$$= \frac{-1}{z\left(1 - \frac{1}{z}\right)} + \frac{1}{-2\left(1 - \frac{z}{2}\right)}$$

$$= \frac{-1}{z}\left(1 - \frac{1}{z}\right)^{-1} - \frac{1}{2}\left(1 - \frac{z}{2}\right)^{-1}$$

$$= \frac{-1}{z}\left(1 + \frac{1}{z} + \frac{1}{z^2} + \frac{1}{z^3} + \ldots\right) - \frac{1}{2}\left(1 + \frac{z}{2} + \frac{z}{2}\right) + \left(\frac{z}{2}\right)$$

$$= -\frac{1}{z} - \frac{1}{z^2} - \frac{1}{z^3} - \frac{1}{z^4} + \ldots - \frac{1}{2} - \frac{z}{4} - \frac{z^2}{8}$$

Coefficient of $\frac{1}{z^2} = -1$

4.4 (b)

Given, $$\frac{1 - e^{2Z}}{Z^4} = \frac{1 - \left[1 + 2Z + \frac{(2Z)^2}{2!} + \frac{(2Z)^3}{3!} + \frac{(2Z)^4}{4!} + \ldots\right]}{Z^4}$$

$$= -\frac{2}{Z^3} - \frac{4}{2Z^2} - \frac{8}{3!Z} - \ldots$$

Residue at $Z = 0$

$$= \frac{-8}{3!} = \frac{-8}{6} = \frac{-4}{3}$$

ADDITIONAL INFORMATION

$$e^x = \sum_{x=0}^{\infty}\frac{x^n}{n!} = 1 + \frac{x}{1!} + \frac{x^2}{2!} + \ldots \frac{x^n}{n!}$$

4.5 (b)

We have, $f\left(re^{i\theta}\right) = u + iv$ ...(i)

Differentiating (i) partially w.r.t. r

$$f'\left(re^{i\theta}\right)e^{i\theta} = \frac{\partial u}{\partial r} + i\frac{\partial v}{\partial r} \quad \text{...(ii)}$$

Differentiating (i) partially w.r.t. θ

$$f'\left(re^{i\theta}\right)ire^{i\theta} = \frac{\partial u}{\partial \theta} + i\frac{\partial v}{\partial \theta} \quad \text{...(iii)}$$

From eq. (ii) and (iii)

$$ir\left[\frac{\partial u}{\partial r} + i\frac{\partial v}{\partial r}\right] = \frac{\partial u}{\partial \theta} + i\frac{\partial v}{\partial \theta}$$

Comparing on both sides, we get,

$$\frac{\partial u}{\partial r} = \frac{1}{r}\frac{\partial v}{\partial \theta}$$

$$\frac{\partial v}{\partial r} = -\frac{1}{r}\frac{\partial u}{\partial \theta}$$

4.6 (d)

If $f(z)$ is a pole of order n at $z = a$

Then,

$$\underset{z=a}{Res}\, f(z) = \lim_{z \to a} \frac{1}{(n-1)!}\frac{d^{n-1}}{dz^{n-1}}\left((z-a)^n f(z)\right)$$

4.7 (c)

General value of log(1 + i) + log (1 – i)

log (1 + i) + log (1 – i) = log (1 + i) (1 – i)

= log ($1^2 + 1^2$) = log 2

= log 2 + log 1

= log 2 + log $e^{i2n\pi}$

= log 2 + $i2n\pi$

4.8 (a)

∵ $f(z) = z^2$ is an analytic function and given contour is an open contour, so integral will be path independent i.e.

A, B, C, P, 0, y, x, or

A = –1 + i

B = 5 + 3i

P = 5 + i

$$\text{Now, } I = \int_C z^2 dz = \int_{AB} z^2 dz = \int_{z=-1+i}^{5+3i} z^2 dz = \left(\frac{z^3}{3}\right)_{-1+i}^{5+3i}$$

$$= \frac{1}{3}\left[(5+3i)^3 - (-1+i)^3\right]$$

$$= \frac{1}{3}\left[\{125 - 27i + 225i - 135\} + \{-2 - 2i\}\right]$$

$$= \frac{1}{3}\left[(-10 + 198i) + (2 - 2i)\right]$$

$$= \frac{1}{3}\left[-12 + 196i\right] = -4 + \frac{196}{3}i$$

4.9 (c)

Given circle $\left|z + \frac{3}{4} + j\right| = \frac{7}{4}$

Put $w = \frac{1}{z} \Rightarrow z = \frac{1}{w}$

$$\left|\frac{1}{w} + \frac{3}{4} + 0\right| = \frac{7}{4}$$

$$\left|1 + \left(\frac{3}{4} + j\right)w\right| = \frac{7}{4}|w|$$

$$\left|1 + \left(\frac{3}{4} + j\right)(4 + jv)\right| = \frac{7}{4}|u + iv|$$

$$\left|1 + \frac{3}{4}4 + j\left(\frac{3}{4}v + u\right) - v\right| = \frac{7}{4}\sqrt{(u^2 + v^2)}$$

$$\left(1 + \frac{3}{4}u - v\right)^2 + \left(\frac{3}{4}v + u\right)^2 = \frac{49}{16}(u^2 + v^2)$$

$$\frac{9}{16}u^2 + (1-v)^2 + 2\left(\frac{3}{4}u\right)(1-v) + \frac{9}{16}v^2 + u^2 + \frac{3}{2}uv$$

$$= \frac{49}{16}u^2 + \frac{49}{16}v^2$$

$$\left(\frac{25}{16} - \frac{49}{16}\right)u^2 + v^2\left[\frac{25}{16} - \frac{49}{16}\right] - 2v + \frac{3}{2}u + 1 = 0$$

$$u^2 + v^2 + u - \frac{4}{3}v + \frac{2}{3} = 0$$

$$2gv = u \Rightarrow 2g = 1 \Rightarrow 9 = \frac{1}{2}$$

$$2fv = -\frac{4}{3}v \Rightarrow f = -\frac{2}{3} \Rightarrow -\frac{2}{3}$$

∴Centre $(-g, -f) = \left(-\frac{1}{2}, +\frac{2}{3}\right)$

Radius =

$$r = \sqrt{g^2 + f^2 - c} = \sqrt{\frac{1}{4} + \frac{4}{9} + \frac{2}{3}} = \frac{7}{6}$$

5.1 (d)

Total number of ways to select 4 person out of 9 persons is ${}^9C_4$.

For the committee to have number of boys less than number of girls means,

the number of girls is 3 and number of boys is 1 (or) number of girls is 4 and number of boys is 0.

This can be done is ${}^4C_3 \times {}^5C_1 + {}^4C_4 \times {}^5C_0$

∴Required probability

$$= \frac{{}^4C_3 \times {}^5C_1 + {}^4C_4 \times {}^5C_0}{{}^9C_4} = \frac{1}{6}$$

5.2 (d)

Bag 1: 4W 2B

Bag 2: 3W 3B

Bag 1: $\begin{pmatrix}\text{Probability to}\\ \text{select bag 1}\end{pmatrix} \times \begin{pmatrix}\text{Probability to select}\\ \text{white ball from bag 1}\end{pmatrix}$

$\frac{1}{2} \times \frac{4}{6} = \frac{4}{12}$

Bag 2: $\begin{pmatrix}\text{Probability to}\\ \text{select bag 2}\end{pmatrix} \times \begin{pmatrix}\text{Probability to select}\\ \text{white ball from bag 2}\end{pmatrix}$

$\frac{1}{2} \times \frac{3}{6} = \frac{3}{12}$

Bag 1: $\frac{1}{2} \times \frac{4}{6} = \frac{4}{12}$

Bag 2: $\frac{1}{2} \times \frac{3}{6} = \frac{3}{12}$

$\frac{4}{12} + \frac{3}{12} = \frac{7}{12}$

$$\text{Total probability} = \frac{4}{12} + \frac{3}{12} = \frac{7}{12}$$

5.3 (b)

$$\int_{-\infty}^{\infty} f(x)dx = 1$$

$$\int_0^2 kx dx + \int_2^4 2k dx + \int_4^6 (-kx + 6k) dx = 1$$

$$k\left(\frac{x^2}{2}\right)\Bigg|_0^2 + 2k(x)\Big|_2^4 + \left(-k\frac{x^2}{2} + 6kx\right)\Bigg|_4^6 = 1$$

$$k(2-0) + 2k(4-2) + (-18k + 36k) - (-8k + 24k) = 1$$

$\Rightarrow \quad 2k + 4k + 18k - 16k = 1$

$\Rightarrow \quad 2k + 4k + 2k = 1$

$8k = 1$

$\therefore \quad k = \frac{1}{8}$

ADDITIONAL INFORMATION

Property of probability density function

$$\int_{-\infty}^{\infty} f(x)\, dx = 1$$

5.4 (a)

The first moment about the origin of Binomial distribution is mean of Binomial distribution $E(x) = np$.

ADDITIONAL INFORMATION

$$r^{th} \text{ moment} = E[X^r] = \int_{-\infty}^{\infty} x^r f(x)\, dx$$

$$\text{Mean} = E[X] = \int_{-\infty}^{\infty} x f(x)\, dx$$

For binomial distribution

$$E(x) = np$$

5.5 (a)

Given that the regression equation are

$$y = 0.516x + 33.73$$

$$x = 0.512y + 32.52$$

$0.516x - y = -33.73$ ...(i)

$x - 0.512y = 32.52$ ...(ii)

The regression lines passes through the mean i.e., the point of intersection of eq. (i) and eq. (ii) are (67.6, 68.6)

5.6 (c)

p(error per page)

$$= \frac{40}{600} = \frac{2}{30} = \frac{1}{15}$$

$$\lambda = np = 10 \times \frac{1}{15} = \frac{2}{3}$$

Let, X = Number of errors per page

$$p[X = 0] = \frac{e^{-\lambda}\lambda^0}{0!} = e^{-\frac{2}{3}}$$

5.7 (b)

Probability of selecting the first capacitor which is within the required tolerance values

$$= \frac{73}{100}$$

Since, this capacitor is replaced before taking out the second capacitor, the probability of getting the second one within the required is also $\frac{73}{100}$.

Thus, the required probability is

$$\frac{73}{100} \times \frac{73}{100} = 0.5329$$

5.8 (a)

In exponential distribution, if λ is the parameter specifying the average rate of occurrences of an event in a given time interval, then mean =

S.D = $\frac{1}{\lambda}$.

5.9 (a)

Consider the event A = {Test is +ve}

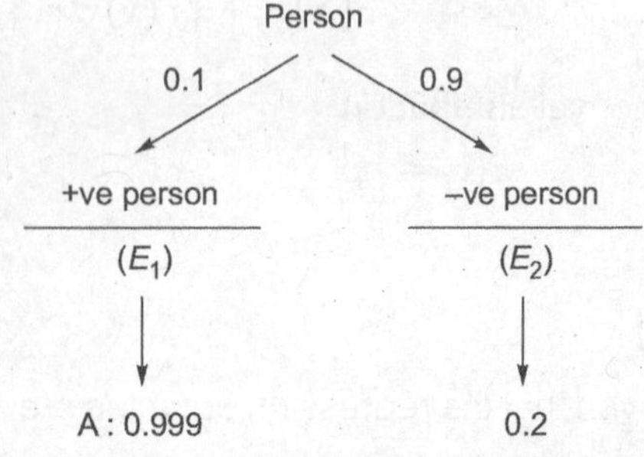

$P(A) = 0.1 \times 0.999 + 0.9 \times 0.2 = 0.2799$

$$P(\text{Actually +ve}) = P\left(\frac{E_1}{A}\right) = \frac{0.1 \times 0.999}{P(A)}$$

$$= \frac{0.1 \times 0.999}{0.2799} = \frac{999}{2799} = 0.356 \approx \frac{1}{3}$$

5.10 (d)

We know that,

$$\int_{-\infty}^{\infty}\int_{-\infty}^{\infty} f_{XY}(x,y)\,dy\,dx = 1$$

$$\therefore \int_{x=0}^{1}\int_{y=x}^{1} c(1-y)\,dy\,dx = 1$$

$$= c\int_0^1 \left[\frac{(1-y)^2}{-2}\right]_x^1 dx = 1$$

$$= \frac{c}{-2}\int_0^1 \left[0-(1-x)^2\right]dx = 1$$

$$= \frac{c}{2}\frac{(1-x)^3}{-3}\Bigg]_0^1 = 1$$

$$\Rightarrow \frac{c}{6}[1-0] = 1 \Rightarrow c = 6$$

So M.D.F. is

$$f_X(x) = \int_{y=x}^{1} 6(1-y)\,dy = 6\left(y-\frac{y^2}{2}\right)_x^1$$

$$= 6\left[\left(\frac{1}{2}\right)-\left(x-\frac{x^2}{2}\right)\right]$$

$$= \frac{6}{2} - 6x + 3x^2 = +6\left(\frac{1}{2} - x + \frac{x^2}{2}\right) \text{ for } 0 \le x \le 1$$

6.1 (c)

$$x = (1468)^{1/3}$$

Let, $x^3 = 1468$

$f(x) = x^3 - 1468 = 0$

By Newton Raphson method

$$x_{n+1} = x_n - \frac{f(x_n)}{f'(x_n)} = x_n - \frac{x_n^3 - 1468}{3x_n^2}$$

$$= \frac{2x_n^3 + 1468}{3x_n^2}$$

Start with $x_0 = 11$

$$x_1 = \frac{2(11)^3 + 1468}{3(11)^2} = 11.377$$

$$x_2 = \frac{2(11.377)^3 + 1468}{3(11.377)^2} = 11.365$$

Alternate solution:

$(11.34)^3 = 1458.27$

$(11.353)^3 = 1463.29$

$(11.365)^3 = 1467.94$

$(11.382)^3 = 1474.53$

$\therefore$ $x = 11.365$ is the nearest value.

6.2 (a)

Method-1:

$x = (1525)^{0.2}$

$x = (1525)^{1/5}$

$x^5 = 1525$

By substitution,

$(4.33)^5 = 1522.08$

$(4.36)^5 = 1575.55$

$(4.38)^5 = 1612.02$

$(4.30)^5 = 1470.08$

$\therefore$ $x = 4.33$ is the nearest value.

Method-2: (Newton Raphson method)

$x = (1525)^{0.2}$

$x^5 = 1525$

$f(x) = x^5 - 1525 = 0$

$f'(x) = 5x^4$

$$x_{n+1} = x_n - \frac{f(x_n)}{f'(x_n)}$$

$$\Rightarrow \quad x_{n+1} = \frac{4x_n^5 + 1525}{5x_n^4}$$

Case-1: $x_0 = 4$

$$x_1 = 4 - \frac{f(4)}{f'(4)} = 4.391$$

Note: This cannot be solution as it is not correct upto 2 decimals.

$$x_2 = (4.391) - \frac{f(4.391)}{f'(4.391)} = 4.333$$

$$x_3 = (4.333) - \frac{f(4.333)}{f'(4.333)} = 4.331$$

Note: This is the correct solution upto 2 decimals.

6.3 (c)

Given, $f(x) = x^3 + x^2 + 3x + 4 = 0$

$f'(x) = 3x^2 + 2x + 3$

$\because \quad f(-1) = 1 > 0$ and $f(-2) = -6 < 0$

$\therefore f(-1) \cdot f(-2) < 0$

$\Rightarrow \exists$ a root lies in $[-1, -2]$

Let, $x_0 = -1$

By Newton Raphson method

First approximation

$$x_1 = x_0 - \frac{f(x_0)}{f'(x_0)}$$

$$= -1 - \frac{1}{4} = -1 - 0.25$$

$$x_1 = -1.25$$

$$x_2 = x_1 - \frac{f(x_1)}{f'(x_1)}$$

$$= -1.25 - \frac{(-1.25)^3 + (-1.25)^2 + 3(-1.25) + 4}{3(-1.25)^2 + 2(-1.25) + 3}$$

$$= -1.2229$$

6.4 (d)

Let, $y = \dfrac{1}{1+x^2}$

	y_0	y_1	y_2	y_3	y_4	y_5	y_6
$x =$	0	1	2	3	4	5	6
	1	$\frac{1}{2}$	$\frac{1}{5}$	$\frac{1}{10}$	$\frac{1}{17}$	$\frac{1}{26}$	$\frac{1}{37}$

$$h = \frac{b-a}{n} = \frac{6-0}{6} = 1$$

$$\int_0^6 \frac{dx}{1+x^2} = \frac{h}{3}[(y_0 + y_6) + 4 \times (y_1 + y_3 + y_5) + 2 \times (y_2 + y_4)]$$

$$= \frac{1}{3}\left[\left(1 + \frac{1}{37}\right) + 4\left(\frac{1}{2} + \frac{1}{10} + \frac{1}{26}\right) + 2\left(\frac{1}{5} + \frac{1}{17}\right)\right]$$

$$= \frac{1}{3}\left(\frac{38}{37} + 4 \times \frac{83}{130} + 2 \times \frac{22}{85}\right)$$

$$= \frac{1}{3}\left(\frac{38}{37} + \frac{166}{65} + \frac{44}{85}\right)$$

$$= 1.36617 \simeq 1.3662$$

6.5 (c)

$$\frac{dy}{dx} = x^2 y - 1, \; y(0) = 1, \; y(0.1) = ?$$

Here, $x_0 = 0, \; y_0 = 1, \; h = 0.1$

So $y(0.1) = y_1 = ?$

Here, $y^{I}(x) = x^2 y - 1$

$\Rightarrow \quad y^{I}(0) = -1$

$y^{II}(x) = 2xy + x^2 y^{I}$

$\Rightarrow \quad y^{II}(0) = 0$

$y^{III}(x) = 2y + 2xy^{I} + 2xy^{I} + x^2 y^{II}$

$\Rightarrow \quad y^{III}(0) = 2$

$y^{IV}(x) = 2y^{I} + 4(xy^{II} + y^{I}) + 2xy^{II} + x^2 y^{III}$

$\Rightarrow \quad y^{IV}(0) = -6$ and so on ....

Hence by Taylor series;

$$y_1 = y_0 + hy^{I}(0) + \frac{h^2}{2!}y^{II}(0) + \frac{h^3}{3!}y^{III}(0) + \frac{h^4}{4!}y^{IV}(0) + \ldots$$

$$= 1 + 0.1(-1) + \frac{(0.1)^2}{2!}(0) + \frac{(0.1)^3}{3!}(2) + \frac{(0.1)^4}{4!}(-6) + \ldots$$

$$= 1 - 0.1 + 0 + 0.00033 + \ldots$$

$$= 0.90031 \approx 0.90033$$

6.6 (d)

Let $x^2 = 2 \Rightarrow x = \dfrac{2}{x}$ so let $f(x) = \dfrac{2}{x}$

To find square root of 2, let's start with $x_o = 1$ i.e.

$$x_1 = f(x_o) = \frac{2}{1} = 2$$

Now, for $x_1 = 2$, $x_2 = f(x_1) = \dfrac{2}{2} = 1 \approx x_o$ i.e. we stuck in a loop b/x x_o and x_2 so we will redefine $f(x)$ using average damping technique i.e.

$$f(x) = \frac{\frac{2}{x}+x}{2} = \text{Average of } \left(\frac{2}{x}\right) \text{ and } x$$

$$f(x) = \frac{1}{2}\left(x+\frac{2}{x}\right)$$ which will converge hence

Answer is (d).

Alternate solution:

Whatever method we apply, the series will converge when

$$x_{n+1} = x_n = \alpha$$

Verify all options with above condition

(d)
$$\alpha = \frac{1}{2}\left(\alpha+\frac{2}{\alpha}\right)$$
$$\alpha = \frac{\alpha}{2}+\frac{1}{\alpha}$$
$$\frac{\alpha}{2} = \frac{1}{\alpha}$$
$$\alpha^2 = 2$$ satisfy the question.

6.7 (b)

Given equations are :

$$-x_2 - x_3 + 4x_4 = 1$$
$$-x_1 + 4x_3 - x_4 = 1$$
$$4x_1 - x_2 - x_3 = 2$$
$$-x_1 + 4x_2 - x_4 = 2$$

It is not diagonally dominant system so using E-operation, we get

$$4x_1 - x_2 - x_3 = 2$$
$$-x_1 + 4x_2 - x_4 = 2$$
$$-x_1 + 4x_3 - x_4 = 1$$
$$-x_2 - x_3 + 4x_4 = 1$$

Now it is diagonally dominant so we can use Gauss-Seidel method as follows

$$x'_1 = \frac{1}{4}(2+x_2+x_3) = 0.5 + 0.25x^0_2 + 0.25x^0_3$$

$$x'_2 = \frac{1}{4}(2+x_1+x_4) = 0.5 + 0.25x'_1 + 0.25x^0_4$$

$$x'_3 = \frac{1}{4}(1+x_1+x_4) = 0.25 + 0.25x'_2 + 0.25x^0_4$$

$$x^1_4 = \frac{1}{4}(1+x_2+x_3) = 0.25 + 0.25x'_3 + 0.25x'_3$$

where $(x^0_1, x^0_2, x^0_3, x^0_4)$ is the initial assumption and (x'_1, x'_2, x'_3, x'_4) is the 1st iteration.

Proceeding in the same manner, option (b) is the right answer.

6.8 (c)

Forward difference table is

x	$f(x)$	$\Delta f(x)$	$\Delta^2 f(x)$	$\Delta^3 f(x)$	$\Delta^4 f(x)$
1	2				
2	5	3	–1		
3	7	2	$\alpha-9$	$\alpha-8$	$56-4\alpha$
4	α	$\alpha-7$	$39-2\alpha$	$48-3\alpha$	
5	32	$32-\alpha$			

∵ Four values of y are given so we can assume $f(x)$ as polynomial of degree 3 and in that situation 4th order differences will be zero.

i.e., $\Delta^4 f(x) = 0 \Rightarrow 56 - 4\alpha = 0$

$\Rightarrow \alpha = 14$

6.9 (d)

Forward difference table is

x	$f(x)$	$\Delta f(x)$	$\Delta^2 f(x)$	$\Delta^3 f(x)$	$\Delta^4 f(x)$	$\Delta^5 f(x)$
0	1	0.16				
0.2	1.16	2.40	2.24			
0.4	3.56	10.40	8.00	5.76		
0.6	13.96	28.00	17.60	9.60	3.84	0
0.8	41.96	59.04	31.04	13.44	3.84	
1.0	101.00					

Here $h = 0.2$

$a = 0$

then $u = \frac{x-a}{h} = \frac{x}{0.2}$

So, at $x = 0.2$

$u = \frac{0.2}{0.2} = 1$

Newton forward difference formula is

$$f(x) = f(a)+u\Delta f(a)+\frac{a(u-1)}{2!}\Delta^2 f(a)+\frac{u(u-1)(u-2)}{3!}\Delta^3 f(a)$$
$$+\frac{u(u-1)(u-2)(u-3)}{4!}\Delta^4 f(a)+............$$

So,

$$f'(x) = \frac{1}{h}\left[\Delta f(a)+\left(\frac{2u-1}{2}\right)\Delta^2 f(a)+\left(\frac{3u^2-6u+2}{6}\right)\Delta^3 f(a)\right]$$
$$+\left(\frac{4u^3-18u^2+22u-6}{24}\right)\Delta^4 f(a)+........$$

Putting $x = 0.1$, $a = 0$, $u = 1$ we get,

$$f'(0.2) = ... = 3.2$$

6.10 (a)

$$a_n = \frac{1}{1}\cdot\frac{5}{2}\cdot\frac{9}{2}\cdot\frac{13}{2}\cdots\frac{4n+1}{2}$$

$$a_0 = \frac{1}{1},\ a_1 = \frac{1}{1}\cdot\frac{5}{2},\ a_2 = \frac{1}{1}\cdot\frac{5}{2}\cdot\frac{9}{2}\cdots$$

Verifying with each options

we get $$a_n = 2^{n+2}\frac{\Gamma\left(n+\frac{5}{4}\right)}{\Gamma\left(\frac{1}{4}\right)}$$

ADDITIONAL INFORMATION

Gamma function, $\Gamma(n+1) = n\Gamma n = n!$

7.1 (c)

Given, $$f(x) = \begin{cases} -2 & -\pi < x < 0 \\ 2 & 0 < x < \pi \end{cases}$$

$$a_n = \frac{2}{\pi}\int_0^{\pi} f(x)\cos\left(\frac{n\pi x}{\pi}\right)dx$$

$$= \frac{2}{\pi}\int_0^{\pi} f(x)\cos nx\, dx = \frac{2}{\pi}\int_0^{\pi} 2\cos nx\, dx$$

$$= \frac{4}{\pi}\left[\frac{\sin nx}{n}\right]_0^{\pi} = \frac{4}{\pi}\left[\frac{\sin n\pi}{n} - \frac{\sin 0}{n}\right]$$

$$= \frac{4}{\pi}(0-0) = 0$$

Alternative:

Since function is odd function.

$\Rightarrow \quad a_n = 0$

7.2 (a)

$$a_5 = \frac{1}{\pi}\int_{-\pi}^{\pi} f(x)\sin 5x\, dx$$

$$= \frac{1}{\pi}\left[\int_{-\pi}^{0} -\pi\sin 5x dx + \int_0^{\pi} \pi\sin 5x dx\right]$$

$$= \frac{1}{\pi}\left[-\pi\left[\frac{-\cos 5x}{5}\right]_{-\pi}^{0} + \pi\left[\frac{-\cos 5x}{5}\right]_0^{\pi}\right]$$

$$= \frac{1}{\pi}\left[-\pi\left(\frac{-2}{5}\right) + \pi\left(\frac{2}{5}\right)\right] = \frac{1}{\pi}\left[\frac{4\pi}{5}\right] = \frac{4}{5}$$

7.3 (a)

Fourier series expansion of $T(x)$ in $[0, L]$ consisting of sine terms is given by

$$T(x) = \sum_{n=1}^{\infty} b_n \sin\frac{n\pi x}{L}$$

$$b_n = \frac{2}{L}\int_0^L T(x)\sin\frac{n\pi x}{L}dx$$

$$= \frac{2}{L}\int_0^L Kx(L-x)\sin\frac{n\pi x}{L}dx$$

$$= \frac{2K}{L}\left[(Lx-x^2)\left(\frac{-\cos\frac{n\pi x}{L}}{\frac{n\pi}{L}}\right) - (L-2x)\left(\frac{-\sin\frac{n\pi x}{L}}{\frac{n^2\pi^2}{L^2}}\right) + (-2)\left(\frac{\cos\frac{n\pi x}{L}}{\frac{n^3\pi^3}{L^3}}\right)\right]_0^L$$

$$= \frac{2K}{L}\left\{\left[\frac{-2L^3}{n^3\pi^3}\cos n\pi + \frac{2L^3}{n^3\pi^3}\right]\right.$$

$$b_n = \frac{4KL^3}{Ln^3\pi^3}[1-(-1)]^n \qquad n \text{ is odd}$$

$$b_{2n-1} = \frac{8KL^2}{(2n-1)^3\pi^3}$$

$$\therefore T(x) = \frac{8KL^2}{\pi^3}\sum_{n=1}^{\infty}\frac{1}{(2n-1)^3}\sin\frac{(2n-1)\pi x}{L}$$

7.4 (b)

$$f(t) = \begin{cases} A\sin\omega t, & 0 \le t \le \frac{\pi}{\omega} \\ 0, & \frac{\pi}{\omega} Mt \le \frac{2\pi}{\omega} \end{cases} \quad \left(0, \frac{\omega\pi}{\omega}\right) \Rightarrow l = \frac{\pi}{\omega}$$

$$a_0 = \frac{1}{l}\int_0^{2l} f(t)dt = \frac{1}{\frac{\pi}{\omega}}\int_0^{\frac{2\pi}{\omega}} f(t)dt = \frac{\omega}{\pi}\left[\int_0^{\frac{\pi}{\omega}} A\sin\omega t\, dt + 0\right]$$

$$= \frac{A\cdot\omega}{\pi}\left(\frac{-\cos\omega t}{\omega}\right)_0^{\frac{\pi}{\omega}} = -\frac{A}{\pi}[\cos\pi - 1] = \frac{2A}{\pi}$$

$$a_n = \frac{1}{l}\int_0^{2l} f(t)\cos\left(\frac{n\pi t}{l}\right)dt = \frac{1}{\frac{\pi}{\omega}}\int_0^{\frac{2\pi}{\omega}} f(t)\cdot\cos(n\omega t)\cdot dt$$

$$= \frac{\omega}{\pi}\left[\int_0^{\frac{\pi}{\omega}} A\sin\omega t\cdot\cos(n\omega t)\cdot dt + 0\right]$$

$$= \frac{\omega A}{2\pi}\int_0^{\frac{\pi}{\omega}}[\sin(1+n)\omega t + \sin(1-n)\omega t]\,dt$$

$$= \frac{\omega A}{2\pi}\left[\frac{-\cos(1+n)\omega t}{(1+n)\omega} - \frac{\cos(1-n)\omega t}{(1-n)\omega}\right]_0^{\frac{\pi}{\omega}}$$

$$= \frac{A}{2\pi}\left[\frac{-\cos(1+n)\pi + 1}{1+n} + \frac{-\cos(1-n)\pi + 1}{1-n}\right]$$

$$= \begin{cases} 0, \; n = \text{odd} \\ \frac{A}{2\pi}\left(\frac{2}{1+n} + \frac{2}{1-n}\right), \; n = \text{even} = \frac{-2A}{(n^2-1)\pi} \quad n = \text{even}\end{cases}$$

$$b_1 = \frac{1}{l}\int_0^{2l} f(t)\sin\left(\frac{\pi t}{l}\right)dt = \frac{1}{\frac{\pi}{\omega}}\int_0^{\frac{2\pi}{\omega}} f(t)\cdot\sin\omega t\,dt$$

$$= \frac{\omega}{\pi}\left[\int_0^{\frac{\pi}{\omega}} A\cdot\sin\omega t\cdot\sin\omega t\,dt + 0\right] = \frac{A\omega}{\pi}\int_0^{\frac{\pi}{\omega}}\left(\frac{1-\cos 2\omega t}{2}\right)dt$$

$$= \frac{A\omega}{2\pi}\left[t - \frac{\sin 2\omega t}{2w}\right]_0^{\frac{\pi}{\omega}} = \frac{A\omega}{2\pi}\left[\left(\frac{\pi}{\omega} - 0\right) - (0-0)\right] = \frac{A}{2}$$

So Fourier series is

$$f(n) = \frac{a_0}{2} + \sum a_n\cos\left(\frac{n\pi t}{l}\right) + \sum b_n\sin\left(\frac{n\pi t}{l}\right)$$

$$f(n) = \frac{A}{\pi} + \frac{A}{2}\sin\omega t - \frac{2A}{\omega\pi}\sum_{n=\text{even}}\left(\frac{\cos n\omega t}{n^2-1}\right)$$

where, $n = 2N$

So answer is option (b).

8.1 (b)

Given, $(z - px - qy)^3 = pq + 2(p^2 + q)^2$

$$z - px - qy = \sqrt[3]{pq + 2(p^2+q)^2}$$

$$z = px + qy + \sqrt[3]{pq + 2(p^2+q)^2}$$

which is in Clairaut form.

$\therefore$ solution is, $z = ax + by + \sqrt[3]{ab + 2(a^2+b)^2}$

ADDITIONAL INFORMATION

Clairaut form : $Z = px + qy + f(p, q)$

Solution, $Z = ax + by + f(a, b)$

Replace p & q by a & b respectively.

8.2 (a)

Verify from options, $u = \sin(3x - y)$

$$u_x = 3\cos(3x - y)$$
$$u_{xx} = -9\sin(3x - y)$$
$$u_y = -\cos(3x - y)$$
$$u_{yy} = -[-\sin(3x - y) \times -1]$$
$$= -\sin(3x - y)$$
$$\therefore \quad u_{xx} = 9u_{yy}$$

8.3 (c)

$$\frac{\partial u}{\partial x} = 2\frac{\partial u}{\partial t} + u \quad \text{...(i)}$$

$$u(x, 0) = 6e^{-3x}$$

Separation of variables

Let, $u(x, t) = X(x)T(t)$

$$u_x = X'T$$
$$u_t = XT'$$

Put in eq. (i), $X'T = 2XT' + XT$

Divide by XT

$$\frac{X'}{X} = \frac{2T'}{T} + 1 = k$$

$$X' = kX$$
$$(D - k)X = 0$$

AE: $m - k = 0$

$$m = k$$
$$X = C_1 e^{kx}$$
$$2T' = (k-1)T$$
$$2T' - (k-1)T = 0$$

AE: $2m - (k - 1) = 0$

$$m = \frac{(k-1)}{2}$$

$$T = C_2 e^{\frac{(k-1)}{2}t}$$

$$u(x, t) = XT$$

$$u(x, t) = C_1C_2 e^{kx + \frac{(k-1)}{2}t}$$

$$u(x, 0) = C_1C_2e^{kx} = 6e^{-3x}$$

$$\therefore \quad C_1C_2 = 6$$
$$k = -3$$
$$u(x, t) = 6e^{-3x-2t}$$
$$u(x, t) = 6e^{-(3x+2t)}$$

Alternate solution:

Given, $\dfrac{\partial u}{\partial x} = 2\dfrac{\partial u}{\partial t} + 4$ ...(i)

$u(x, 0) = 6e^{-3x}$ ...(ii)

Verify equations (i) and (ii) with options only (c) satisfy these two conditions.

9.1 (a)

Given : $L^{-1}\left\{\dfrac{5s^2+8s-1}{(s+3)(s^2+1)}\right\}$

$$\frac{5s^2+8s-1}{(s+3)(s^2+1)} = \frac{A}{(s+3)} + \frac{Bs+C}{(s^2+1)}$$

$\Rightarrow$ $A = 2$

$\Rightarrow$ $B = 3$

$\Rightarrow$ $C = -1$

$$= L^{-1}\left\{\frac{2}{s+3} + \frac{3s-1}{s^2+1}\right\}$$

$$= L^{-1}\left(\frac{2}{s+3}\right) + L^{-1}\left\{\frac{3s}{s^2+1}\right\} - L^{-1}\left\{\frac{1}{s^2+1}\right\}$$

$= 2e^{-3t} + 3\cos t - \sin t$

9.2 (d)

$L(4\cos 2t - 5\sin 2t) = 4L[(\cos 2t) - 5L(\sin 2t)]$

$$= 4\left(\frac{s}{s^2+4}\right) - 5\left(\frac{2}{s^2+4}\right) = \frac{4s-10}{s^2+4}$$

Using frequency shifting theorem,

$e^{-at}\, x(t) \Longleftrightarrow X(s+a)$

$L[2e^{3t}(4\cos 2t - 5\sin 2t)] = 2L[e^{3t}(4\cos 2t - 5\sin 2t)]$

$$= 2\left[\frac{4(s-3)-10}{(s-3)^2+4}\right] = 2\left[\frac{4s-12-10}{s^2-6s+9+4}\right]$$

$$= \frac{8s-44}{s^2-6s+13}$$

9.3 (c)

Fourier series of $f(x) = x + 1, -1 \le x \le 1$ is given by

$$f(x) = \frac{a_0}{2} + \sum_{n=1}^{\infty} a_n \cos\frac{n\pi x}{l} + \sum_{n=1}^{\infty} b_n \sin\frac{n\pi x}{l} \quad ...(1)$$

given $l = 1$

$$a_0 = \frac{1}{l}\int_{-l}^{l} f(x)dx = \frac{1}{1}\int_{-1}^{1}(x+1)dx = 0 + (x)_{-1}^{1} = 2$$

$$a_n = \frac{1}{l}\int_{-l}^{l} f(x)\cos\frac{n\pi x}{l} = \frac{1}{1}\int_{-1}^{1}(1+x)\cos\frac{n\pi x}{1}dx$$

$$= \left(\frac{\sin\frac{n\pi x}{1}}{n\pi}\right)_{-1}^{1} = 0$$

$$b_n = \frac{1}{l}\int_{-l}^{l} f(x)\sin\frac{n\pi x}{l}dx$$

$$= \frac{1}{1}\int_{-1}^{1}(1+x)\sin(n\pi x)dx = 0 + \frac{2}{1}\int_{0}^{1} x\sin\frac{n\pi x}{1}dx$$

$$b_n = 2\left[x\left(\frac{-\cos n\pi x}{n\pi}\right) - \left(\frac{-\sin n\pi x}{n^2\pi^2}\right)\right]_0^1$$

$$= \frac{2}{n\pi}\left[(-1)^{n+1} + 0\right]$$

$\therefore$ Fourier series of $f(x)$ is

$$f(x) = 1 + \sum_{n=1}^{\infty}\frac{2}{n\pi}(-1)^{n+1}\sin n\pi x$$

9.4 (d)

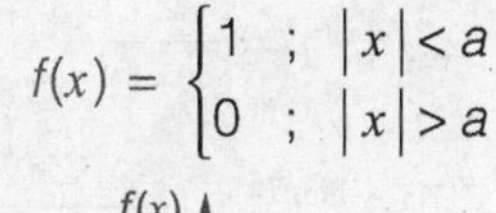

$$f(x) = \begin{cases} 1 \ ; & |x| < a \\ 0 \ ; & |x| > a \end{cases}$$

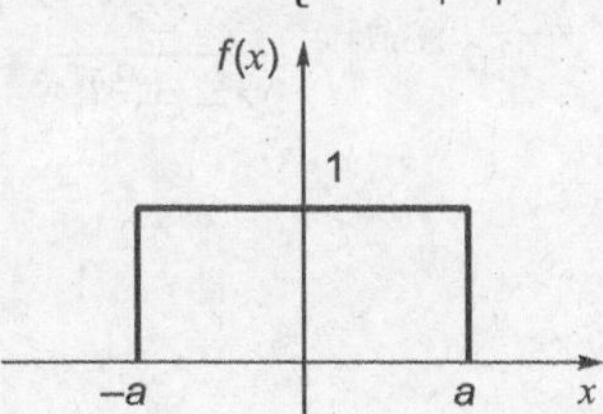

$$g(x) = \begin{cases} 1 \ ; & 0 < x < a \\ 0 \ ; & \text{otherwise} \end{cases}$$

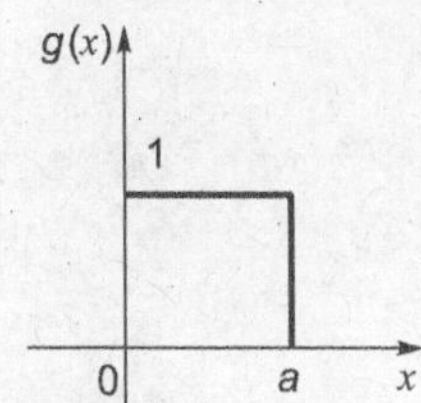

Fourier transform :

$$X(\omega) = \frac{1}{\sqrt{2\pi}}\int_{-\infty}^{\infty} x(t)\cdot e^{-j\omega t}dt$$

$F\{3f(x) - 2g(x)\}$

$$= \sqrt{\frac{1}{2\pi}}\left[\int_{-\infty}^{+\infty} e^{-i\omega n}(3f(x))dx - \int_{-\infty}^{+\infty} e^{-i\omega n}(2g(x))dx\right]$$

$$= \sqrt{\frac{1}{2\pi}}\left[3\int_{-a}^{a} e^{-iWx}(1)dx - 2\int_{0}^{a} e^{-iWx}(1)dx\right]$$

$$= \sqrt{\frac{1}{2\pi}}\left[3\left(\frac{e^{-i\omega x}}{-i\omega}\right)_{-a}^{a} - 2\left(\frac{e^{-i\omega x}}{-i\omega}\right)_{0}^{a}\right]$$

$$= \sqrt{\frac{1}{2\pi}}\left[-\frac{3}{\omega}\left(\frac{e^{-i\omega a} - e^{-i\omega a}}{i}\right) + \frac{2}{i\omega}(e^{-i\omega a} - 1)\right]$$

$$= \sqrt{\frac{1}{2\pi}}\left[+\frac{3}{\omega}(2\sin\omega) + \frac{2}{i\omega}(e^{-i\omega a} - 1)\right]$$

$$= \sqrt{\frac{2}{\pi}}\left[+\frac{3\sin\omega a}{\omega} - \left(\frac{1 - e^{-i\omega a}}{i\omega}\right)\right]$$

9.5 (a)

The discrete-time signal f(n) obtained after sampling at an interval of T is given by

$$f(n) = e^{-2\omega nT}$$

$$f(t) = e^{-2\omega t}$$

$$Z.T.\{a^n\} = \frac{z}{z-a} \quad \text{We know that}$$

$$\text{Hence, } Z.T.[e^{-2\omega nT}] = \frac{z}{z - e^{-2\omega T}}$$

9.6 (b)

$$z^{-1}(Y(z)) = z^{-1}\left(\frac{z}{z^2 - z + 1}\right)$$

$$Y(z) = \frac{z^{-1}}{1 - z^{-1} + z^{-2}} = \frac{z^{-1}}{z^{-2} - z^{-1} + 1}$$

$$Y(z) = \left(\frac{2}{\sqrt{3}} \cdot \frac{\frac{\sqrt{3}}{2} \cdot z^{-1}}{z^{-2} - z^{-1} + 1}\right)$$

$$Y(z) = \left(\frac{2}{\sqrt{3}} \frac{\sin\frac{\pi}{3} z^{-1}}{1 - 2\left(\cos\frac{\pi}{3}\right)z^{-1} + z^{-2}}\right)$$

Taking inverse z-transform, we get

$$y(n) = \frac{2}{\sqrt{3}}\sin\frac{\pi n}{3}u(n) = \frac{2}{\sqrt{3}}\sin\frac{\pi n}{3}$$

Formula:

$$Z^{-1}\left[\frac{\sin(\omega_0)z^{-1}}{1 - 2\cos\omega_0 z^{-1} + z^{-2}}\right] = \sin(n\omega_0)u(n)$$

■■■■

General Principles of Design, Drawing, Importance of Safety

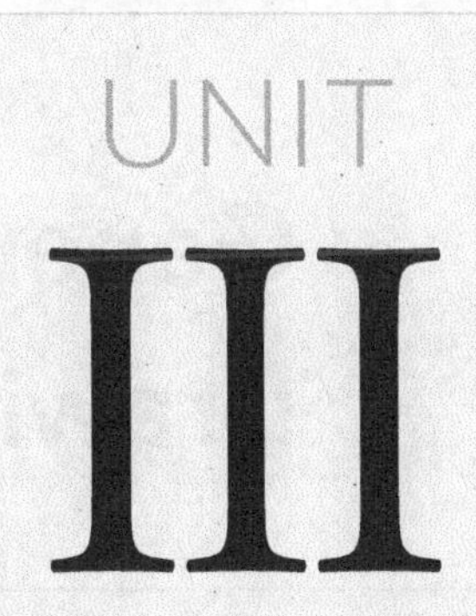

CONTENTS

General Principles of Design, Drawing, Importance of Safety

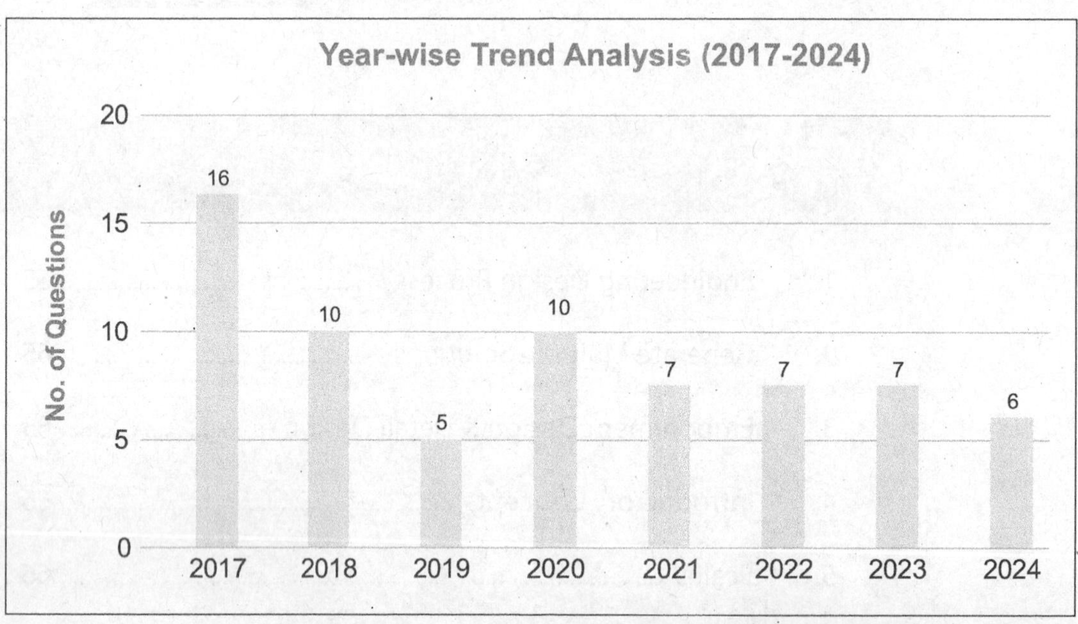

TOPIC-WISE TREND ANALYSIS (2017-2024)		
Sl. No.	**Topics**	**Total Number of Questions**
1.	Engineering Design Process	3
2.	Generate Multiple Solutions	2
3.	Embodiment Design & Detail Design	3
4.	Introductory Concepts	8
5.	Scales	1
6.	Engineering Curves	2
7.	Conic Sections	4
8.	Projection of Straight Lines	3
9.	Projection of Solids and Isometric Projection	8
10.	Section of Solids and Intersection of Surfaces	4
11.	Development of Surfaces	4
12.	Auto CAD	2
13.	Safety and its Importance	4
14.	Accidents, Hazards and Risks	3
15.	Fire Safety	1
16.	Safety in Other Industries	9
17.	Miscellaneous	8

General Principles of Design, Drawing, Importance of Safety

1. ENGINEERING DESIGN PROCESS

1.1 Concurrent engineering is implemented by involving a cross-functional team for design, production, testing and operational work
(a) During the project execution
(b) Long before the project execution
(c) Towards the end of the project execution
(d) After completion of the project execution

[ESE Prelims : 2017]

1.2 The design of highway interchanges involves the application of the geometry of
(a) circle arcs (b) semi ellipse
(c) hyperbola (d) semi-circle

[ESE Prelims : 2021]

1.3 It is important to carefully identify the needs and expectations of the customer prior to beginning the design of mechanical device. One of the step in formulating usually quantitative statements of expected performance levels, environmental conditions in which the device must operate, limitations on space or weight, or available materials and components that may be used. These are part of which one of the following elements of design?
(a) Functions
(b) Evaluation criterias
(c) Design requirements
(d) Drawings

[ESE Prelims : 2024]

2. GENERATE MULTIPLE SOLUTIONS

2.1 With reference to problem solving-fixation refers to
(a) Focused approach to problem solving
(b) Planned approach to problem solving
(c) Inability to see a problem from fresh perspective
(d) Inability to comprehend the goal to be achieved

[ESE Prelims : 2017]

2.2 The creative design routes are practised by adopting following steps:
1. Concentration 2. Illumination
3. Preparation 4. Verification
5. Incubation

Arrange the above steps in correct sequence:
(a) 3, 1, 5, 2, 4 (b) 3, 5, 2, 1, 4
(c) 3, 2, 1, 5, 4 (d) 3, 1, 2, 5, 4

[ESE Prelims : 2021]

3. EMBODIMENT DESIGN & DETAIL DESIGN

3.1 Consider the following statements regarding Golden Ratio for positive integers:
1. It is the ratio of difference of two numbers and the smaller number
2. It is the ratio of sum of two numbers and the smaller number
3. It is the ratio of the sum of two numbers and the larger number

Which of the above statements is/are correct?
(a) 1, 2 and 3 (b) 3 only
(c) 2 only (d) 1 only

[ESE Prelims : 2017]

3.2 Consider the following statements regarding Ergonomic Design:
1. Reducing the stress on the spinal cord and providing for lesser fatigue-causing sitting arrangements.
2. Arrangements of keys on the computer keyboard towards optimizing finger stress level.

3. Catering to increasing demand to produce more pleasing objects.

Which of the above statements are correct?

(a) 1 and 2 only (b) 2 and 3 only
(c) 1 and 3 only (d) 1, 2 and 3

[ESE Prelims : 2018]

3.3 Which one of the following makes the design, assembly and operation of complex systems feasible and practical?

(a) System Architecture
(b) Modularization
(c) Standardization
(d) Composition

[ESE Prelims : 2019]

4. INTRODUCTORY CONCEPTS

4.1 Dimensions in series may be placed in any one of the following ways except

(a) Progressive dimensioning
(b) Proportional dimensioning
(c) Continuous dimensioning
(d) Chain dimensioning

[ESE Prelims : 2020]

4.2 Points to be remembered while dimensioning:

1. Dimensions are to be placed on the view which clearly express the relevant features.
2. Once dimension is marked in one view, it should not be repeated in another view.
3. Dimensions are to be drawn from hidden lines.
4. Dimensions should be given from the base line or centre line of a hole.

Which of the above statements are correct?

(a) 1, 2 and 3 only (b) 1, 2 and 4 only
(c) 1, 3 and 4 only (d) 2, 3 and 4 only

[ESE Prelims : 2021]

4.3 On a multi view drawing a visible or invisible line represents the following:

1. Intersection of two surfaces
2. Edge view of surface
3. Limiting elements of a surface

Which of the above points are correct?

(a) 1 and 2 only (b) 2 and 3 only
(c) 1 and 3 only (d) 1, 2 and 3

[ESE Prelims : 2021]

4.4 On a view showing assembled parts, section line on adjacent pieces are drawn in

(a) same directions usually at an angle of 90°
(b) opposite directions usually at an angle of 45°
(c) opposite directions usually at an angle of 30°
(d) same directions usually at an angle of 60°

[ESE Prelims : 2021]

4.5 Continuous thin (narrow) with zigzags (straight) lines are generally used to represent

(a) long-break line (b) hidden outline
(c) visible outline (d) reference line

[ESE Prelims : 2022]

4.6 Which one of the following is used when components of same shape but different dimensions are to be manufactured?

(a) Drawing for installation
(b) Tabular drawing
(c) Schematic assembly drawing
(d) Patent drawing

[ESE Prelims : 2023]

4.7 Which one of the following lines is used to represent the outlines of adjacent parts or alternative and extreme positions of movable parts?

(a) Continuous thick line
(b) Continuous thin line
(c) Chain thin double-dashed line
(d) Dashed thin line

[ESE Prelims : 2023]

4.8 Which one of the following is/are used for drawing curves which cannot be drawn with a compass?

(a) Scale (b) Protractor
(c) French curves (d) Set square

[ESE Prelims : 2024]

5. SCALES

5.1 In a plain scale, if 1.5 inches = 1 foot and it can measure upto 4 feet, what is the representative factor of the scale?

(a) $\frac{1}{8}$ (b) $\frac{1}{4}$
(c) $\frac{1}{1.5}$ (d) $\frac{2}{1.5}$

[ESE Prelims : 2023]

6. ENGINEERING CURVES

6.1 The locus traced by a point moving along a pendulum from one end to another, when the pendulum oscillates, is
(a) A spiral (b) An involute
(c) A cycloid (d) A helix
[ESE Prelims : 2017]

6.2 If the radius of a generating circle which is moving inside the directing circle is half of the radius of the directing circle, the curve generated by a point on the circumference of the generating circle is
(a) A circle (b) An ellipse
(c) A straight line (d) A spiral
[ESE Prelims : 2017]

7. CONIC SECTIONS

7.1 A cone is resting with its base on HP. A section plane parallel to VP cuts the cone. The section plane is some distance away from the centre and does not pass through the apex. The true shape of the section is
(a) Hyperbola
(b) Rectangular Hyperbola
(c) Parabola
(d) Rectangular Parabola.
[ESE Prelims : 2017]

7.2 A cone resting on its base in horizontal plane (H.P.) is cut by a plane inclined to the axis and parallel to one of its generators, the sectional view will be
(a) Ellipse (b) Parabola
(c) Hyperbola (d) Circle
[ESE Prelims : 2019)]

7.3 The line passing through the focus and perpendicular to the directrix is called
(a) Axis (b) Vertex
(c) Eccentricity (d) Conic
[ESE Prelims : 2020]

7.4 In a rectangular hyperbola, if a curve is traced out by a point moving in such a way that the product of its distances from two fixed lines at right angles to each other is a constant, then those fixed lines are called
(a) Asymptotes (b) Intercepts
(c) Holes (d) Limits
[ESE Prelims : 2020]

8. PROJECTION OF STRAIGHT LINES

8.1 If a line is inclined to Vertical Plane and parallel to Horizontal Plane, then which of the following statements is always correct?
(a) True length = Plan length
(b) True length = Elevation length
(c) True length < Plan length
(d) Vertical Trace of the line is above the XY plane
[ESE Prelims : 2017]

8.2 If a line is perpendicular to the V.P. and its V.T. coincides with its front view which is a point, then
(a) V.T. is a point on H.P.
(b) H.T. is a point on V.P.
(c) it has no V.T.
(d) it has no H.T. [ESE Prelims : 2022]

8.3 If a line is inclined to the H.P. and parallel to the V.P., then it has
(a) No trace
(b) Only V.T. but no H.T.
(c) Both H.T. and V.T.
(d) Only H.T. but no V.T.
[ESE Prelims : 2023]

9. PROJECTION OF SOLIDS AND ISOMETRIC PROJECTION

9.1 If a square pyramid is freely suspended from one of the corners of its base, then the imaginary line joining that corner with the centre of gravity of the pyramid will be
(a) Inclined at 60° with the vertical
(b) Inclined at 45° with the vertical
(c) Inclined at 30° with the vertical
(d) Vertical
[ESE Prelims : 2017]

9.2 In which one of the following projection types, the object is kept in such a way that its three mutual perpendicular edges make equal angles with the plane of projection and the object stands on one of its corners?
(a) Non Isometric projection
(b) Oblique projection
(c) Isometric projection
(d) Point projection
[ESE Prelims : 2021]

9.3 Oblique drawing has the following advantage over isometric drawing:
(a) Distortion can be increased by foreshortening measurements along the receding axis
(b) A greater choice is permitted in orthographic top view
(c) Circular or irregular outlines on the front face show in their true shape
(d) Oblique drawing is often less flexible
[ESE Prelims : 2021]

9.4 When the receding lines are drawn to full size scale and the projectors inclined at an angle of 30° or 45° or 60° to the plane of projection, such oblique projection is known as
(a) Cabinet projection
(b) Vertical projection
(c) Cavalier projection
(d) Horizontal projection [ESE Prelims : 2022]

9.5 Which one of the following methods is used when the non-isometric lines or their ends lie in isometric planes?
(a) Intersection method
(b) Box method
(c) Co-ordinate method
(d) Offset method [ESE Prelims : 2022]

9.6 A triangular prism, base 40 mm side and axis 50 mm long is resting on one of its bases on the H.P. with a vertical face perpendicular to the V.P. What is the front view of the prism?
(a) a triangle
(b) a rectangle
(c) combination of two rectangles
(d) combination of triangle and rectangle
[ESE Prelims : 2023]

9.7 Consider the following points while drawing the isometric view of any solid:
1. The isometric view should be drawn according to the given views and in such a way that maximum possible details are visible.
2. At every point for the corner of a solid, at least three lines for the edges must converge. Of these, at least two must be for visible edges.
3. Two lines (for visible edges) will never cross each other.

Which of the above statements are correct?
(a) 1 and 2 only (b) 2 and 3 only
(c) 1 and 3 only (d) 1, 2 and 3
[ESE Prelims : 2023]

9.8 Which one of the following methods is used when the non-isometric lines or their ends lie in isometric planes?
(a) Co-ordinate method
(b) Box method
(c) Offset method
(d) Visual-ray method [ESE Prelims : 2023]

10. SECTION OF SOLIDS AND INTERSECTION OF SURFACES

10.1 A pentagonal prism is lying on H.P. on one of its rectangular faces. When it is cut by a section plane, the largest possible section thereof has
(a) Five edges (b) Six edges
(c) Seven edges (d) Eight edges
[ESE Prelims : 2017]

10.2 The plan view at just below window-sill level, but not showing door openings is shown of an outpost building of, say, the forest department. Section AA extending just a little above ground level GL, and fully below ground level is shown. The wall and first footing are of random-rubble masonry in cement mortar; and the lowest part of the foundation is of mass rubble in cement mortar. The total volume of 40 cm deep footing for the whole building is nearly

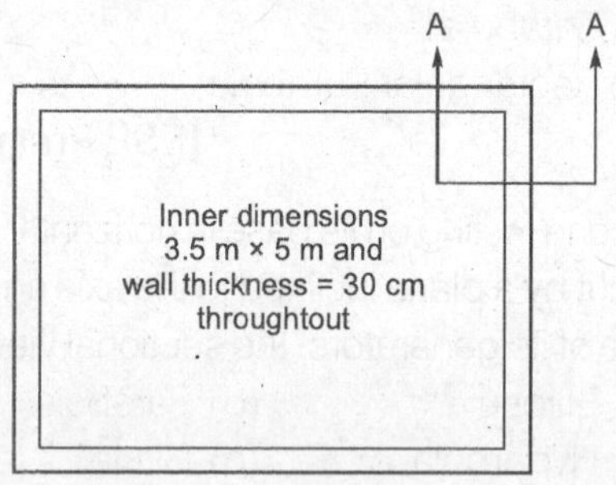

NOT TO SCALE

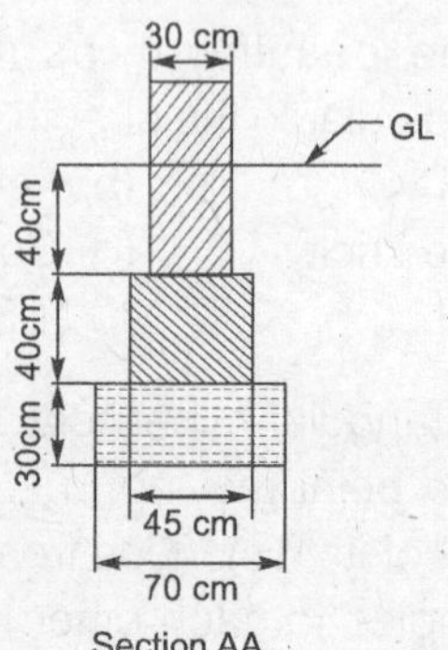

Section AA

(a) 3.8 m^3 (b) 3.3 m^3
(c) 2.8 m^3 (d) 2.3 m^3
[ESE Prelims : 2018]

10.3 Which one of the following statements is correct about oblique projection?
(a) The object is drawn with the reduced (about 82%) dimensions.
(b) All the faces of the object are distorted in the shape and size.
(c) Projectors from an object are parallel to each other and perpendicular to the plane of picture.
(d) The faces of object which are perpendicular to the plane of projection will be distorted in the shape and size. [ESE Prelims : 2022]

10.4 Which of the following are the methods for determining the line of intersection between surfaces of two interpenetrating solids?
(a) Line method and cutting plane method
(b) Line method and box method
(c) Co-ordinate method and cutting plane method
(d) Co-ordinate method and box method
[ESE Prelims : 2024]

11. DEVELOPMENT OF SURFACES

11.1 If the development of the lateral surface of a cone is semicircle, then
(a) The slant height of the cone < diameter of the base of the cone
(b) The slant height of the cone > diameter of the base of the cone
(c) The slant height of the cone = diameter of the base of the cone
(d) The slant height of the cone = radius of the base of the cone
[ESE Prelims : 2017]

11.2 Which one of the following statements is correct in the development of lateral surfaces of solids?
(a) The development of a right cone is a triangle
(b) Triangulation is the recommended method in the development of a prism
(c) The development of the lateral surface of a right circular cylinder is a rectangle
(d) The elements of an elliptical cone are equal in length
[ESE Prelims : 2017]

11.3 If a thread is wound around a cone, starting from a point on the base, and brought back to the same point, then the shortest possible length of the thread is equal to the
(a) Diameter of the base of the cone
(b) Slant height of the cone
(c) Largest chord of the development sector
(d) Length of the perpendicular from a corner of the development sector to the opposite edge
[ESE Prelims : 2017]

11.4 Which one of the following is used for pyramids and cones in which the true length of the slant edge or the generator is used as radius?
(a) Parallel-line development
(b) Radial-line development
(c) Triangulation development
(d) Approximate method
[ESE Prelims : 2024]

12. AUTO CAD

12.1 The insert command is used in Auto CAD to insert
(a) Objects in current file
(b) Objects in any file
(c) Blocks in any drawing file
(d) Blocks and wblocks in the current drawing
[ESE Prelims : 2019]

12.2 **Statement (I)** : The drawings and machining processes can be automated using CAD/CAM change the primary function of these drawings and processes.
Statement (II) : The primary function is to provide information about the product to the designer and production people.
(a) Both Statement (I) and Statement (II) are individually true and Statement (II) is the correct explanation of Statement (I)
(b) Both Statement (I) and Statement (II) are individually true, but Statement (II) is not the correct explanation of Statement (I)
(c) Statement (I) is true, but Statement (II) is false
(d) Statement (I) is false, but Statement (II) is true
[ESE Prelims : 2022]

13. SAFETY AND ITS IMPORTANCE

13.1 For which of the following sectors do comprehensive safety and health statutes, for regulating Occupational Safety and Health (OSH) at work places, exist at present in India?
(a) Mining, Factories, Ports and Agriculture
(b) Factories, Mining, Agriculture and Construction

(c) Mining, Factories, Ports and Construction
(d) Factories, Ports, Agriculture and Construction

[ESE Prelims : 2017]

13.2 Consider the goals of the safety policy of a company, at once designating also the responsibilities and authority for their achievement:

1. Safety of employees and the public at large
2. Efforts to be made to involve all managers, supervisors and employees in the development and implementation of safety procedures.
3. Clearing of all caveats that may tell on keeping the morale of the employees high

Which of the above statements are correct?

(a) 1, 2 and 3 (b) 2 and 3 only
(c) 1 and 2 only (d) 1 and 3 only

[ESE Prelims : 2017]

13.3 Trial runs are recommended for which of the following reasons?

1. Trial runs provide an opportunity to remedy the situation during the experiment.
2. Trial runs provide a final chance to fine-tune levels of a factor.
3. Trial runs provide a chance to make any needed changes in the experimental plan during experiment.
4. Trial runs can help considerably in estimating the time to complete a run, the logistical support required for level changes and total time needed to complete an experiment.

Select the correct answer using the code given below:

(a) 1, 2, 3 and 4 (b) 1 and 3 only
(c) 2 and 4 only (d) 2, 3 and 4 only

[ESE Prelims : 2022]

14. ACCIDENTS, HAZARDS AND RISKS

14.1 Human engineering approach is followed to prevent accidents by giving due consideration to physical and mental limitations of the workers by

(a) giving constant attention to how a worker is likely to react rather than how supervisor would like him to react
(b) imposing too many safety rules
(c) not permitting to make safety device or guard inoperative
(d) giving first-aid attention in case of injury

[ESE Prelims : 2020]

14.2 Which one of the following is not an operator error through triggers leading to an accident which confuses and traps into making mistake?

(a) Faulty design or construction of machine tool
(b) Poor housekeeping and cleanliness
(c) Standard operating safety practice
(d) Lack of standardization and identification

[ESE Prelims : 2020]

14.3 Which one of the following is not included in the safety program for achieving good results during the prevention of accidents?

(a) Development of safe working conditions
(b) Promotion of employees participation in safety
(c) Compensation and medical payment
(d) Corrective action when safety rules are ignored

[ESE Prelims : 2020]

15. FIRE SAFETY

15.1 Statement (I): Lifts and external staircases are provided with access from the lobby area of each floor in multistory blocks. The external staircase must be accessible through self-closing, 180°-swing unlocked doors (with provision for locking at appropriate conditions).

Statement (II): Such staircases should not be inadvertently subjected to spreading of smoke, but must yet provide unhindered exit from the lobby of each floor.

(a) Both Statement (I) and Statement (II) are individually true; and Statement (II) is the correct explanation of Statement (I)
(b) Both Statement (I) and Statement (II) are individually true; but Statement (II) is NOT the correct explanation of Statement (I)
(c) Statement (I) is true; but Statement (II) is false
(d) Statement (I) is false; but Statement (II) is true

[ESE Prelims : 2018]

16. SAFETY IN OTHER INDUSTRIES

16.1 Consider the following statements regarding safety during demolition of a multistory building:
1. Demolition need not proceed story by story.
2. Floor openings, not used for material chutes, should be enclosed with adequate guard rails.
3. Before demolitions starts all lath and loose plaster shall be stripped off throughout the building.
4. Adequate and well-stiffened lateral bracing shall be provided for walls.

Which of the above statements are correct?
(a) 1, 2 and 3 only (b) 1 and 4 only
(c) 2, 3 and 4 only (d) 1, 2, 3 and 4

[ESE Prelims : 2017]

16.2 Consider the following statements:
(i) The failure of many structures
(ii) Accidents on major highways
(iii) Discharge of effluents which pollute rivers
(iv) Dangerous gas emissions

These are attributable to:
1. Improper design
2. Not adhering to proper maintenance practices
3. Lack of proper inspection and quality control during construction
4. Corruption amongst several concerned individuals
5. Lack of public concern

Select the most appropriate answer using the codes given below:
(a) 1, 2 and 3 only (b) 1, 2 and 4 only
(c) 3, 4 and 5 only (d) 1, 2, 3, 4 and 5

[ESE Prelims : 2017]

16.3 Consider the following provisions regarding safety on highways, where major improvement works may also be in progress:
1. Highly visible barricades to avoid falling of vehicles in deep interspaces ahead (including drainages).
2. Wire-net provisions to ward off road slippages.
3. Signages for wild-animals crossing (like deer, elephant, etc.)
4. Signages on minor gradients

Which of the above are relevant?
(a) 1, 2, 3 and 4 (b) 1, 2 and 4
(c) 1, 3 and 4 (d) 1, 2 and 3

[ESE Prelims : 2018]

16.4 Consider the following statements:
1. Mobile cranes are sophisticated machines which are designed for lifting efficiently
2. Mobile cranes are a versatile and reliable means of lifting on site

Which of the above statements is/are correct?
(a) 1 only (b) 2 only
(c) Both 1 and 2 (d) Neither 1 nor 2

[ESE Prelims : 2019]

16.5 Which of the following statements are correct for portable step-ladders?
1. Used on working platforms to gain height above the protected edge
2. Used in the fully opened position
3. Should be of a length that ensures a person's feet are not positioned any higher than the second top rung

Select the correct answer using the codes given below:
(a) 2 and 3 only (b) 1 and 3 only
(c) 1 and 2 only (d) 1, 2 and 3

[ESE Prelims : 2019]

16.6 Which of the following measures is/are correct for using Mobile Equipment Working Platform (MEWP)?
1. Tyres are properly inflated and air filled.
2. SWL to be marked in platforms as identification for carrying loads.

(a) 1 only (b) 2 only
(c) Both 1 and 2 (d) Neither 1 nor 2

[ESE Prelims : 2019]

16.7 Which one of the following is not the way of estimating the statistical parameters and is integral part of analysis or synthesis tasks when probability of failure is involved?
(a) Propagation of error
(b) Propagation of uncertainty
(c) Propagation of weight
(d) Propagation of dispersion

[ESE Prelims : 2020]

16.8 In order to limit the seriousness of an accident, emergency controls should be provided with which of the following as determining factors in the location of emergency stops?
(a) Speed and ease of operation
(b) Common sense of workers
(c) Nearest exits and checkpoints
(d) Supervisors and decision-makers

[ESE Prelims : 2020]

16.9 The sound level L is

(a) $\log_{10}\dfrac{Q_0}{Q}$ (bels)

(b) $20\log_{10}\dfrac{Q}{Q_0}$ (bels)

(c) $\log_{10}\dfrac{Q}{Q_0}$ (bels)

(d) $20\dfrac{Q}{Q_0}+\log_{10}\dfrac{Q}{Q_0}$ (bels)

where, Q = Measured quantity of sound pressure or sound intensity
Q_0 = Reference standard quantity of sound pressure

[ESE Prelims : 2020]

16.10 What are the functions of axles?
(a) Support the weight of the mower. Permit easy, rolling movement. Provide for mounting on an axle. Ensure safe operation on flat or sloped lawn surfaces.
(b) Support, safely enclose, and protect operating components, including the blade and motor. Accommodate the attachment of two axles and a handle. Permit cut grass to exit the cutting area.
(c) Cut blades of grass and weeds while rotating at high speed. Facilitate connection to motor shaft. Operate safely when foreign objects are encountered, such as stones, sticks, or metal pieces.
(d) Transfer the weight of mower from the housing to the wheels. Allow rotation of the wheels. Maintain location of the wheels relative to the housing.

[ESE Prelims : 2024]

17. MISCELLANEOUS

17.1 A weight of 240 N is dropped on to a close-coiled helical spring made up of 18 mm spring steel wire. The spring consists of 22 coils wound to a diameter of 180 mm. If the instantaneous compression is 120 mm, what is the height of drop of the weight, given G = 88 × 103 N/mm2?

(a) 450 mm (b) 300 mm
(c) 250 mm (d) 150 mm

[ESE Prelims : 2018]

17.2 For a vehicle travelling at 24 km/hr having a wheel radius of 0.305 m with overall gear ratio G = 19.915, and when torque transmitted is 203.6 N.m, the engine speed and power are, nearly
(a) 4155 rpm and 88.6 kW
(b) 4500 rpm and 88.6 kW
(c) 4155 rpm and 95.4 kW
(d) 4500 rpm and 95.4 kW

[ESE Prelims : 2018]

17.3 A vehicle moving at a speed of 88 km/hr weighs 62293.5 N and its rolling resistance coefficient is 0.018. The rolling resistance of the vehicle is

(a) 1121.3 N (b) 1000.4 N
(c) 975.7 N (d) 845.6 N

[ESE Prelims : 2018]

17.4 A cantilever beam ABC is shown to a highly exaggerated vertical scale. Horizontally, AB is 2 m long and BC is 0.6 m long. Loads act only in the region AB, and there are no loads in the region BC. Under this load system, the deflection at B is 0.24 cm and the slope of the beam at B is **θ**, where sin**θ** = 0.038. What is the deflection at D, which is midway between B and C?

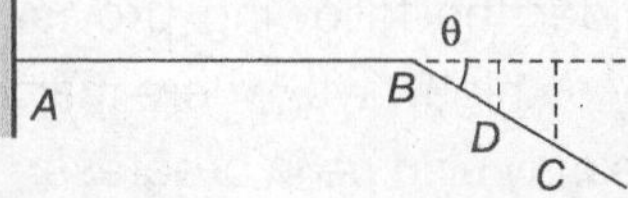

(a) 0.2406 cm (b) 0.2514 cm
(c) 0.2530 cm (d) 0.2452 cm

[ESE Prelims : 2018]

17.5 Consider the following statements regarding a Grillage Foundation:
1. It is provided for heavily loaded isolated columns.
2. It is treated as a spread foundation.

3. It consists of two sets of perpendicularly placed steel columns.

Which of the above statements are correct?

(a) 1 and 2 only (b) 1 and 3 only
(c) 2 and 3 only (d) 1, 2 and 3

[ESE Prelims : 2018]

17.6 A rod of length L, cross-section area A1 and modulus of elasticity E1, has been placed inside a tube of length L, of cross-section area A2 and modulus of elasticity E2, and the two are firmly held by end plates. The portion of the load P applied on the end plates shared by rod and tube, respectively, are

(a) $\dfrac{PA_1E_1}{A_1E_1 + A_2E_2}$ and $\dfrac{PA_2E_2}{A_1E_1 + A_2E_2}$

(b) $\dfrac{PA_2E_2}{A_1E_1 + A_2E_2}$ and $\dfrac{PA_1E_1}{A_1E_1 + A_2E_2}$

(c) $\dfrac{PA_1E_1}{A_1E_1 + A_2E_1}$ and $\dfrac{PA_2E_2}{A_1E_2 + A_2E_2}$

(d) $\dfrac{PA_1E_1}{A_1 + A_2}$ and $\dfrac{PA_2E_2}{A_1 + A_2}$

[ESE Prelims : 2018]

17.7 Among the effects of design specifications on costs, which one of the following is the most significant that influences the producibility of end product?

(a) Standard size (b) Large tolerance
(c) Breakeven point (d) Cost estimate

[ESE Prelims : 2020]

17.8 A plane, extended if necessary, will meet the reference planes in lines, unless it is parallel to any one of them. These lines are called

(a) Projection lines (b) Traces of the plane
(c) Dimension lines (d) Imaginary lines

[ESE Prelims : 2024]

■■■■

General Principles of Design, Drawing, Importance of Safety : Answer Key

1.1	(b)	**1.2**	(a)	**1.3**	(c)	**2.1**	(c)	**2.2**	(a)	**3.1**	(b)	**3.2**	(a)	**3.3**	(b)
4.1	(b)	**4.2**	(b)	**4.3**	(d)	**4.4**	(b)	**4.5**	(a)	**4.6**	(b)	**4.7**	(c)	**4.8**	(c)
5.1.	(a)	**6.1**	(a)	**6.2**	(c)	**7.1**	(b)	**7.2**	(b)	**7.3**	(a)	**7.4**	(a)	**8.1**	(a)
8.2	(d)	**8.3**	(d)	**9.1**	(d)	**9.2**	(c)	**9.3**	(c)	**9.4**	(c)	**9.5**	(b)	**9.6**	(b)
9.7	(d)	**9.8**	(b)	**10.1**	(c)	**10.2**	(b)	**10.3**	(d)	**10.4**	(a)	**11.1**	(c)	**11.2**	(c)
11.3	(c)	**11.4**	(b)	**12.1**	(d)	**12.2**	(d)	**13.1**	(c)	**13.2**	(a)	**13.3**	(a)	**14.1**	(a)
14.2	(a)	**14.3**	(c)	**15.1**	(a)	**16.1**	(c)	**16.2**	(d)	**16.3**	(d)	**16.4**	(c)	**16.5**	(a)
16.6	(c)	**16.7**	(c)	**16.8**	(a)	**16.9**	(c)	**16.10**	(a)	**17.1**	(d)	**17.2**	(a)	**17.3**	(a)
17.4	(*)	**17.5**	(d)	**17.6**	(a)	**17.7**	(b)	**17.8**	(b)						

EXPLANATIONS

1.1 (b)

Concurrent engineering brings together multidisciplinary teams, in which product developers from different functions work together and in parallel with the intention of getting things right as quickly as possible and as early as possible. It decreases product development time and leads to improve the productivity and reduces cost.

ADDITIONAL INFORMATION

Some times, only design engineers and manufacturing engineers are involved in concurrent engineering. In other cases, the cross-functional teams include representatives from purchasing, marketing, production, quality assurance, the field and other functional groups. In concurrent engineering approach to development, input is obtained from as many functional areas as possible before the specifications are finalized. It is greatly facilitated by the use of computer aided engineering.

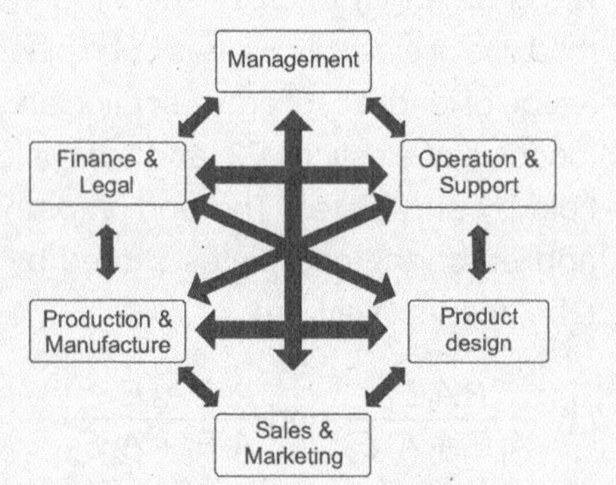

Concurrent Engineering

1.2 (a)

The design of highway interchanges involves the application of the geometry of circle, arcs. The setting of highway networks necessitated interchanges where two highways intersect. The grade separation of interchanges allows continuous traffic without interruptions such as stops or traffic lights, increasing capacity.

ADDITIONAL INFORMATION

There are wide variety of interchange designs with the common including :

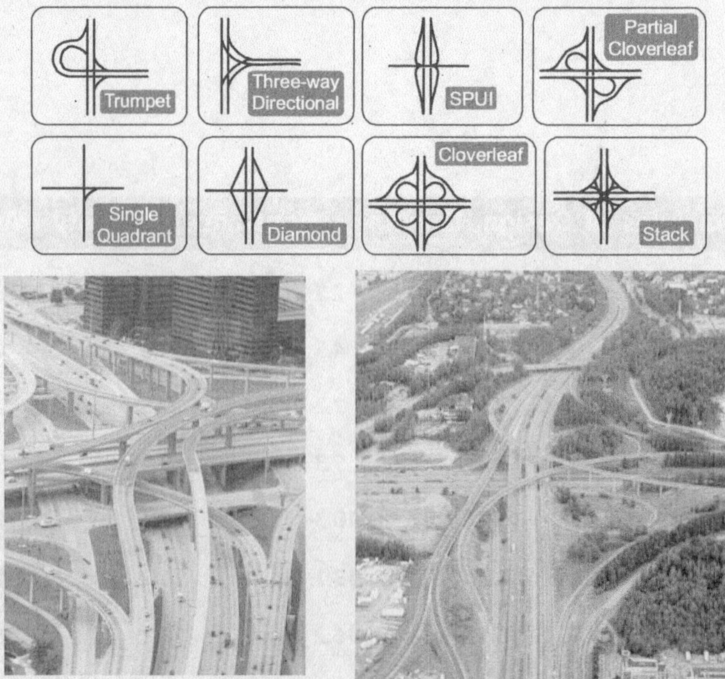

From the above images, we can say that the design of highway interchanges involves the application of circular arcs.

2.1 (c)

Fixation is the human tendency to approach a given problem in a set way that limits one's ability to shift to a new approach to that problem. As such fixation impairs the ideation for designers and results in inpasses. Two examples of fixation are mental set and functional fixedness.

ADDITIONAL INFORMATION

Perpetual blocks have to do with not properly defining the problem and not recognizing the information needed to solve it. These are stereotyping, information overload, limiting the problem unnecessarily, fixation and priming or provision of cues. Psychologists identified two common obstacles to successful problem solving:

1. **Fixation :** It refers to the tendency to set the mind into a routine approach to problem solving. Thinkers who adopt an automatic or mindless approach to problems often overlook ways of representing and searching the problem space that are ideal.
2. **Functional Fixedness :** It refers to the tendency to see the objects as having only a single typical use. Thinkers prematurely categories the elements of a problem in accordance with their typical use, thus overlooking novel and useful alternatives.

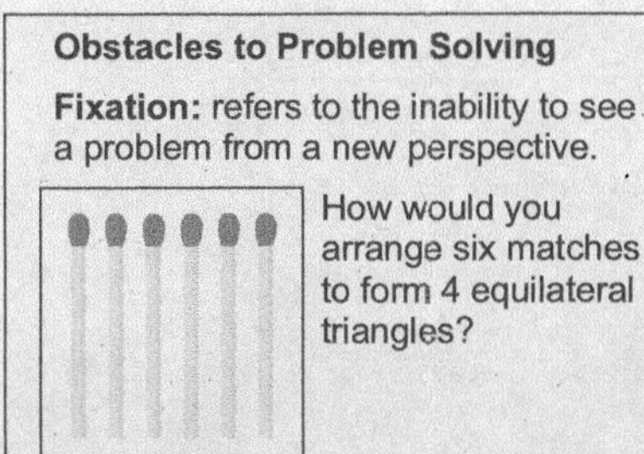

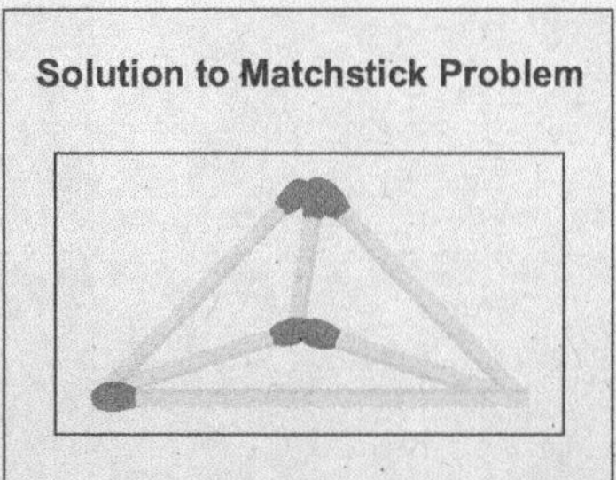

2.2 (a)

The creative design routes are practised by adopting following steps :

1. **Preparation :** It is the first step in the creative process. In this stage, all parts of the problem are thoroughly investigated.
2. **Concentration :** In this stage, energy and resources of the person or of the organisation are focused on solving the problem.
3. **Incubation :** This stage is a largely unconscious phase of the creative process. It is in essence the black box of creativity.
4. **Illumination :** This is the moment of insight or discovery when the answer simply seems to arrive in person's conscious mind from his or her subconcious mind.
5. **Verification :** The final stage of the creative process is verification. This involves testing and verifying the idea or insight as viable.

ADDITIONAL INFORMATION

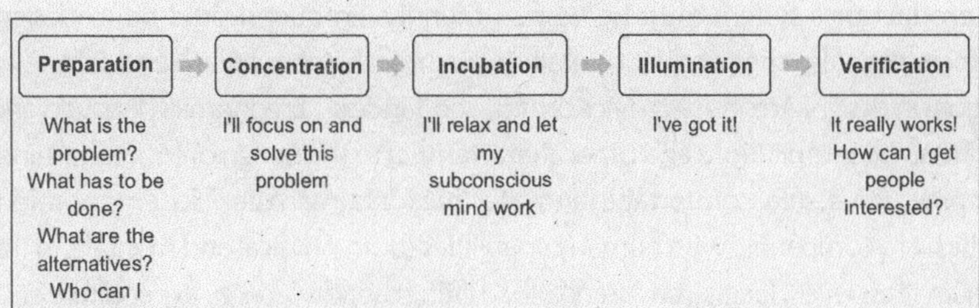

3.1 **(b)**

Two numbers are said to be in the golden ratio if their ratio is the same as the ratio of the sum of the two numbers to the larger number. Let the two numbers be *a* and *b* such that a > b > 0

So, golden ratio :

$$\frac{a}{b} = \frac{a+b}{a}$$

$$\Rightarrow \quad a^2 - b^2 - ab = 0$$

Assuming $b = 1$, we get, $a = \dfrac{1+\sqrt{5}}{2} = 1.618$

ADDITIONAL INFORMATION

Golden ratio has been used in design of structures from ancient times and is used till today in designing buildings with modern architecture.

The **Golden Rectangle** was discovered by Pythagoras in ancient Greece. The golden rectangle has proportion of its sides such that when a square is cut from the rectangle, the remaining rectangle has the same proportions.

Use of golden rectangle

Use of golden spiral

3.2 **(a)**

Ergonomic chairs for instance reduce stress posed on the lumbar region of spinal cord and provide less fatiguing sitting arrangement along with the sitting comfort. Provision of power steering in the design of heavy vehicles is an ergonomic consideration reducing lots of driving fatigue. Even the arrangement of keys on a computer keyboard which is not in continuation (*A, B, C, D, ... Z*), and rather seem to be haphazardly arranged is attributed to ergonomic considerations. This ergonomic design reduces and optimizes the fingers' stress level and also facilitates speedy typing.

ADDITIONAL INFORMATION

A sound design of a product or a system must be human friendly. Product should be so designed that it is easy to understand, use and maintain. Also the product operation ought to be convenient and safe. All these are ensured in ergonomics considerations in design. According to McKormick and Sanders "Ergonomics discovers and applies information about human behaviour, abilities, limitations and other characteristics to the design of tools, machines, systems, tasks, jobs and environments for productive, safe, comfortable and effective for human use." So, ergonomics helps the designer to create the products which are compatible with human capabilities (both physical and mental). In the domain of physical capacities come maximum stress levels, fatigue, endurance, load lifting capacity, energy expenditure etc. These considerations in the product design help in reducing operator or user's physical stress level and avoid injuries and accidents.

3.3 (b)

Modularisation is a design approach that subdivides a system into smaller parts called modules that can be independently created and then used in different systems. A modular system can be characterised by functional partitioning into discrete scalable, reusable modules; rigorous use of well defined modular interfaces; and making use of industry standards for interfaces.

ADDITIONAL INFORMATION

There are two entirely opposite styles of product architecture :

1. **Integral Architecture :** In this type, the implement of functions is accompanied by only one or a few modules. In integral product architectures, components perform multiple functions. This reduces the number of components, generally decreasing cost unless the integral architecture is obtained at the expense of extreme part complexity.
2. **Modular Architecture :** A modular architecture makes it easier to evolve a design over time. The product can be adapted to the needs of different customers by adding or deleting modules. Modular design may even be carried to the point of using the same set of basic components in multiple products, creating a product family.

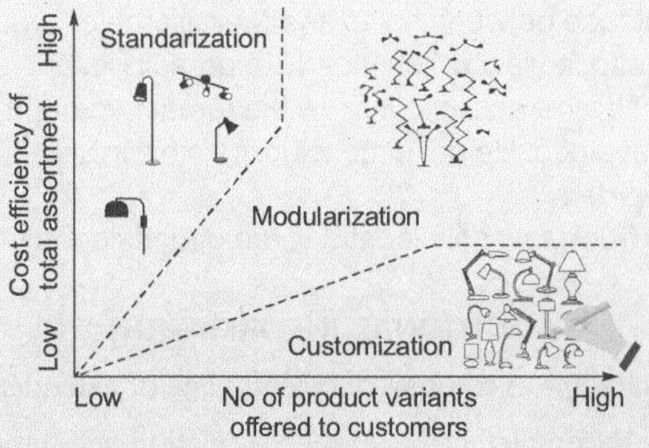

4.1 (b)

Dimension in series may be placed in any one of the following two ways :

1. **Chain Dimensioning/Continuous Dimensioning :**

   When successive dimensions are arranged in a straight line, the method is known as chain dimensioning. Note that the chain of dimensions shall be arranged in a continuous straight line.

   Chain of single dimension should be used only where the possible accumulation of tolerances dos not impinge on the functional requirements of the part.

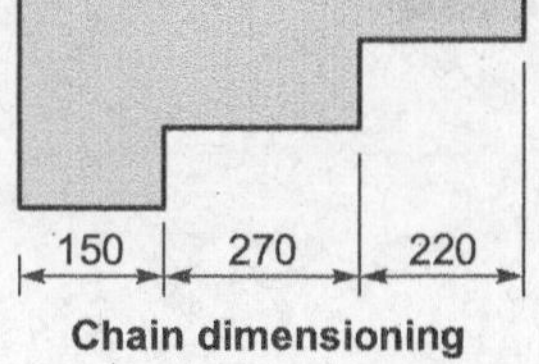

Chain dimensioning

2. **Parallel Dimensioning/Progressive Dimensioning :**

   Parallel dimensioning is the placement of a number of single dimension lines parallel to one another from a common origin. It is used where a number of dimensions have a common origin. Cumulative error is avoided by this method.

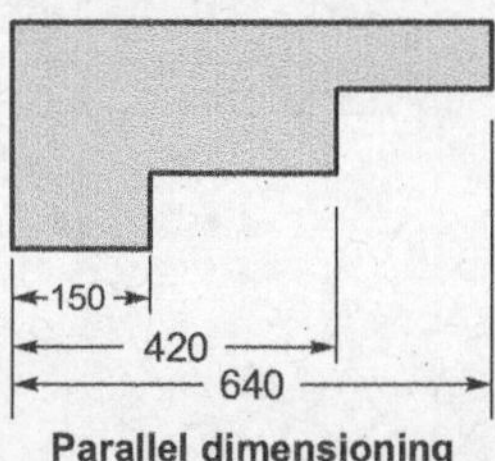

Parallel dimensioning

ADDITIONAL INFORMATION

Dimensional lines should be drawn at least 8 mm away from the outlines and from each other. The mutual crossing of dimension lines and dimensioning between hidden lines should be avoided. Dimension lines should not cross any other line of the drawing. Aligned system of dimensioning is recommended.

4.2 (b)

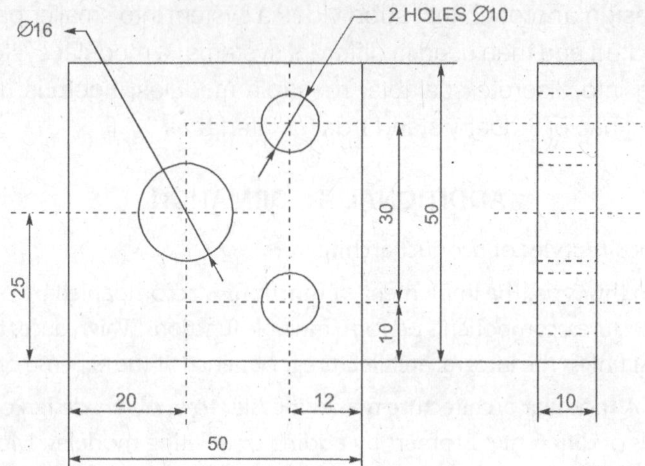

You can observe that dimensioning between hidden lines is avoided. In side view holes are appearing as hidden lines and hence distance between holes and their diameter is not dimensioned in side view.

Dimensions are always placed in view which clearly express relevant features. In front view all width and height dimensions are placed whereas in side view the remaining depth dimension is marked.

Repetition of dimensions is avoided. Height dimension is visible in side view also but since it is dimensioned in front view repetition is avoided.

You can observe that in the front view hole locations are dimensioned from their center lines.

ADDITIONAL INFORMATION

1. Every dimension lines are a thick line as compared to centre line or extended line.
2. Dimensioning should be done so completely that further calculation or assumption of any dimension, or direct measurement from the drawing is not necessary.
3. Dimensions should be placed outside the views, unless they are clearer and more easily read inside.

4.3 (d)

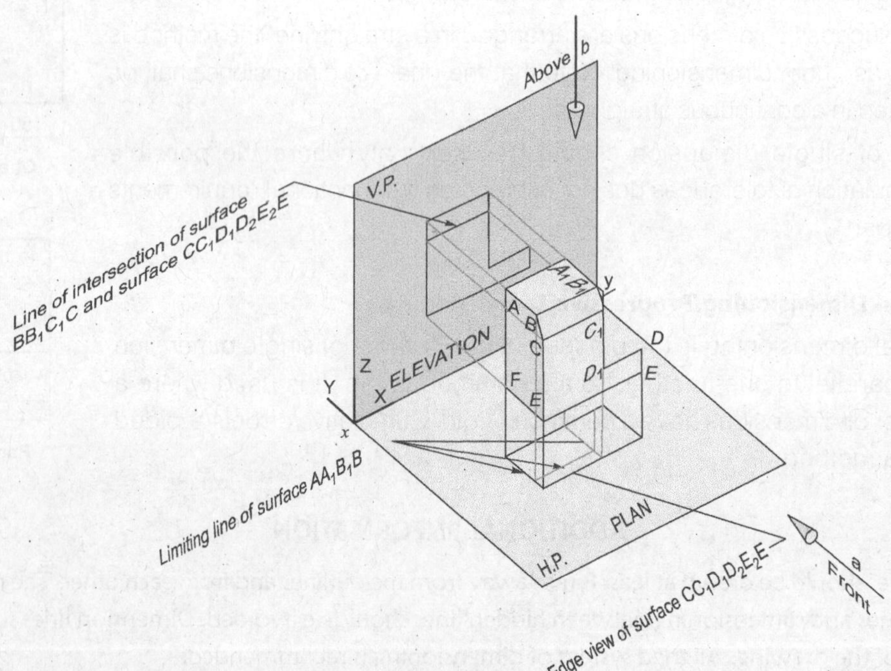

Figure above shows all three conditions.

ADDITIONAL INFORMATION

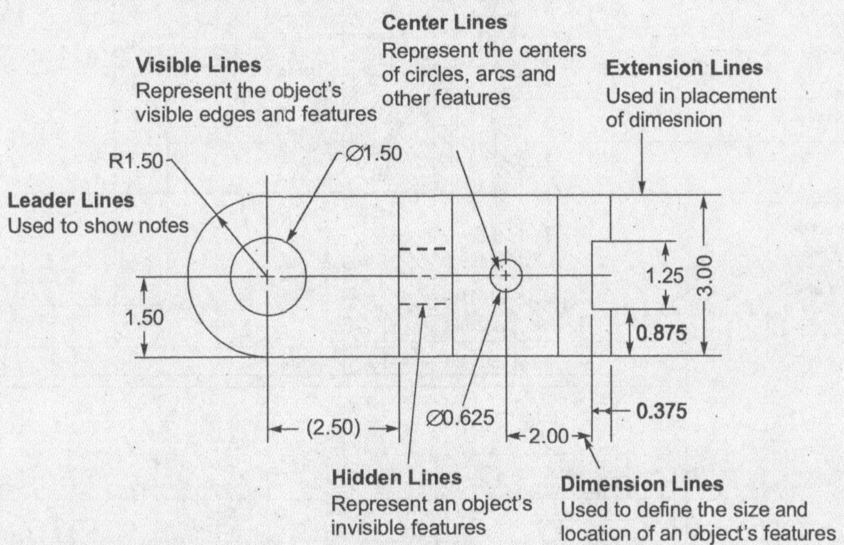

- Visible lines represent the objects visible edges and features.
- Invisible/hidden lines represent the objects invisible features.

4.4 **(b)**

Section or Hatching lines are drawn to make the section evident. They are continuous thin lines and are drawn generally at an angle of 45° to the main outline of the section. They are uniformly spaced about 1 mm to 2 mm apart.

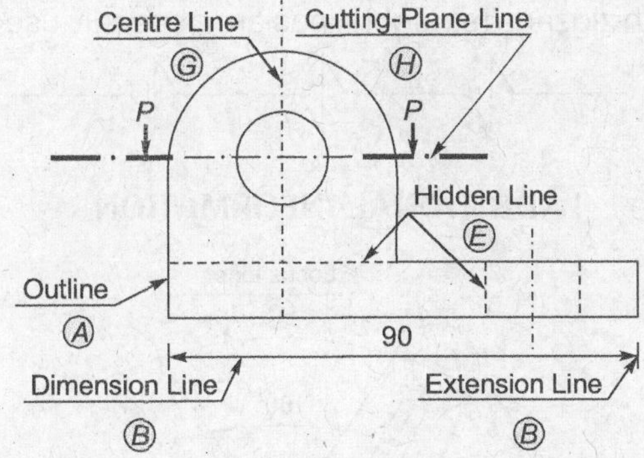

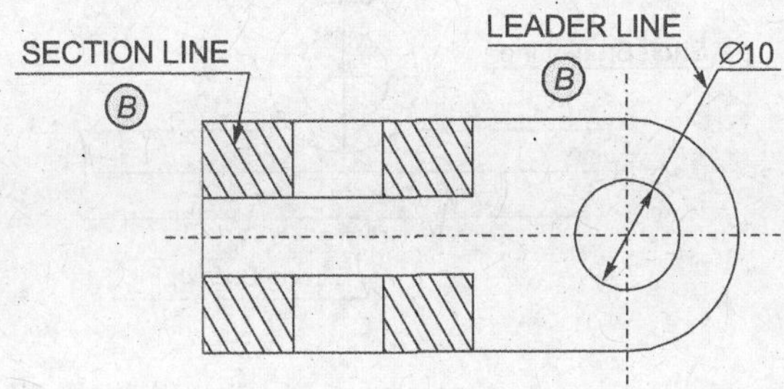

ADDITIONAL INFORMATION

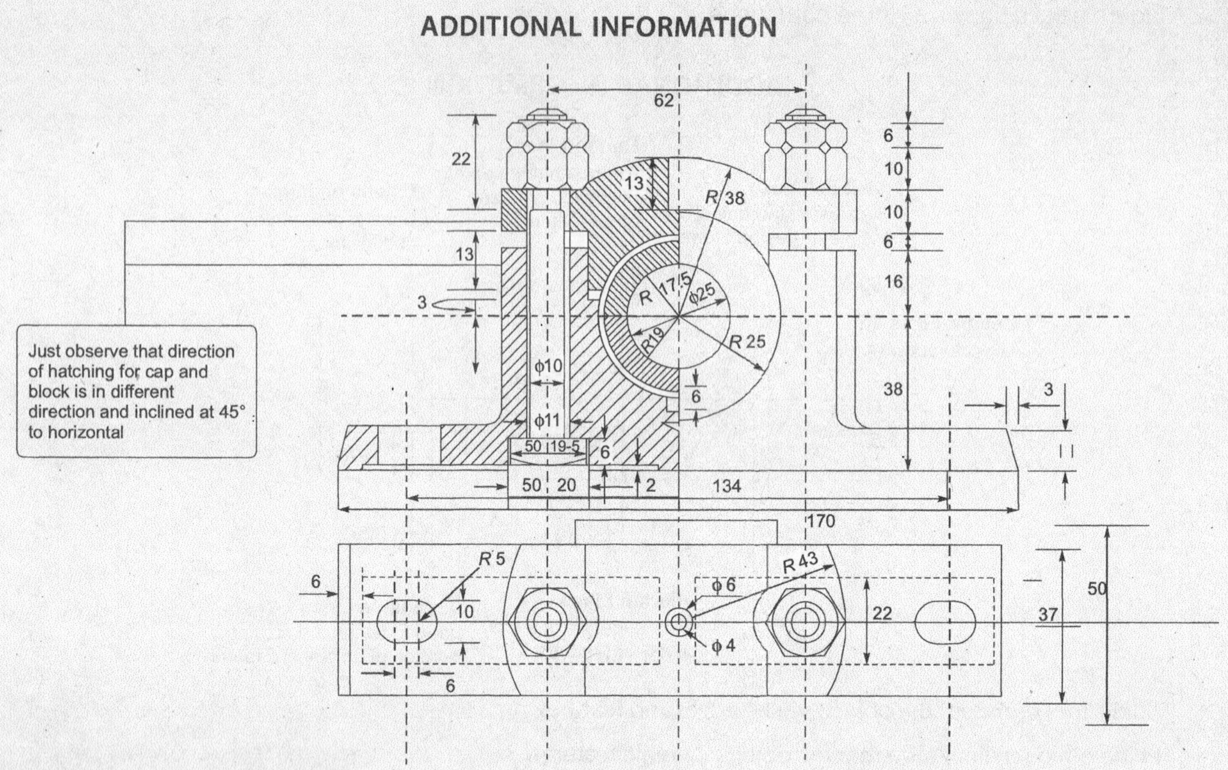

Figure above clearly shows that in front view left half in section of the Plummer block, Cap and Block which are in contact, are hatched in mutually perpendicular direction and at 45° to horizontal.

4.5 (a)

Continuous thin (narrow) with zigzags (straight) lines are generally used to represent long-break line.

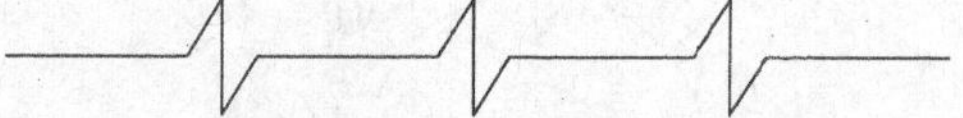

ADDITIONAL INFORMATION

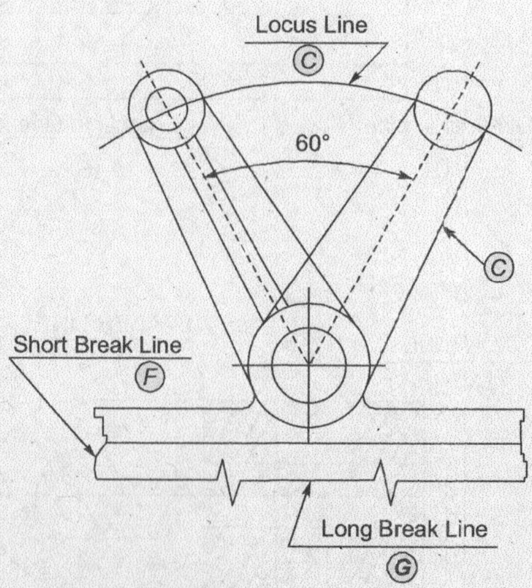

As shown in figure above, long break lines are continuous narrow lines with Zig Zag. Hidden line are represented by dashed wide line in general. Visible outline is generally represented by Continuous wide lines. Reference line is generally represented by continuous narrow line.

4.6 (b)

Tabular drawings are used for parts that have different dimension but same shape. When several parts that differ only a few dimensions are required, only one drawing with a table of sizes is made. In other words, a series of objects having like features but varying in dimensions may be represented by one drawing as shown in figure. Letters are substituted for dimensions value on the drawing, and the varying dimensions are given in the tabular form.

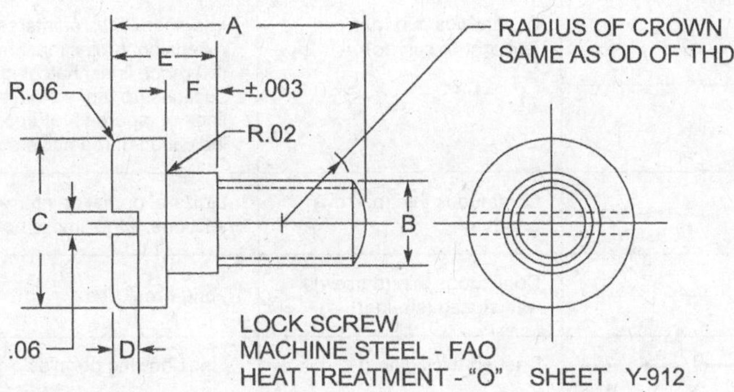

DETAILS	A	B	C	D	E	F	UNC THD	STOCK	LBS
1	.62	.38	.62	.06	.25	.135	.312 – 18	ø.75	.09
2	.88	.38	.62	.09	.38	.197	.312 – 18	ø.75	.12
3	1.00	.44	.75	.12	.38	.197	.375 – 16	ø.875	.19
4	1.25	.50	.88	.12	.50	.260	.437 – 14	ø.1	.30
5	1.50	.56	1.00	.16	.62	.323	.5 – 13	ø1.125	.46

ADDITIONAL INFORMATION

- A patent application or patent may contain drawings, illustrating the invention, some of its embodiments or the prior art are called patent drawings.
- A schematic drawing is intended to explain how something works, a drawing showing the relation between the parts.
- Installation drawings means a drawing based on the detailed design or coordination drawing with the primary purpose of defining that information needed by the tradesmen on site to install the works.

4.7 (c)

Chain thin double-dashed or long-dashed double-dotted (narrow) lines are used to represent outlines of adjacent parts, alternative and extreme positions of movable parts, centroidal lines, initial outlines prior to forming parts situated in front of the cutting plane.

______ . . ______ . . ______ . . ______ . . ______

Chain thin double-dashed line

ADDITIONAL INFORMATION

Line	Description	General applications
A ———	Continuous thick or Continuous wide	Visible outlines, visible edges; crests of screw threads; limits of length of full depth thread, lines of cuts and section arrows; parting lines of moulds in views; main representations in diagrams, maps, flow charts; system lines (structural metal engg.)
B ———	Continuous thin (narrow) (straight or curved)	Imaginary lines of intersection ; grid dimension, extension, projection, short centre, leader, reference lines; hatching; outlines of revolved sections; root of screw threads; interpretation lines of tapered features; framing of details; indication of repetitive details.
C	Continuous thin (narrow) freehand	Limits of partial or interrupted views and sections, if the limit is not a chain thin line.
D	Continuous thin (narrow) with zigzag (straight)	Long-break line.
E	Dashed thick (wide)	Line showing permissible of surface treatment.
F	Dashed thin narrow)	Hidden outlines; hidden edges.
G	Chain thins Long-dashed dotted (narrow)	Centre line; lines of symmetry; trajectories; pitch circle of gears, pitch circle of holes.
H THICK THIN THICK	Chain thin (narrow) with thick (wide) at the ends and at changing of position	Cutting planes.
J	Chain thick or Long-dashed dotted (wide)	Indication of lines or surfaces to which a special requirement applies.
K	Chain thins double-dashed or long-dhased double-dotted (narrow)	Outlines of adjacent parts : Alternative and extreme positions of movable parts. Centroidal lines : Initial outlines prior to forming parts situated in front of the cutting plane.

4.8 **(c)**

In manual drafting, French curves are used for drawing curves which cannot be drawn by compass. Faint freehand curve is first drawn through the known points. Longest possible curves exactly coinciding with the freehand curve are then found out from the French curve. Finally, neat continuous curve is drawn with the aid of the French curve. Care should be taken to see that no corner is formed anywhere within the drawn curve.

5.1 **(a)**

- 1.5 inches length on drawing is represented by actual length of 12 inches.
- Conceptually representative fraction of a drawing indicates the ratio by which linear distance in drawing are changed compared to actual length.

- The formula for R.F. is

$$\text{R.F.} = \frac{\text{Distance between points A and B on drawing}}{\text{Actual distance between the points A and B}} = \frac{1.5}{12} = \frac{1}{8}$$

Hence the correct answer is (a).

ADDITIONAL INFORMATION

If we have to construct this scale, the following steps should be followed :

(i) Determine the R.F. of the scale.

$$\text{R.F.} = \frac{1.5 \text{ inches}}{1 \times 12 \text{ inches}} = \frac{1}{8}$$

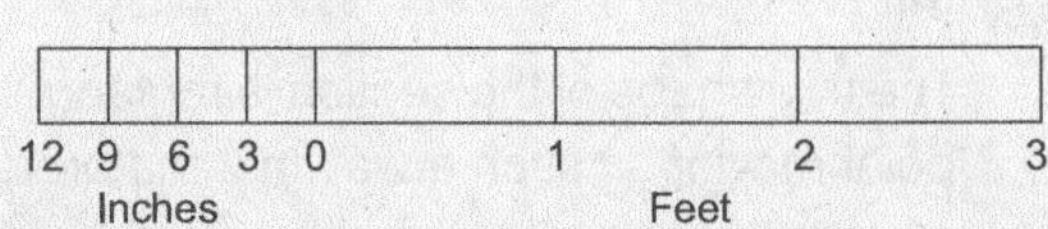

(ii) Draw a line, 1.5 × 4 = 6 inches long.

(iii) Divide it into four equal parts, each part representing one foot.

(iv) Divide the first division into 12 equal parts, each representing 1". Complete the scale.

6.1 **(a)**

"A spiral"

AB is pendulum

P, P_1, P_2, P_3, P_4 are points on the curve and that is spiral.

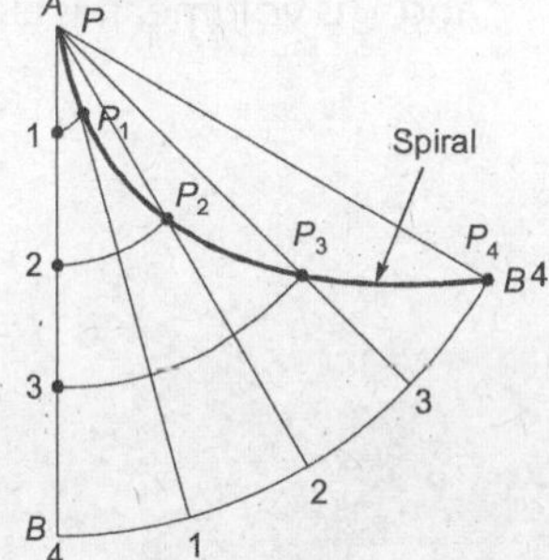

ADDITIONAL INFORMATION

- If the line rotates in a plane about one of its ends and at the same time if a point moves along the line continuously in one direction the traced out curve by the moving point is called **spiral.**
- A **helix** is formed when a point moves along and around the outermost portion or along the surface of a cylinder or cone.
- A **cycloid** is formed when a rolling circle rolls and you mark a point on the parameter of the rolling circle then the locus of that point forms cycloid.
- An **involute** is is a curve traced out by an end of a piece of thread unwound from a circle or a polygon, the thread being kept tight.

6.2 **(c)**

A straight line.

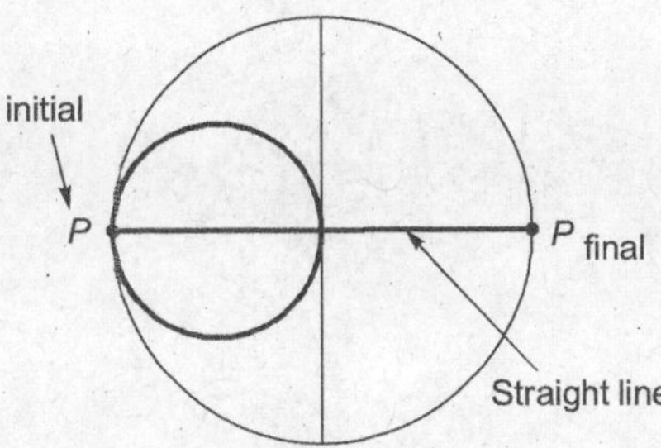

"When the diameter of directing circle is twice the diameter of generating circle then hypocycloid is a straight line".

ADDITIONAL INFORMATION

- The curve generated by a point on the circumference of a circle, which rolls without slipping along another circle inside it, is called a hypocycloid.
- When the diameter of the rolling circle is half the diameter of the directing circle, the hypocycloid is a straight line and is a diameter of the directing circle.

7.1 (b)

The true shape of the section is rectangular hyperbola. Rectangular hyperbola is a curve traced out by a point moving in such a way that the product of its distances from two fixed lines at right angles to each other is constant. The fixed lines are called asymptotes.

ADDITIONAL INFORMATION

The rectangular hyperbola curve graphically represents the Boyle's law i.e., PV = constant, where P is pressure and V is volume. It is also useful in design of water channels.

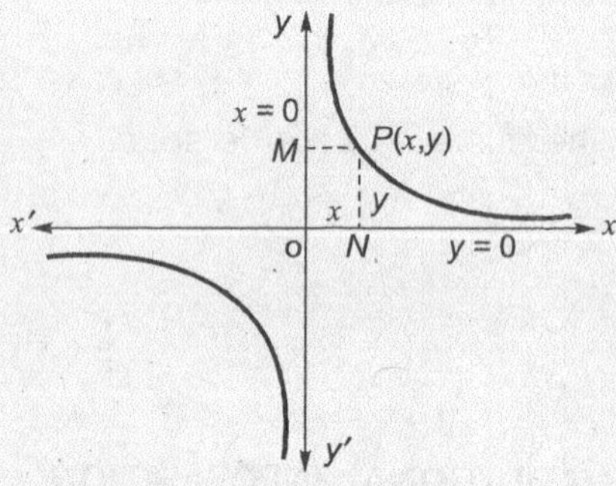

7.2 (b)

When the section plane is inclined to the axis and is parallel to one of the generators, the section is a **parabola**.

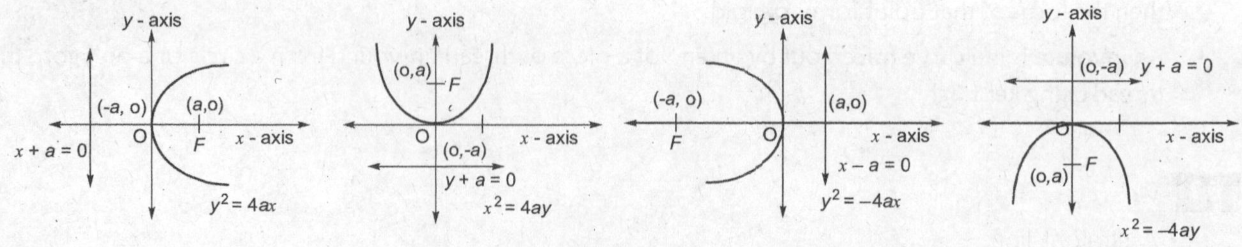

ADDITIONAL INFORMATION

The ratio $\frac{\text{Distance of the point from the focus}}{\text{Distance of the point from the directrix}}$ is called eccentricity and is denoted by *e*. It is always equal to 1 for parabola.

The sections obtained by the intersection of a right circular cone by a plane in different positions relative to axis of the cone are called conics. The various conic sections are :

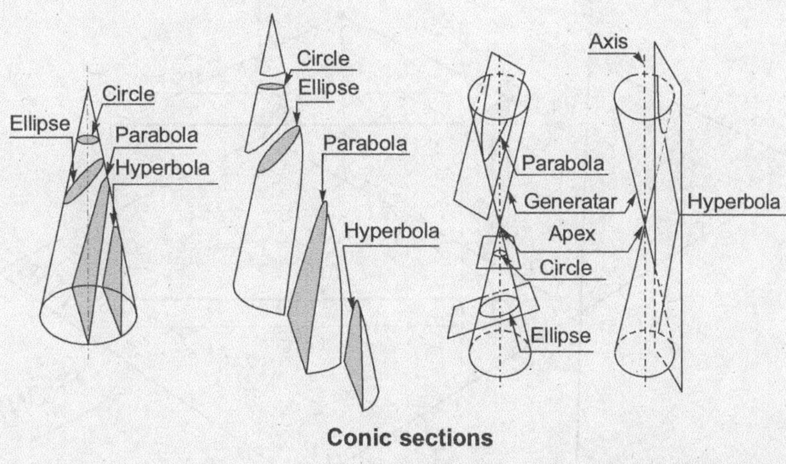

Conic sections

7.3 (a)

The line passing the focus and perpendicular to the directrix is called the axis.

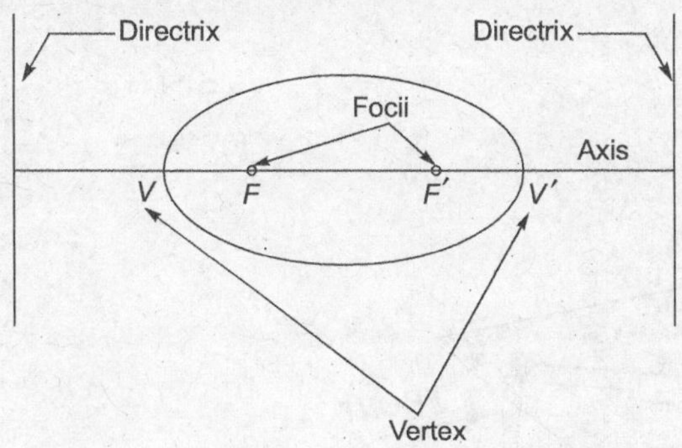

ADDITIONAL INFORMATION

- The point at which the conic cuts its axis is called vertex.
- The ratio $\frac{\text{Distance of the point from the focus}}{\text{Distance of the point from the directrix}}$ is called **eccentricity**.
- The sections obtained by intersection of a right circular cone by a plane in different positions relative to the axis of the cone are called **conics**.

7.4 (a)

Rectangular hyperbola is a curve traced out by a point moving in such a way that the product of its distances from two fixed lines at right angles to each other is constant. The fixed lines are called asymptotes.

ADDITIONAL INFORMATION

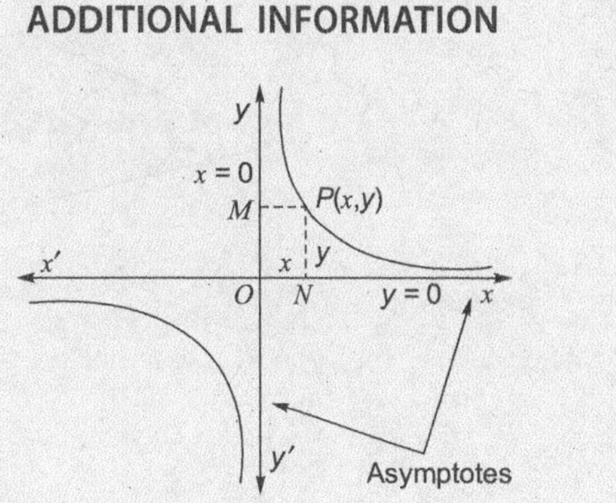

The curve graphically represents the Boyle's law i.e. PV = constant.

8.1 (a)

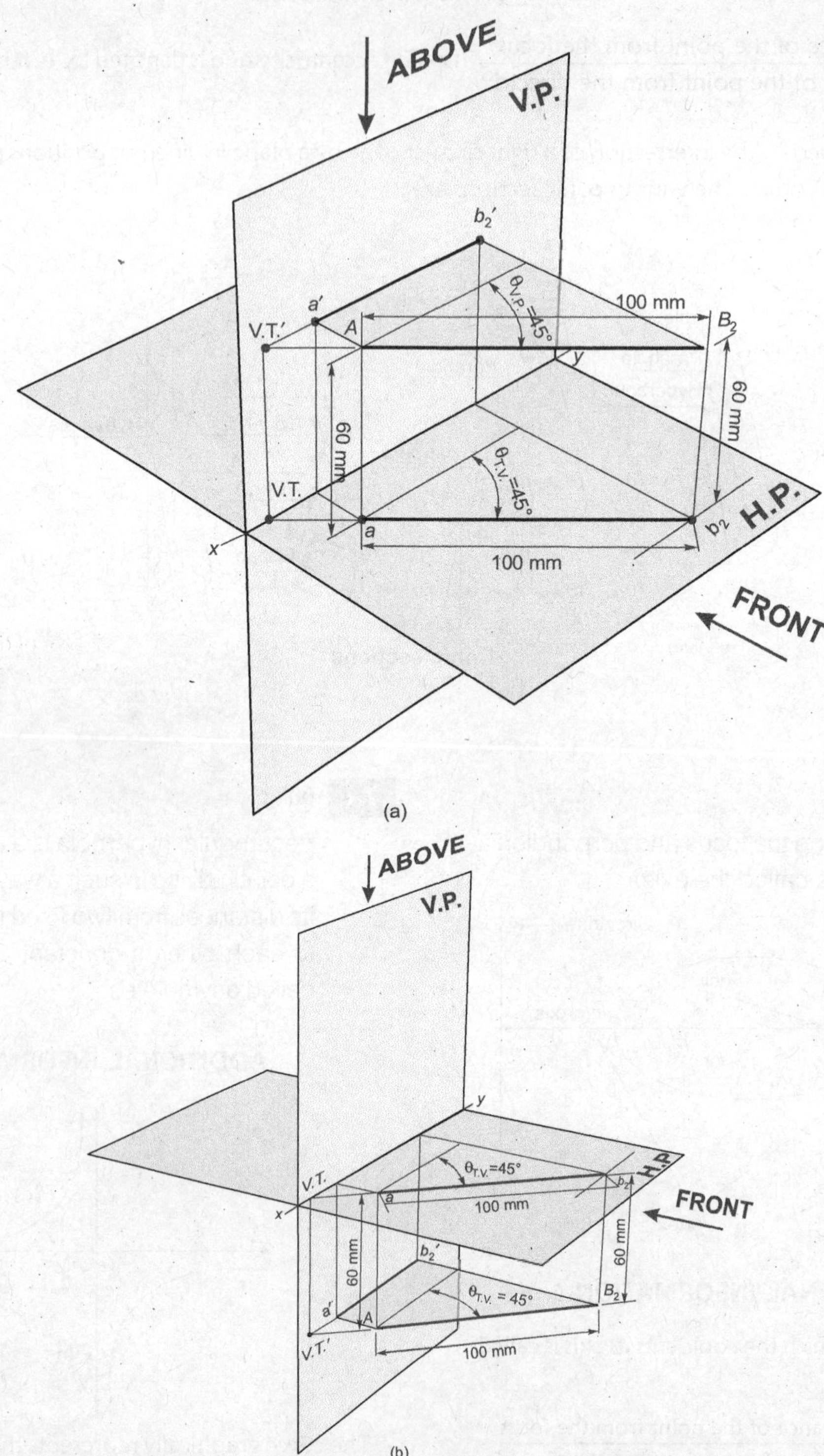

(a)

(b)

Pictorial view in figure (a) and figure (b) clearly shows that when a line is parallel to H.P. and inclined to V.P. then

- In 3 dimensions a rectangle AB_2b_2a is formed between line and its top view. Since opposite sides of a rectangle are equal, length of top view ab_2 is equal to true length of line AB_2.
- As far as vertical trace is concerned as shown in figure (a) if line is above H.P. then V.T. is above H.P. and as shown in figure (b) if line is below H.P. then V.T. is below H.P.

ADDITIONAL INFORMATION

When a line is inclined to one plane and parallel to the other, its projection on the plane to which it is inclined, is a line shorter than its true length but parallel to the reference line. Its projection on the plane to which it is parallel, is a line equal to its true length and inclined to the reference line at its true inclination. In other words, the inclination of a line with the H.P. is seen in the front view and that with the V.P. is seen in the top view.

8.2 (d)

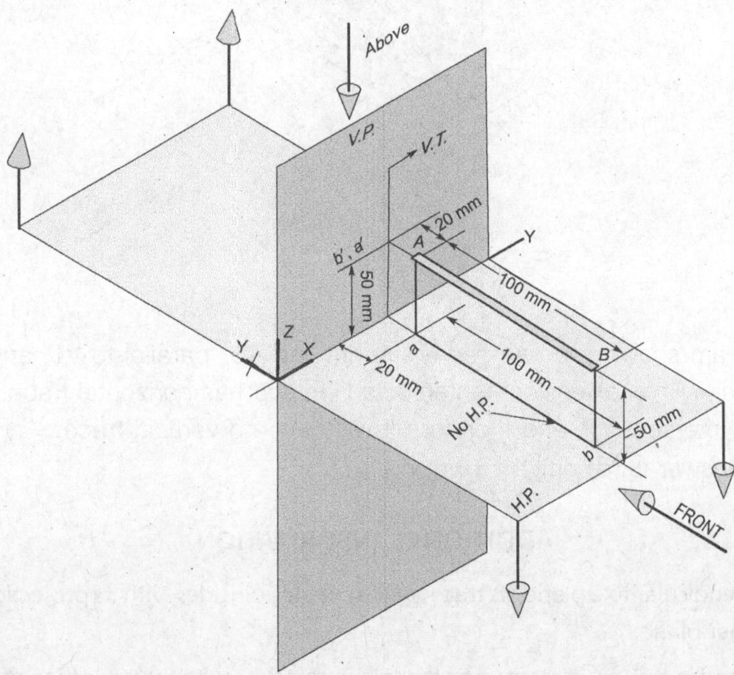

Figure above shows a line perpendicular to V.P. From 3-D model. It is clear that Vertical trace is on V.P. hence statement (a) & (c) are wrong. A line perpendicular to V.P. is always parallel to H.P. and it never intersects H.P. & has no H.T. hence statement (d) is correct.

ADDITIONAL INFORMATION

- When a line is perpendicular to a plane, its trace on that plane coincides with its projection on that plane. It has no trace on the other plane.
- When a line has an end in a plane, its trace upon that plane coincides with the projection of that end on that plane.

8.3 **(d)**

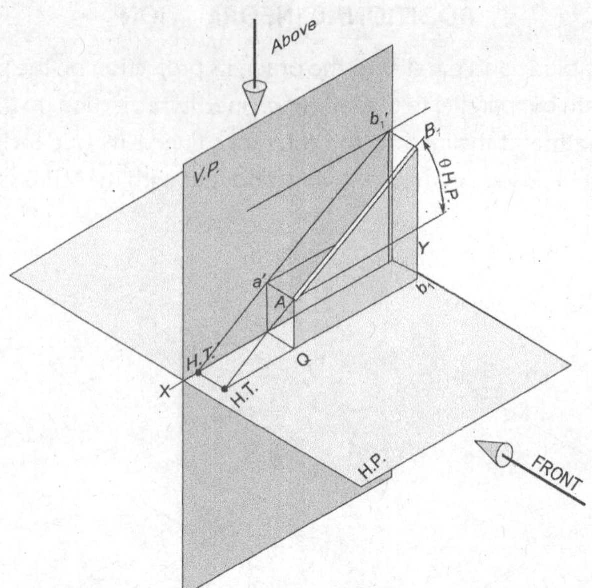

In pictorial diagram shown you can see a straight line AB_1 parallel to V.P. and inclined to H.P. Since line is inclined to H.P. on extension it intersects H.P. and has horizontal trace. Since line is parallel to V.P. it will never intersect V.P. and therefore it will have no vertical trace.

Hence correct answer is (d) only H.T. but no V.T.

ADDITIONAL INFORMATION

- When a line is perpendicular to a plane, its trace on that plane coincides with its projection on that plane. It has no trace on the other plane.
- When a line has an end in a plane, its trace upon that plane coincides with the projection of that end on that plane.

9.1 **(d)**

Whenever a solid is suspended freely then it means that solid is free to rotate about point of suspension. When the solid attains a state of rotational equilibrium then net external torque on it becomes zero. When line joining point of suspension and centre of gravity becomes vertical, line of action of downward gravitational attraction force through CG and upward reaction force offered by support at point of suspension becomes same. Hence Net external torque on the body becomes zero. Hence for a freely suspended body line joining point of suspension and centre of gravity becomes vertical.

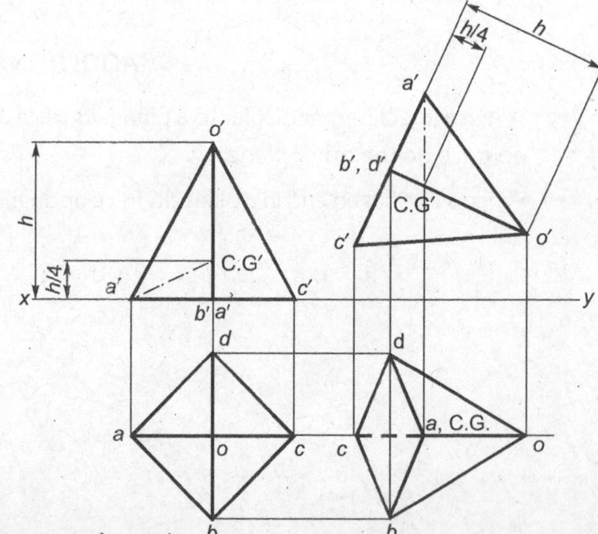

a'CG' is the imaginary line joining the corner and center of gravity.

ADDITIONAL INFORMATION

When a solid is freely suspended from a corner then the line joining point of contact and centre of gravity becomes vertical.

9.2 (c)

In case of isometric projection, the object is kept in such a way that its three mutually perpendicular edges make equal angles with the plane of projection and the object stands on one of its corners. When all three coordinate axes of solid are equally inclined to plane of projection, then the single view orthographic projection obtained is called isometric projection.

ADDITIONAL INFORMATION

Isometric projection is a type of pictorial projection in which the three dimensions of a solid are not only shown in one view, but their actual sizes can be measured directly from it.

The three lines *CB*, *CD* and *CG* meeting at the point *C* and making 120° angles with each other are termed isometric axes. The lines parallel to these axes are called isometric lines. The planes representing the faces of the cube as well as other planes parallel to these planes are called isometric planes.

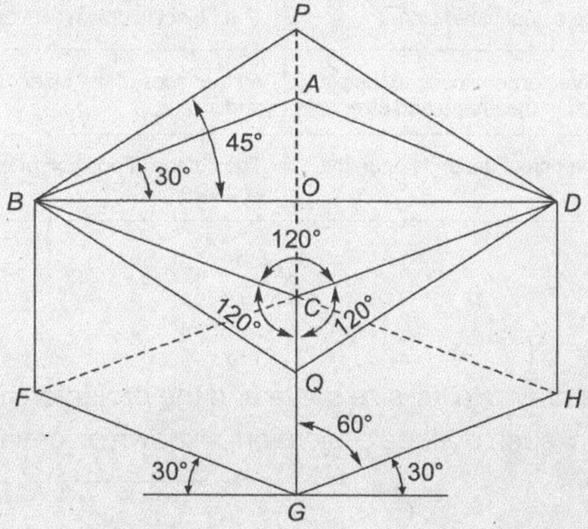

9.3 (c)

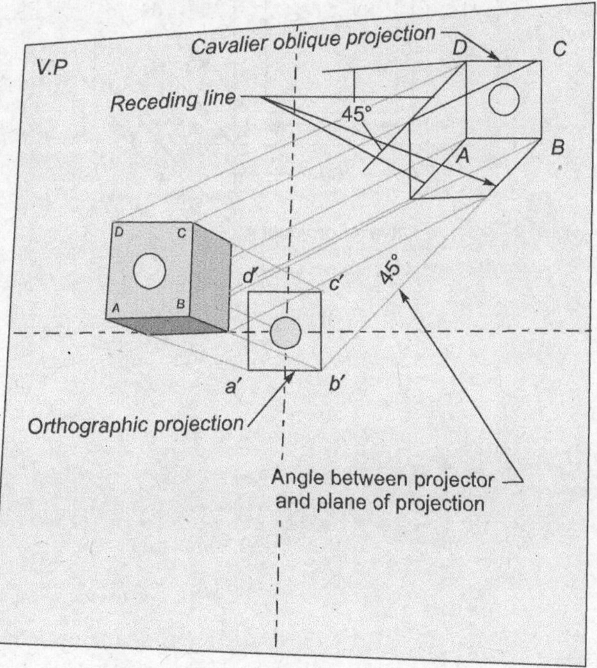

Single view parallel projection when projectors are inclined to the projection plane at angle other than 90° is called oblique projection. While obtaining oblique projection face containing maximum information is kept parallel to plane of projection. Figure above clearly shows that since front face is kept parallel to plane of projection & circle on front face is appearing of true shape in oblique projection.

ADDITIONAL INFORMATION

Sl. No.	Oblique projection	Isometric projection
(i)	Projectors from an object are parallel to each other and inclined to the plane of projection.	Projectors from an object are parallel to each other and perpendicular to the plane of picture.
(ii)	The object is placed in such a way that one of its prominent faces remains parallel to the plane of projection. The object rests on one of its faces.	The object is kept in such a way that its three mutually perpendicular edges (axes) make equal angles with the plane of projection. The object stands on one of its corners.
(iii)	The object is drawn with the actual dimensions.	The object is drawn with the reduced (about 82%) dimensions.
(iv)	The faces of object which are perpendicular to the plane of projection will be distorted in the shape and size.	All the faces of the object are distorted in the shape and size.
(v)	The choice of position of the object depends upon the shape and size.	The choice of position of the object depends upon the shape and size.

9.4 **(c)**

When the receding lines are drawn to full size scale and the projectors inclined at an angle of 30° or 45° or 60° to the plane of projection, such oblique projection is known as cavalier projection.

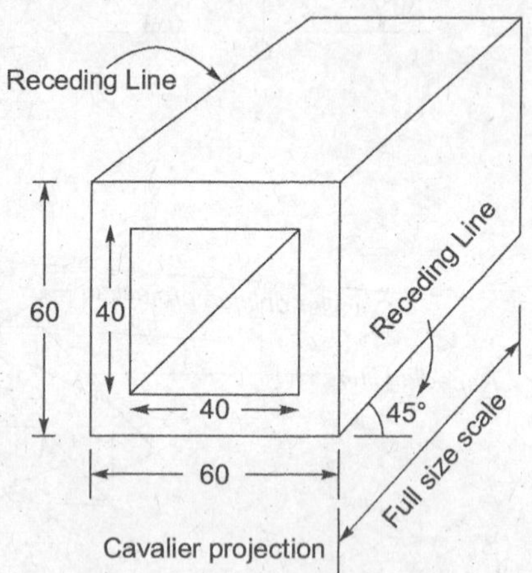

Cavalier projection

ADDITIONAL INFORMATION

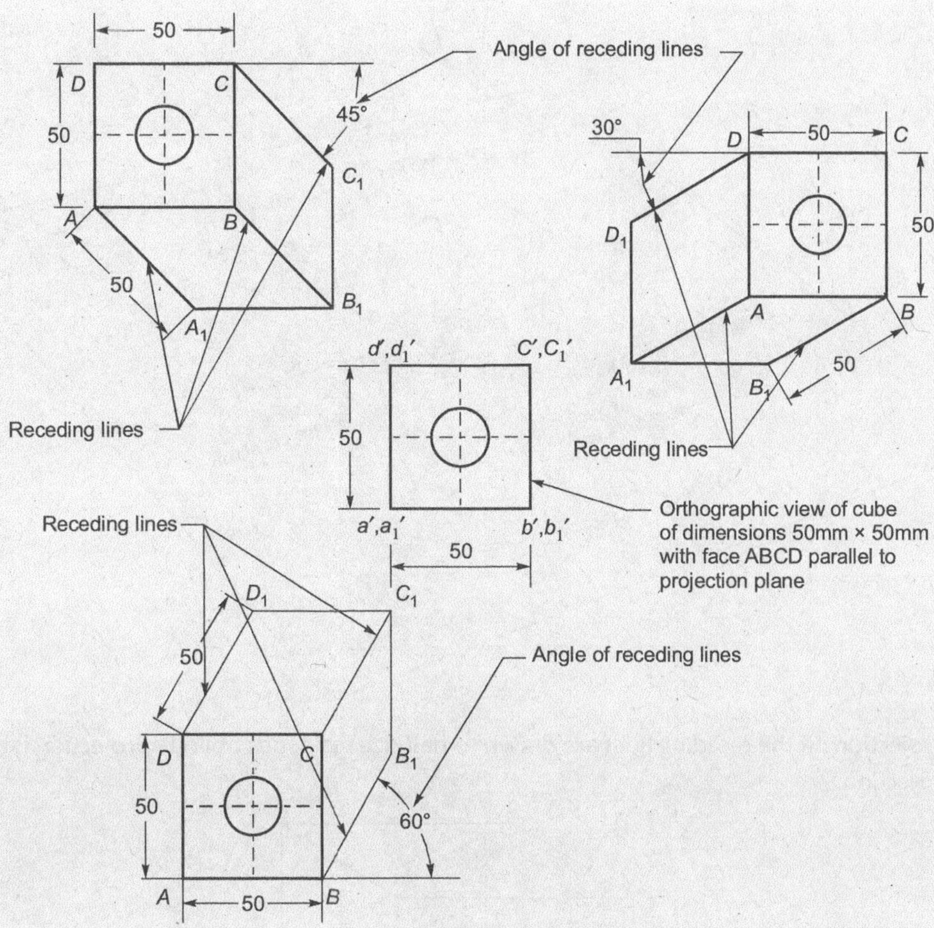

Cavalier Oblique Projection of Cube

As shown in above figure, cavalier oblique projection receding lines are drawn to full size scale. Figure below gives the 3-Dimensional concept of cavalier oblique projection. In 3-dimensions when obtaining cavalier oblique projection most important face of solid is kept parallel to projection plane and parallel projectors dropped from corner point of solid make 45° angle with plane of projection. In 3-dimensions lines AA_1, BB_1, CC_1 & DD_1 of cube are perpendicular to projection plane and their oblique projection is called receding lines. As angle between projectors and plane of projection is 45° in Cavalier oblique projection, by geometry we reach the conclusion that receding lines will be of true length.

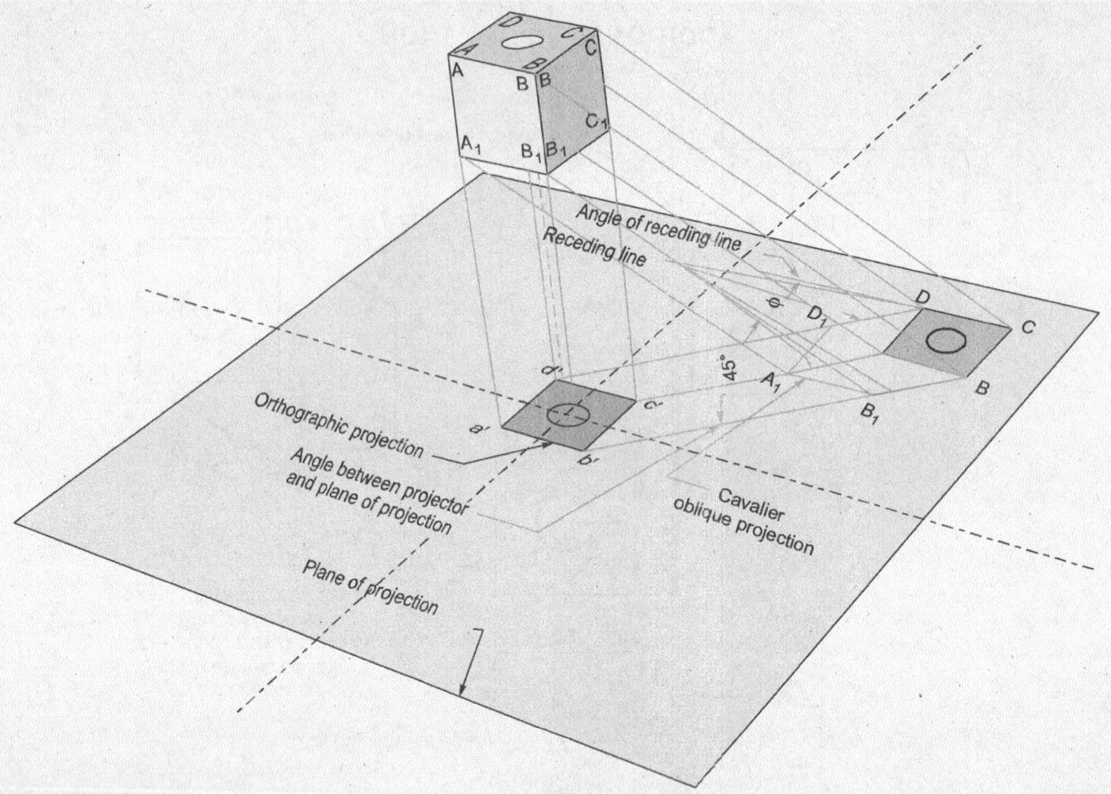

- **Cabinet Projection :** If the receding lines are drawn to half size scale such oblique projection is known as the cabinet projection.

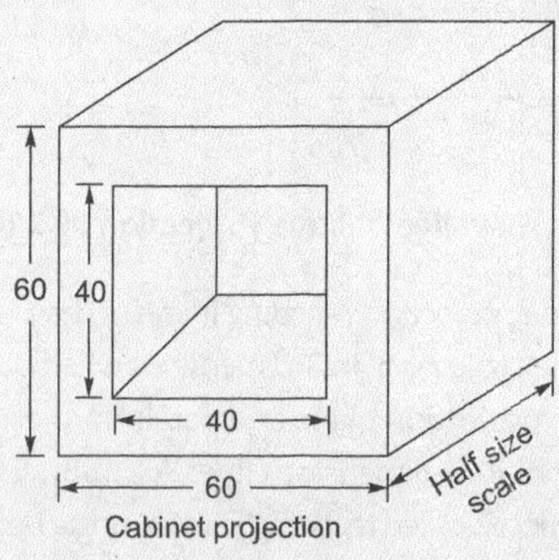

Cabinet projection

9.5 (b)

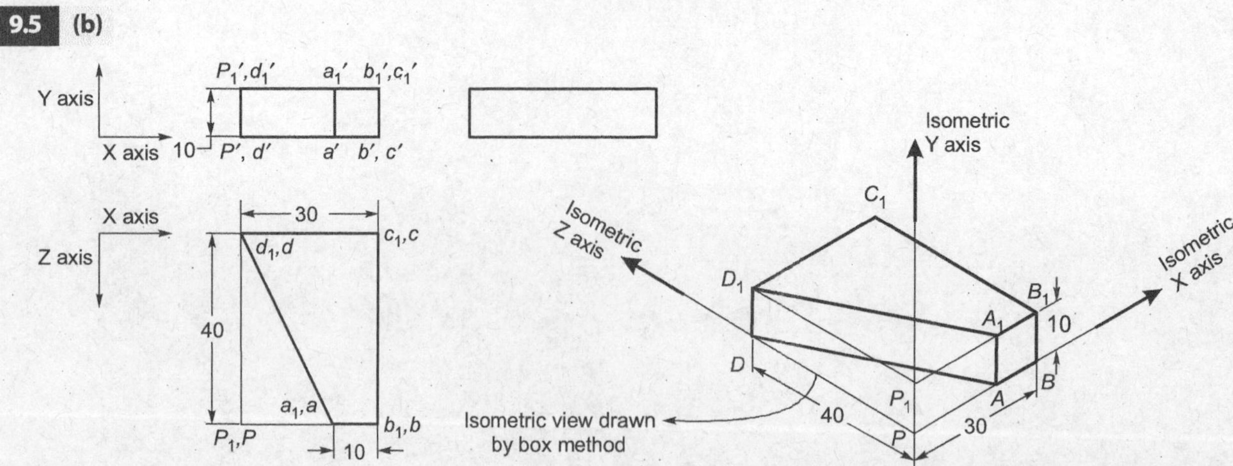

In above drawing AD is a non-isometric line lying in Isometric plane XZ. A_1D_1 is also a non-isometric line lying in Isometric plane parallel to XZ plane and 10 mm above it. To draw the non-isometric lines AD and A_1D_1 lying in isometric planes you enclose the solid in a box with upper rectangle $P_1B_1C_1D_1$ &lower rectangle PBCD. You can draw the isometric view of box easily because all its lines of box are along one of the 3-dimensional axis i.e. X axis or Y axis or Z axis. Point A_1 is located on isometric line P_1B_1 and joined to D_1 to obtain the non-isometric line A_1D_1. Point A is located on isometric line PB and joined to D to obtain the non-isometric line AD.

ADDITIONAL INFORMATION

When an object contains inclined edges which in the isometric view would be shown by non-isometric lines, the view may be drawn by using any one of the following methods:

(i) **Box method :** This method is used when the non-isometric lines or their ends lie in isometric planes.

(ii) **Co-ordinate or offset method :** This method is adopted for objects in which neither non-isometric lines nor their ends lie in isometric planes.

9.6 **(b)**

In pictorial diagram shown you can see a right triangular prism of edge of base 40 mm and height 50 mm. The prism is having its base ABC in H.P. and vertical face A_1B_1BA is perpendicular to V.P.. You can observe that when you look at the prism from front direction you see the rectangle B_1C_1CB. Hence front view of the prism is a rectangle.

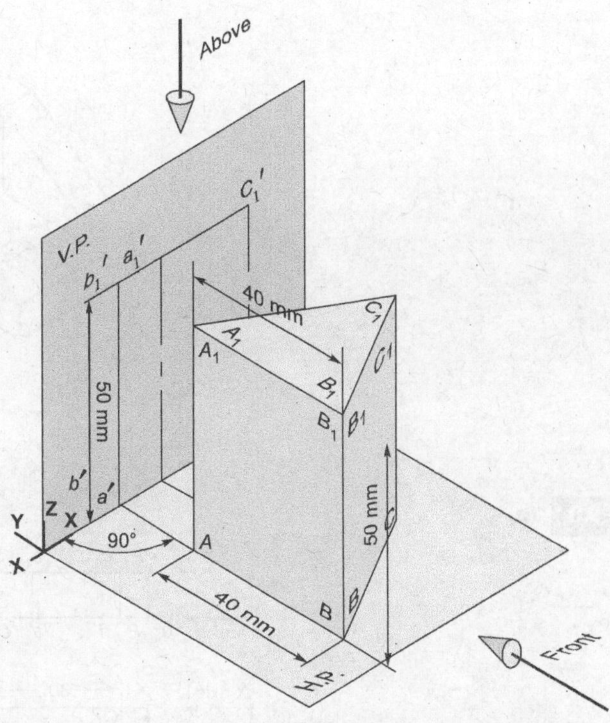

ADDITIONAL INFORMATION

The top view of the prism is a triangle.

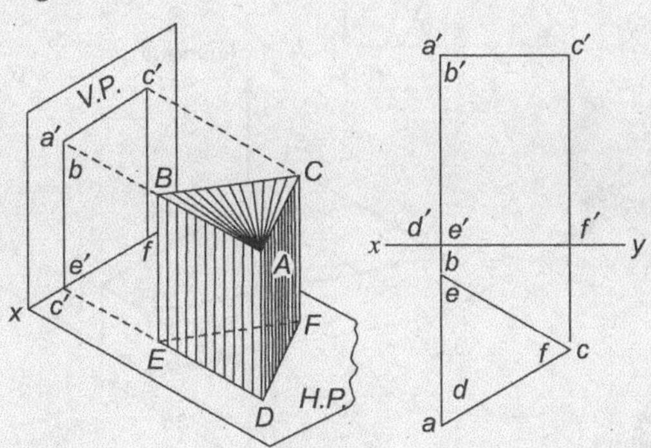

9.7 **(d)**

While drawing the isometric view of any solid, the following important points should be carefully noted:

(i) The isometric view should be drawn according to the given views and in such a way that maximum possible details are visible.

(ii) At every point for the corner of a solid, at least three lines for the edges must converge. Of these, at least two must be for visible edges. Lines for the hidden edges need not be shown, but it is advisable to check up every corner so that no line for a visible edge is left out.

(iii) Two lines (for visible edges) will never cross each other.

ADDITIONAL INFORMATION

The view drawn with the true scale is called isometric drawing or isometric view, while that drawn with the use of isometric scale is called isometric projection.

An isometric graph as shown below which facilitates the drawing of isometric view of an object.

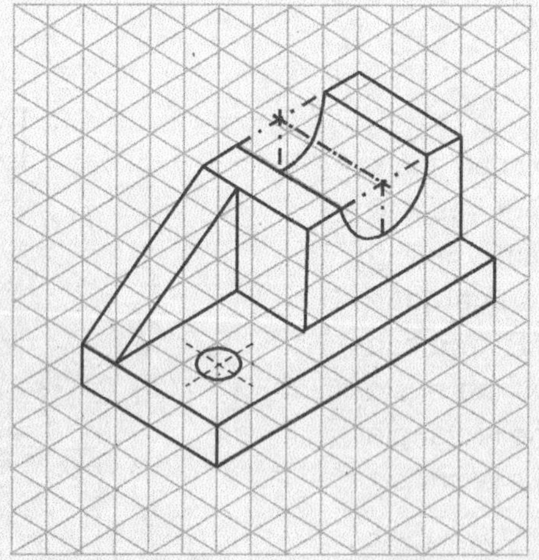

9.8 **(b)**

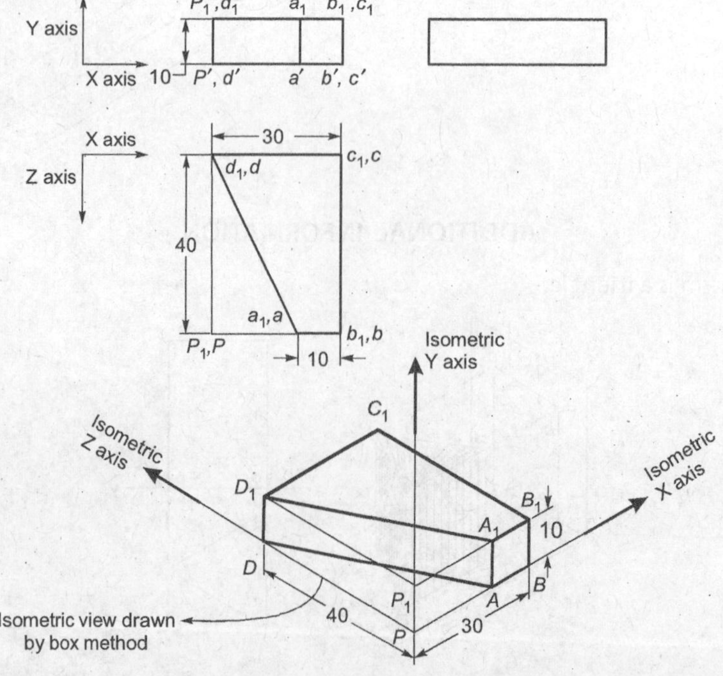

In above drawing AD is a non-isometric line lying in Isometric plane XZ. A_1D_1 is also a non-isometric line lying in Isometric plane parallel to XZ plane and 10 mm above it. To draw the non-isometric lines AD and A_1D_1 lying in isometric planes you enclose the solid in a box with upper rectangle $P_1B_1C_1D_1$ &lower rectangle PBCD. You can draw the isometric view of box easily because all its lines of box are along one of the 3-dimensional axis i.e. X axis or Y axis or Z axis. Point A_1 is located on isometric line P_1B_1 and joined to D_1 to obtain the non-isometric line A_1D_1. Point A is located on isometric line PB and joined to D to obtain the non-isometric line AD.

ADDITIONAL INFORMATION

When an object contains inclined edges which in the isometric view would be shown by non-isometric lines, the view may be drawn by using any one of the following methods:

(i) **Box method :** This method is used when the non-isometric lines or their ends lie in isometric planes.

(ii) **Co-ordinate or offset method :** This method is adopted for objects in which neither non-isometric lines or their ends lie in isometric planes.

10.1 (c)

When we cut a right regular pentagonal prism by a cutting plane in different orientations then out of the alternative given above, we can get section surface of five edges, six edges, and seven edges only. After experimentation it is found that when the section plane cuts seven edges then for a particular orientation of cutting plane area of section surface is maximum.

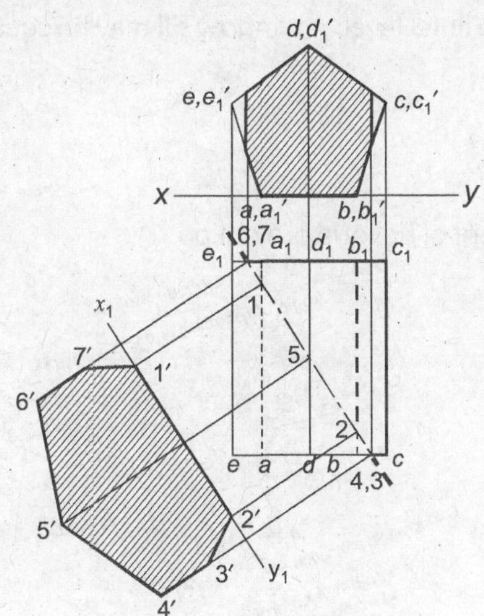

ADDITIONAL INFORMATION

1. When a polyhedra is cut by a section plane the section surface boundary is made up of straight lines.
2. When the cutting plane passes through mid point of six sides of cube then section produced is a regular hexagon.
3. When a right prism is cut parallel to axis the section surface produced is a rectangle.

10.2 (b)

After interpreting orthographic drawing, the visualization of sectional isometric is as follows:

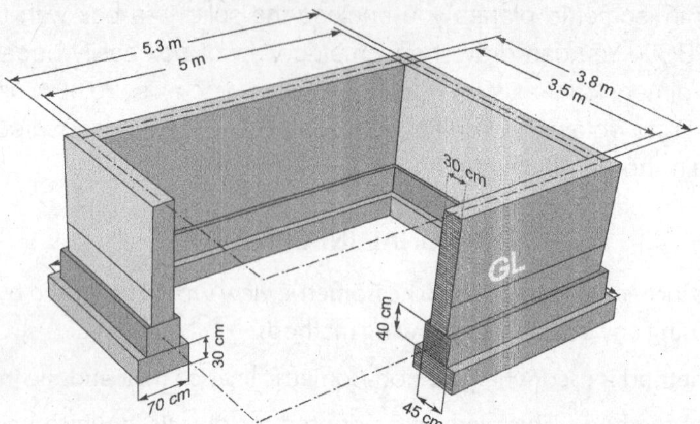

Volume of first footing = Cross-sectional area × Perimeter of path followed by centroid of cross-sectional area

From above diagram it is clear that

Perimeter of path followed by centroid of cross-sectional area

$$= \{5.3 \text{ m} + 3.8 \text{ m} + 5.3 \text{ m} + 3.8 \text{ m}\} = 18.2 \text{ m}$$

Volume of first footing $= 0.45 \text{ m} \times 0.4 \text{ m} \times 18.2 \text{ m} = 3.26 \text{ m}^3$

ADDITIONAL INFORMATION

A **level** between the building's window base and floor **level** above ground **level** is known as the **sill level.** The concrete bed or mortar bed is placed at the window base **level**. Granite is also used at the base **level** of the window. **Sill level** height varies from room to room and depends on the room type.

The window finishes at the top at the **lintel** level. A window sill may be equal to or greater than the width of the brickwork beneath it.

10.3 (d)

Diagram below gives 3D concept of oblique projection

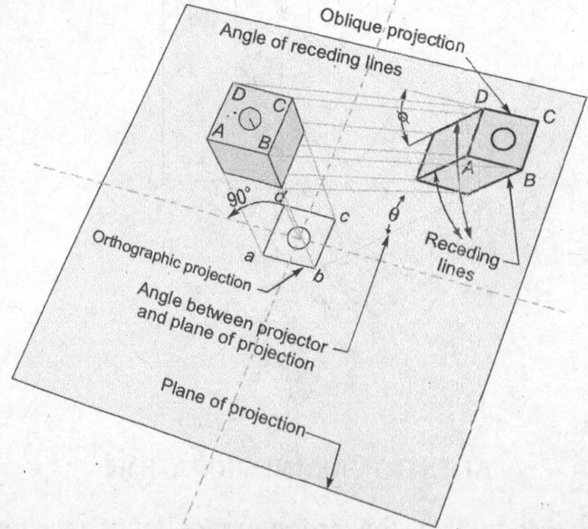

While obtaining oblique projection, most important face of solid is kept parallel to plane of projection and appears to be of true shape and size in the oblique drawing. Hence option (a) and (b) are wrong. Oblique projection is a type of parallel projection, hence when obtaining oblique projection of a solid, projectors are parallel to each other but oblique/inclined to plane of projection. Hence, option (c) is wrong. From figure below is clear that face of solid perpendicular to projection plane get distorted in oblique projection. Hence option (d) is the right answer.

ADDITIONAL INFORMATION

Sl. No.	Oblique projection	Isometric projection
(i)	Projectors from an object are parallel to each other and inclined to the plane of projection.	Projectors from an object are parallel to each other and perpendicular to the plane of picture.
(ii)	The object is placed in such a way that one of its prominent faces remains parallel to the plane of projection. The object rests on one of its faces.	The object is kept in such a way that its three mutually perpendicular edges (axes) make equal angles with the plane of projection. The object stands on one of its corners.
(iii)	The object is drawn with the actual dimensions.	The object is drawn with the reduced (about 82%) dimensions.
(iv)	The faces of object which are perpendicular to the plane of projection will be distorted in the shape and size.	All the faces of the object are distorted in the shape and size.
(v)	The choice of position of the object depends upon the shape and size.	The choice of position of the object depends upon the shape and size.

10.4 (a)

There are two methods of determining the line or curve of intersection between surfaces of two interpenetrating solids

Line method: A number of lines are drawn on the lateral surface of one of the solids and in the region of the line of intersection. Points of intersection of these lines with the surface of the other solid are then located. The line or curve drawn through these points will be the line or curve of intersection

Cutting plane method: The two solids are assumed to be cut by a series of cutting planes. The intersection points of sections of two solids and lying on surface of solid are the required points. These intersection points when joined in proper sequence give the line or curve of intersection.

11.1 (c)

We Know, $$\theta = \frac{360° \times r}{S.H.}$$

Where θ = Angle of development of lateral surface of right cone; r = radius of base circle of right cone; S.H. = Slant height of right cone

When development of lateral surface of right cone is semicircle it means that θ = 180°

Hence we get, $$180° = \frac{360° \times r}{S.H.}$$

⇒ $$S.H. = 2r = \text{diameter of base circle}$$

ADDITIONAL INFORMATION

The development of the curved surface of a cone is a sector of a circle, the radius and the length of the arc of which are respectively equal to the slant height and the circumference of the base-circle of the cone.

11.2 (c)

The development of cylinder is Rectangle.

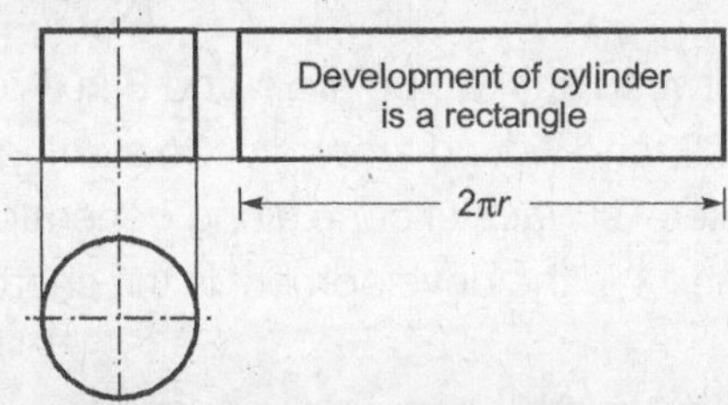

ADDITIONAL INFORMATION

- The development of the curved surface of a cone is a sector of a circle, the radius and the length of the arc of which are respectively equal to the slant height and the circumference of the base-circle of the cone.

Right circular cone unrolled lateral area

- Triangulation development method is used to develop transition pieces. This is simply a method of dividing a surface into a number of triangles and transferring them into the development.
- The elements of an elliptical cone are not equal in length.

11.3 (c)

If a thread is wound around a cone, starting from a point on the base, and brought back to the same point, then the shortest possible length of the thread is equal to the largest chord of the development sector.

ADDITIONAL INFORMATION

We all know that shortest distance between any two points A and B in a plane is the straight line segment AB. To find out shortest possible length of thread wound around cone starting and ending at point 1 on circumference of base circle we develop the lateral surface of cone along generator O1 and join 1 to 1 by a straight line as shown in diagram. This straight line 11 in the development is the shortest path.

11.4 (b)

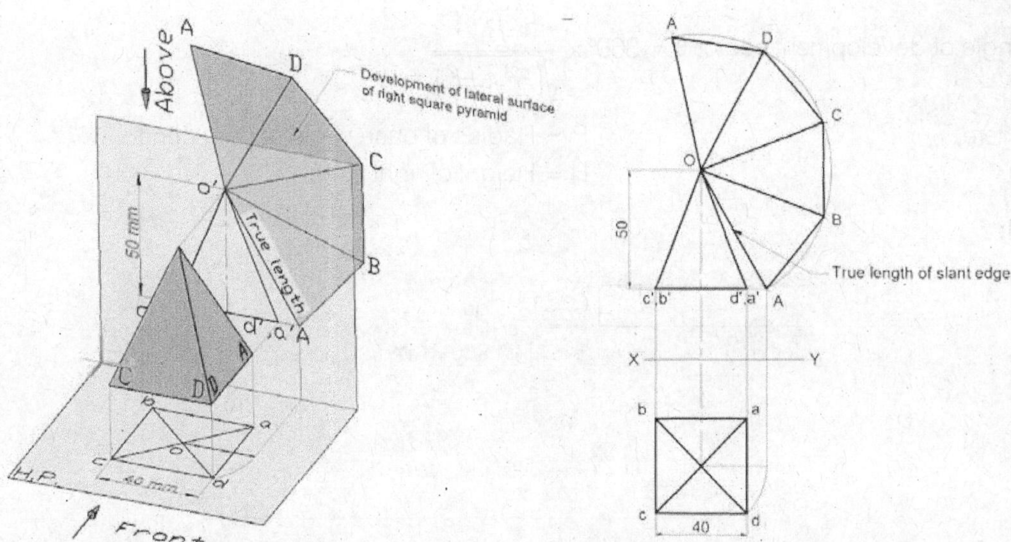

As shown in figure, Radial line development method is used for development of lateral surface of right pyramids in which the true length of slant edge is used as radius.

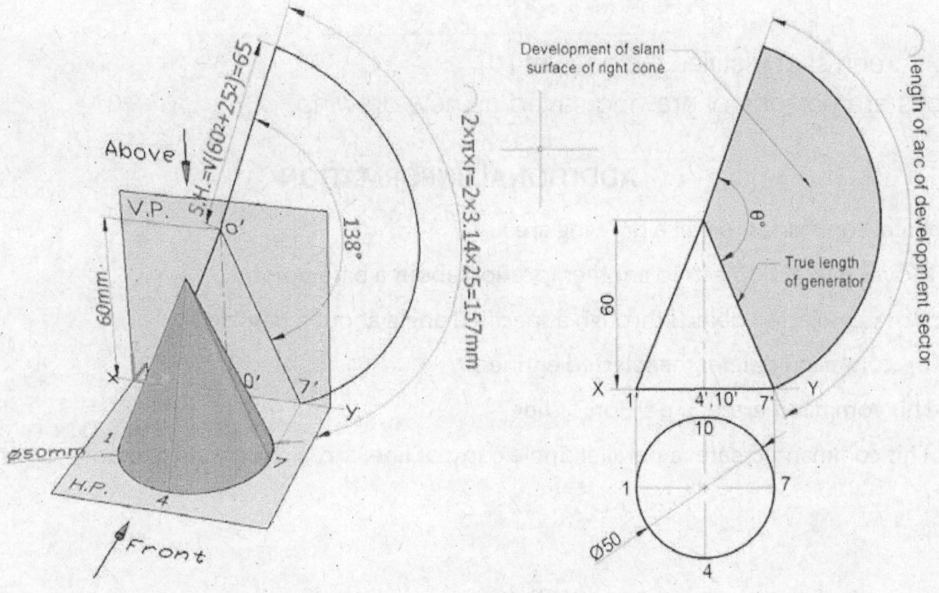

As shown in figure, Radial line development method is used for development of lateral surface of right cone in which the true length of generator is used as radius of development sector

Length of arc of development sector = Circumference of base circle of cone

$$\text{Angle of development sector } \theta = \frac{\text{Length of arc}}{\text{Radius of development sector}}$$

$$\text{Angle of development sector } \theta = \frac{\text{Circumference of base circle of cone}}{\text{Slant height of cone}}$$

$$\text{Angle of development sector } \theta = \frac{2\pi R}{\text{S.H.}} \text{ radians}$$

Angle of development sector $\theta = 360° \times \frac{\text{Radius of base circle}}{\text{Slant height}}$

Angle of development sector $\theta = 360° \times \frac{R}{\sqrt{(R^2 + H^2)}}$

where, R = Radius of base circle of right cone

H = Height of right cone

12.1 (d)

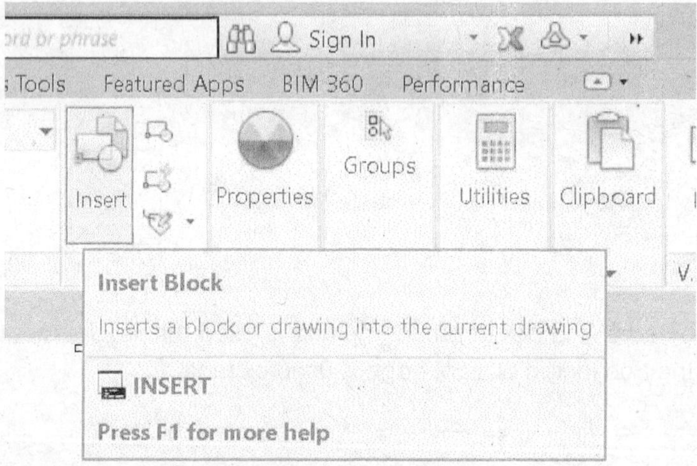

Above screen shot justifies the answer (d).

W blocks are portions of drawing saved as new drawing.

ADDITIONAL INFORMATION

The fundamental commands to edit a drawing are :

- **Move :** Moves selected objects to another location about a base point.
- **Rotate :** Rotates selected objects through a specified angle about a base point.
- **Erase :** This command deletes the selected entities.
- **Break :** This command erases a portion of line.
- **Offset :** This command creates a parallel single copy of line, arc, circle, rectangle, polygon at a given offset distance.

12.2 (d)

- Once the automation starts, the primary function can not be changed. Hence statement (I) is wrong.
- To provide information about the product to the designer and production people is the primary function.

ADDITIONAL INFORMATION

Computer aided drawing and drafting system uses the computer to assist in generation of blueprint data. CAD systems are essential in design and a large number of computer based systems are commercially available. 2D drawing systems correspond directly to traditional engineering drawings, and they are developed to substitute manual drafting.

13.1 (c)

The constitution of India contains specific provisions on the occupational health and safety of workers, including a duty on the government to direct its policies towards ensuring "the health and strength of workers, men and women, and the tender age of children, are not abused". However, despite this health and safety laws currently only cover four industry sectors:

1. Factories 2. Mining 3. Ports 4. Construction

ADDITIONAL INFORMATION

The basic statutory requirements of OHS have been embedded in the constitution of India which every organization is legally bound to follow. Apart from this there are also a large number of statuary acts, rules and regulations which spell out the mandatory requirements to maintain a minimum level of OHS standards.

13.2 (a)

A safety policy should convey at least the following messages :

- The company and its top managers are committed to safety and health.
- Employees are expected to perform their duties in a safe and health manner.
- The company's commitment extends beyond the walls of its plant to include customers and the community.
- It clears all the caveat that may tell on keeping the moral of the employees high.

ADDITIONAL INFORMATION

The policy should make it clear that safe work practices are expected of all employees at all levels at all times. The safety policy serves as the foundation upon which all other promotional efforts are built.

13.3 (a)

Trial runs are recommended for a number of reasons. Some of these are :

1. They afford an opportunity to evaluate potential problem treatment combinations before proceeding with an experiment.
2. They provide a final chance to fine-tune levels of a factor.
3. As the word suggests, a trial is a test of the system that will be engaged during the course of an experiment. This is important in a production environment where the experimenter may only have one chance to obtain the needed data. Trial runs provide a chance to make any needed changes in the experimental plan before an experiment begins.
4. Trial runs can help considerably in estimating: the time to complete a run, the logistical support required for level changes, and the total time needed to complete an experiment.
5. Finally, a trial run is also an excellent communications test. any breakdown in communications will likely be apparent during trial runs. Trial runs also provide an opportunity to remedy the situation before proceeding with the experiment.

ADDITIONAL INFORMATION

Trial runs are very useful because it helps to identify problems in the production system which is not identified during design. During trial runs, several useful learnings are discovered in a real environment.

14.1 (a)

A well-designed and implemented ergonomics reduces the strains and plays an important role in performance improvement. This is achieved by designing tasks, work spaces, controls, displays, tools, lighting and equipments fit to employee's physical capabilities and limitations.

ADDITIONAL INFORMATION

By improvement of workplace layout and by reduction of worker strain, engineering ergonomics (human engineering) reduces waste, improves process flow and production metrics, reduces injury and risk exposure.

14.2 (a)

Faulty design or construction of machine tool is not an operator error.

ADDITIONAL INFORMATION

Poor housekeeping and cleanliness, standard operating safety practice and lack of standardization and identification all comes under operator's error.

14.3 (c)

Compensation and medical payment is not included in the safety program for achieving good results during the prevention of accidents.

ADDITIONAL INFORMATION

Important aspects of safety programmes :

1. Using a qualified safety person.
2. Develop safe working conditions.
3. Promote participation of workers for safety.
4. Assigning a competent person.
5. Enforcing safety rules and using appropriate discipline.

15.1 (a)

Every lift lobby should have access, without any obstruction and lockable door, to an exit route. Such access should be available at all times to any person who may come out from a lift car to the lift lobby. The provision of a direct intercom link connecting a lift lobby with the management office of the building will be accepted as an adequate alternative.

Another escape route may be an external staircase, provided that there is atleast one internal staircase from every part of each storey.

ADDITIONAL INFORMATION

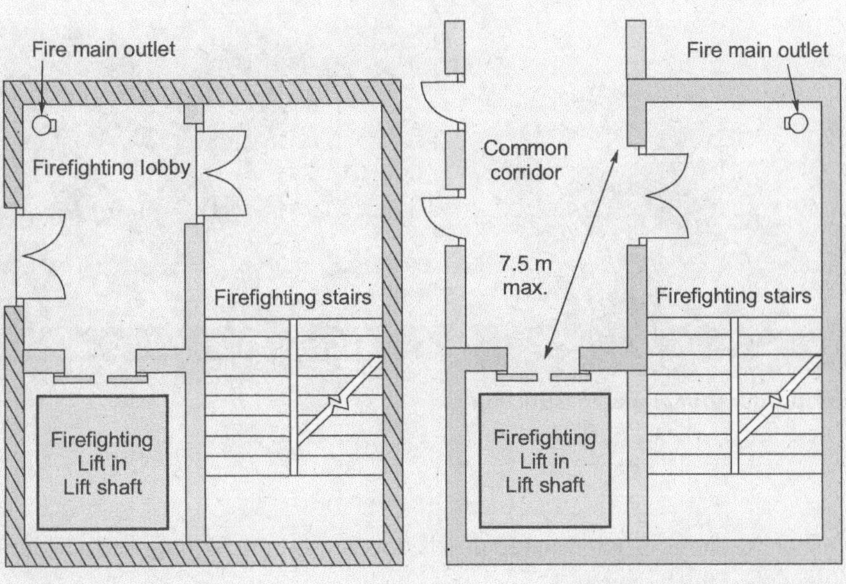

16.1 (c)

Precautions before starting demolition work :

- All doors and openings giving access to the structure shall be kept barricaded or manned except during the actual passage of workmen or equipment.
- If a structure to a demolished has been partially wrecked by fire, explosion or other catastrophe, the walls and damaged roofs shall be shored or braced suitably.
- Before demolition work is started, glazed doors and windows shall be removed. All fragile and loose fixtures shall be removed.

ADDITIONAL INFORMATION

The demolition shall always proceed systematically storey by storey in descending order and the work on the upper floors shall be completely over before any of the supporting members or other important portion on the lower floor is disturbed.

16.2 (d)

- Improper designs, manufacturing errors and not adhering to proper maintenance practices are the reasons of failure of many structures.
- Lack of proper inspection and lack of public concern results accidents on major highways.
- Corruption amongst several concerned individuals and lack of public concern also results accident on major highways, discharge of effluents which pollute rivers and dangerous gas emissions.

ADDITIONAL INFORMATION

Building collapse due to improper construction

Current scenario of water pollution in India

16.3 (d)

Signages on minor gradients is not required as it is not dangerous. Signage is provided for a steep gradient during improvement on the highway.

ADDITIONAL INFORMATION

- Highly visible barricades are used to alert to drivers so that falling of vehicles are prevented.
- To avoid accidents, signages are provided for wild animals where the roads pass through jungle areas.

16.4 (c)

Mobile cranes are highly sophisticated machines that are designed to efficiently lift heavy loads and move them from one location to another. They are also designed to work with different types of loads and cargo with little or no set up or assembly, which makes mobile cranes a very versatile and reliable means of lifting on site.

ADDITIONAL INFORMATION

16.5 (a)

Step ladders shall always be used in fully open position and fixed with a safety locking device. These shall not be used where strenuous action on the part of the workman is called for. Step ladders should not be used on top of scaffolds, platforms, or other surfaces above the ground. Top most rung of a step ladder should not be used.

ADDITIONAL INFORMATION

Unprotected edge means any side or edge of a walking/working place. Therefore, for working platform to gain height above the protected edge e.g. roof, floor, ramp etc., we have to use extension ladder in place of step ladder.

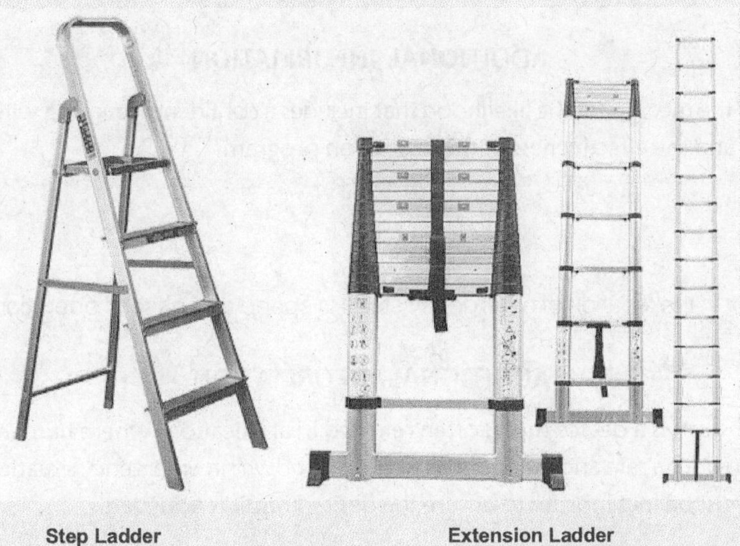

Step Ladder   Extension Ladder

16.6 (c)

Tyres should be properly inflated to provide proper balancing and safety of the workers and equipments on the platform when MEWP is moving.

Safe Working Load (SWL) is the maximum safe force that a piece of lifting equipment can exert to lift without breaking. It is generally marked in platforms.

ADDITIONAL INFORMATION

Requirements to carry out work at height :

- Ensure the safe working load (SWL) allowance in the MEWP basket is clearly displayed and is never exceeded.
- Check tyres for correct inflation, steering linkages, wheel nuts etc. for damage and security.

16.7 **(c)**

If bearing balls are produced by a manufacturing process in which a diameter distribution is created, we can say upon choosing a ball that there is uncertainty as to size. If we wish to consider weight or moment of inertia in rolling, this size uncertainty can be considered to be propagated to our knowledge of weight or inertia. There are ways of estimating the statistical parameters describing weight and inertia from those describing size and density. These methods are variously called *propagation of error, propagation of uncertainty, propagation of dispersion*. These methods are integral part of analysis or synthesis tasks when probability of failure is involved.

ADDITIONAL INFORMATION

Probability of failure is the measure of the likelihood that includes a confidence levelling with regard to the damage rates and mechanisms and the effectiveness of the inspection program.

16.8 **(a)**

Determining factor in the location of emergency stops is speed and ease of operation of the emergency stops.

ADDITIONAL INFORMATION

An emergency stop (E-stop) is a device that is often referred in applications where human and machine safety is required. An E-stop provides a safe and quick disconnection of power in emergency situations. However, rigorous codes and standards must be met inorder to ensure the desired result is achieved.

16.9 **(c)**

The sound level, $$L = \log_{10}\frac{Q}{Q_0} \text{ (bels)}$$

where, Q = Measured quantity of sound pressure or sound intensity

Q_0 = Reference standard quantity of sound pressure

ADDITIONAL INFORMATION

Sound/noise level meter equipment measures noise pollution. It consists of several parts, mainly microphone, preamplifier, display system, power supply etc. A "decibel" is a unit in noise measurement.

17.1 (d)

$$\text{Loss of P.E.} = \text{Spring energy stored}$$

$$Mg(H + \Delta) = \frac{1}{2}K \cdot \Delta^2$$

$$W(H + \Delta) = \frac{1}{2}K\Delta^2$$

$$W = 240 \text{ N}$$

$$D = 120 \text{ mm}$$

$$K = \frac{G \cdot d^4}{64R^3 n}$$

where, Δ = Comp. in spring ; d = diameter of wire = 18 mm ; n = number of coils = 22

G = shear modulus = 88 × 10³ N/mm² ; R = Radius of spring $\frac{180}{2} = 90$ mm

$$K = \frac{88 \times 10^3 \times 18^4}{64 \times 90^3 \times 22} \text{ N/mm} = 9 \text{ N/mm}$$

$$W(H + \Delta) = \frac{1}{2} \cdot K \cdot \Delta^2$$

$$240(H + 120) = \frac{1}{2} \times 9 \times 120^2$$

$$H + 120 = 270$$

$$H = 150 \text{ mm}$$

ADDITIONAL INFORMATION

The two main types of impact testing using a pendulum are Izod and Charpy :

1. The Izod impact testing holds the sample vertically with the notch facing the pendulum.
2. The Charpy impact testing holds the sample horizontally with the notch facing away from the pendulum.

17.2 (a)

$$\text{Engine rpm} = \text{Overall gear ratio} \times \text{Wheel rpm} = 19.915 \times \text{Wheel rpm}$$

$$\text{Wheel rpm} = \frac{\text{Vehicle Velocity}}{\text{Circumference of wheel}} = \frac{24 \text{ km/hr}}{2 \times \pi \times 0.305 \text{ m}} = \frac{24 \times \frac{1000}{60} \text{ m/min}}{2\pi \times 0.305 \text{ m}}$$

$$= 208.7 \text{ rpm}$$

$$\text{Engine rpm} = 19.915 \times 208.7 = 4156.8 \text{ rpm}$$

$$\text{Engine power} = \text{Torque} \times \text{Angular displacement/sec}$$

$$= \frac{203.6 \text{ Nm} \times 4156.8 \times 2\pi \text{ rad}}{60 \text{ sec}} = 88626 \text{ Watt} = 88.6 \text{ kW}$$

ADDITIONAL INFORMATION

In the context of automotive performance, torque is the amount of twisting force applied to the engine crankshaft by the pistons. The higher the torque, the more force the engine can generate to turn the wheels. Speed, on the other hand measures how fast the engine crankshaft is rotating.

17.3 **(a)**

$$\text{Rolling resistance} = C \times W$$

Where, C = Rolling resistance coefficient; W = Weight of the vehicle

$\Rightarrow$ Rolling resistance = 0.018×62293.5 N = 1121.31 N

ADDITIONAL INFORMATION

Rolling resistance is the force resisting the motion when a body rolls on a surface. The coefficient of rolling resistance is much smaller than the coefficient of sliding friction.

17.4 **(*)**

$$L_{BD} = 0.3 \text{ m} = 30 \text{ cm}$$

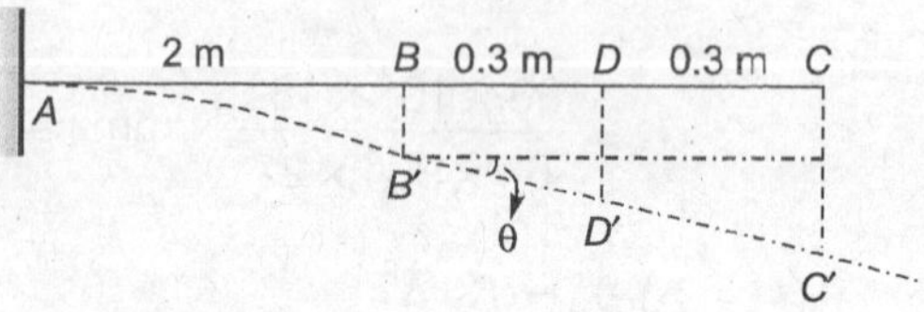

2 m B 0.3 m D 0.3 m C
A
B′ θ D′ C′

$$BB' = \Delta_B = 0.24 \text{ cm}$$

$$DD' = \Delta_D = \Delta_B + \tan\theta \times L_{DB}$$

$$\Delta_D = \Delta_B + \sin\theta \times L_{DB}$$

(For small angle $\tan\theta \approx \sin\theta$)

$$= 0.24 + 0.038 \times 30 \text{ cm} = 0.24 + 1.14$$

$$\Delta_D = 1.38 \text{ cm}$$

Hence none is correct. However, as per UPSC answer is option (b).

ADDITIONAL INFORMATION

Some direct results for cantilever beams in standard cases are :

Beam	Slope	Deflection
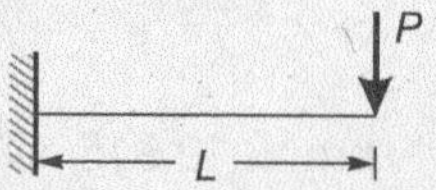 P, L	$\theta = \dfrac{PL^2}{2EI}$	$\Delta = \dfrac{PL^3}{3EI}$
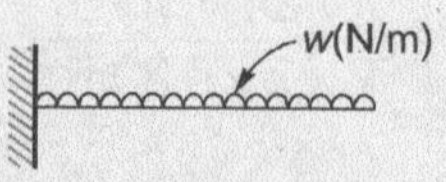 w(N/m)	$\theta = \dfrac{wL^3}{6EI}$	$\Delta = \dfrac{wL^4}{9EI}$
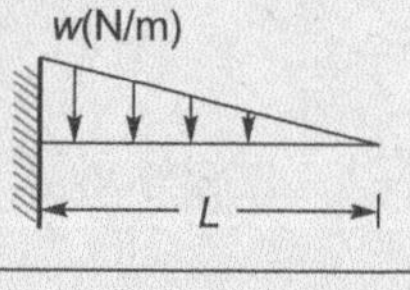 w(N/m), L	$\theta = \dfrac{wL^3}{24EI}$	$\Delta = \dfrac{wL^4}{30EI}$

17.5 (d)

Grillage foundation consists of one or more tiers of beams (steel or timber) superimposed at right angles to each other on a concrete layer to disperse load over an extensive area. This type of foundation is generally used for heavy structural columns, piers and scalfolds where two foundation is expected to disperse heavy loads over large areas. With the use of a grillage foundation, deep excavations can be avoided and the depth of the foundation can be limited from 1 m to 1.5 m.

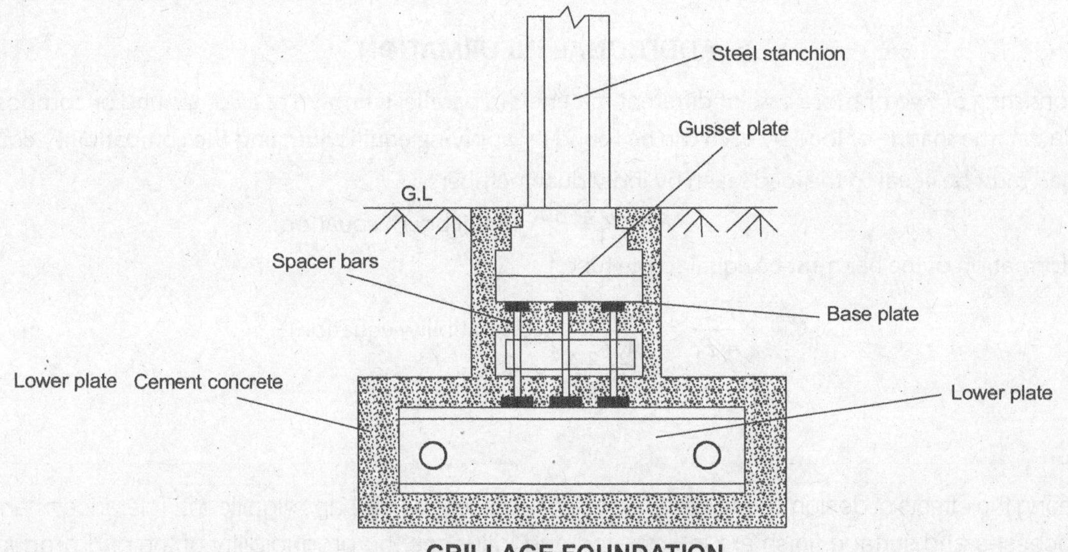

GRILLAGE FOUNDATION

ADDITIONAL INFORMATION

- The time required for the installation and casting of grillage foundation is comparatively less than that of conventional foundation.
- Grillage foundation is capable of successfully transforming heavy loads over a larger area.
- The major types of grillage foundation are
  1. Steel grillage foundation 2. Timber grillage foundation

17.6 (a)

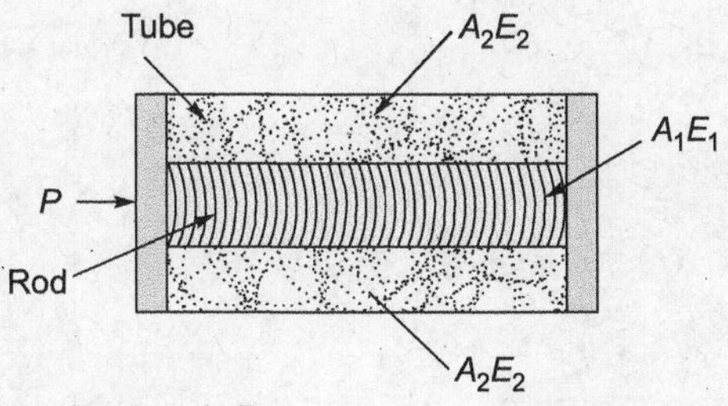

Let Rod takes load P_1 and Tube take load P_2

$$P_1 + P_2 = P \quad ...(1)$$

Since rigid plates are not placed at both the ends hence deflection will be equal.

$$\Delta_{rod} = \Delta_{tube}$$

$$\frac{P_1L}{A_1E_1} = \frac{P_2L}{A_2E_2}$$

$$\frac{P_1}{P_2} = \frac{A_1E_1}{A_2E_2} \quad ...(2)$$

Solving (1) and (2)

$$P_1 + P_2 = P$$

$$P_2 \cdot \frac{A_1E_1}{A_2E_2} + P_2 = P$$

$$\Rightarrow \quad P_2 = \frac{PA_2E_2}{A_1E_1 + A_2E_2} \quad \text{and} \quad P_1 = \frac{P \cdot A_1E_1}{A_1E_1 + A_2E_2}$$

ADDITIONAL INFORMATION

A bar consisting of two or more base of different materials in parallel is known as a compound or composite bar. In such a bar, the sharing of load by each can be found by applying equilibrium and the compatibility equation. Total load must be equal to the load taken by individual members.

$$P = P_1 + P_2 \quad \text{(Equilibrium equation)}$$

The deformation of the bar must be equal to the tube.

$$\frac{P_1 L}{A_1 E_1} = \frac{P_2 L}{A_2 E_2} \quad \text{(compatibility equation)}$$

17.7 (b)

Among the effects of design specification on cost, tolerance are perhaps significant. Tolerance, manufacturing processes and surface finish are interrelated and influence the producibility of the end product in many ways. Tolerance cover dimensional variation and surface roughness range and also the variation in mechanical properties resulting from heat treatment and other processing operations.

ADDITIONAL INFORMATION

The product design specification (PDS) is the essential definition of what the product is required to provide. The PDS is a statement of what the customer wants the product to achieve. In same cases the customer is external and in some cases the customer is internal. In every case, a PDS must be prepared to act as a reference for the objectives of the design.

17.8 (b)

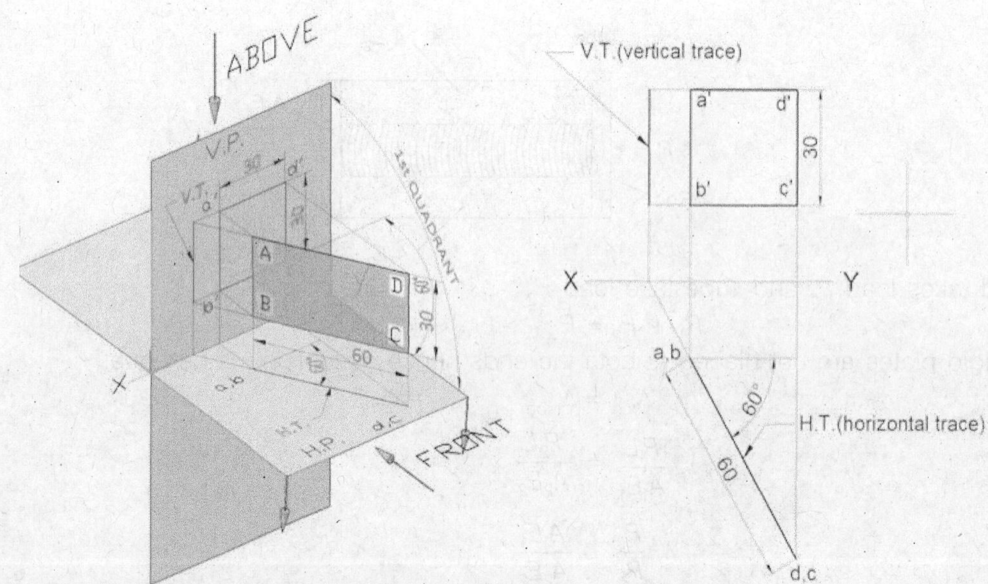

Traces of a plane are defined as line of intersection of plane or plane extended with reference planes or planes of projection.

In figure, you have a rectangular plane ABCD perpendicular to H.P. and inclined to V.P. at an angle of 60°. To obtain the vertical trace of the plane ABCD, you simply extend it in until it intersects the V.P. V.T. (vertical trace) as shown in pictorial diagram is the line of intersection of plane extended and vertical reference plane.

To obtain the Horizontal trace of the plane ABCD, you simply extend it in until it intersects the H.P. H.T. (horizontal trace) as shown in pictorial diagram is the line of intersection of plane extended and Horizontal reference plane.

■■■■

Standards and Quality Practices in Production

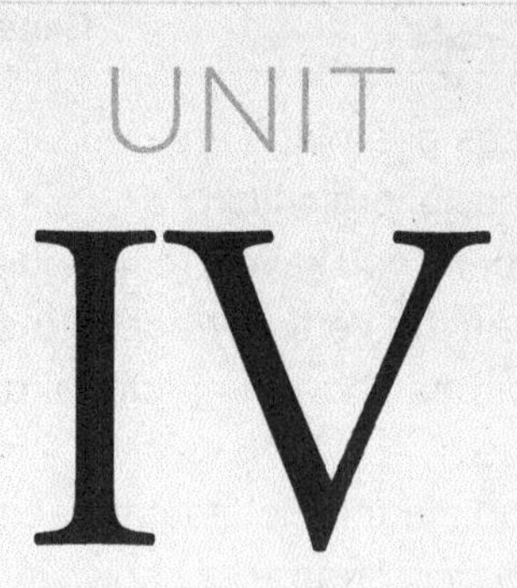

CONTENTS

Standards and Quality Practices in Production

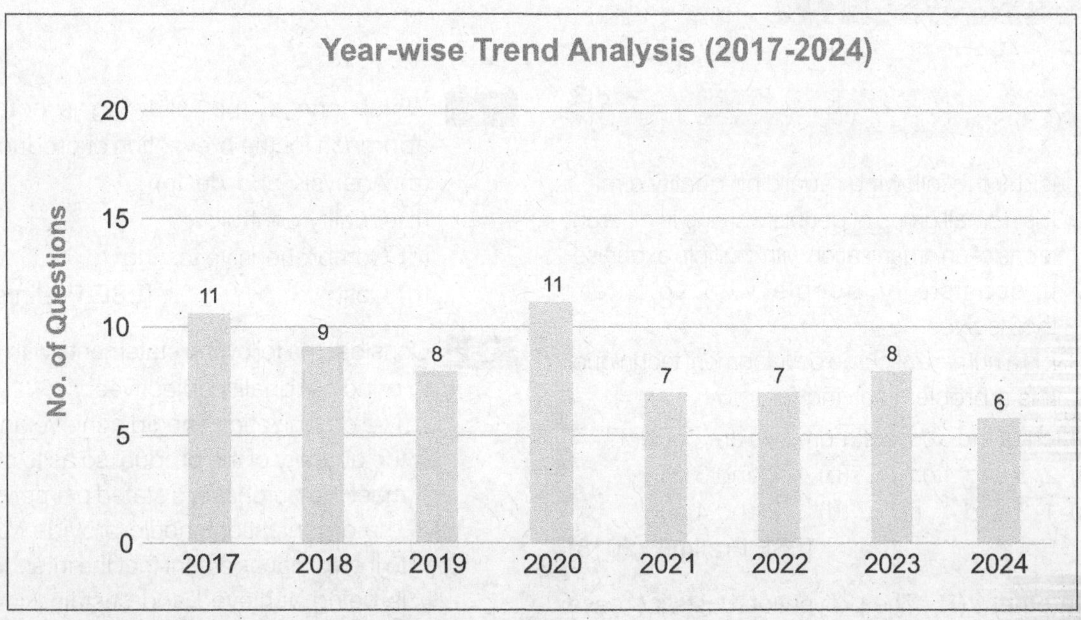

TOPIC-WISE TREND ANALYSIS (2017-2024)

Sl. No.	Topics	Total Number of Questions
1.	Quality	8
2.	Different Approaches Regarding Quality	8
3.	Quality Control Tools OR, Statistical Tools & Techniques in TQM Practice	11
4.	Sampling	4
5.	Total Quality Management	3
6.	ISO Standards	2
7.	Six Sigma	7
8.	Inventory	5
9.	Quality in Construction	7
10.	Quality Practices in Services	2
11.	Reliability and Maintenance	6
12.	Miscellaneous	4

UNIT

IV

Standards and Quality Practices in Production

1. QUALITY

1.1 Consider the following regarding quality circle:

1. It is a small group of people working in different areas of an organization with multiple expertise.
2. It consists of people who volunteer themselves.
3. It is a human resource development technique.
4. It is a problem solving forum.

Which of the above statements are correct?

(a) 2, 3 and 4 only (b) 1, 2 and 3 only
(c) 1, 3 and 4 only (d) 1, 2 and 4 only

[ESE Prelims : 2018]

1.2 **Statement (I)** : The prevention costs increase with the introduction of a quality system and may be a significant proportion of the total quality costs.

Statement (II) : Costs associated with education and training are not included in prevention costs.

(a) Both Statement (I) and Statement (II) are individually true; and Statement (II) is the correct explanation of Statement (I)
(b) Both Statement (I) and Statement (II) are individually true; but Statement (II) is NOT the correct explanation of Statement (I)
(c) Statement (I) is true; but Statement (II) is false
(d) Statement (I) is false; but Statement (II) is true

[ESE Prelims : 2019]

1.3 Which of the following approaches are correct regarding total quality?

1. Opportunity to improve
2. Adoption requires little change
3. React to competitive threats

Select the correct answer using the codes given below:

(a) 1 and 2 only (b) 1 and 3 only
(c) 2 and 3 only (d) 1, 2 and 3

[ESE Prelims : 2019]

1.4 Which one of the following is not the best approach for the prevention of product liability?

(a) Analysis and design
(b) Quality control
(c) Comprehensive testing
(d) Cost

[ESE Prelims : 2020]

1.5 Consider the following statements with reference to principal quality objectives:

1. The organization should achieve and sustain the quantity of the product so as to continually meet the purchaser's stated or implied needs.
2. The organization should provide confidence to it own management that the intended quality is being achieved and sustained.
3. The organization should provide confidence to the purchaser that the intended quality is being, or will be, achieved in the delivered product.

Which of the above statements are correct?

(a) 1 and 2 only (b) 2 and 3 only
(c) 1 and 3 only (d) 1, 2 and 3

[ESE Prelims : 2022]

1.6 The quality characteristics can be categorized in which of the following groupings?

1. Sensory characteristics
2. Structural characteristics
3. Statistical characteristics
4. Time oriented characteristics

Select the correct answer using the code given below:

(a) 1, 2, 3 and 4 (b) 1 and 4 only
(c) 2 and 3 only (d) 1, 2 and 4 only

[ESE Prelims : 2022]

1.7 What are the major categories for quality costs?

1. Prevention costs 2. Appraisal costs
3. Production costs 4. Internal failure costs

Select the correct answer using the code given below:
(a) 1, 2 and 3 only
(b) 1, 2 and 4 only
(c) 3 and 4 only
(d) 1, 2, 3 and 4 [ESE Prelims : 2022]

1.8 Consider the following statements regarding production:
The major aspects of production that may lead to sickness are
1. Increase in the cost of production.
2. Decrease in the quantity of production.
3. Quality of product not meeting the standards/ customer expectation.
4. Producing more quantity than can be sold, leading to accumulation of stock.

Which of the above statements are correct?
(a) 1 and 2 only (b) 1, 2, 3 and 4
(c) 2 and 3 only (d) 1, 3 and 4 only
[ESE Prelims : 2022]

2. DIFFERENT APPROACHES REGARDING QUALITY

2.1 **Statement-I:** Quality circle is a method to bring organization improvement through indulging the workers.
Statement-II: The main aim of quality circle is self-development and mutual development of grass root level employees.
(a) Both Statement (I) and Statement (II) are individually true and Statement (II) is the correct explanation of Statement (I)
(b) Both Statement (I) and Statement (II) are individually true, but Statement (II) is not the correct explanation of Statement (I)
(c) Statement (I) is true, but Statement (II) is false
(d) Statement (I) is false, but Statement (II) is true
[ESE Prelims : 2017]

2.2 Concurrent engineering is implements by involving a cross-functional team for design, production, testing and operational work.
(a) During the project execution.
(b) Long before the project execution.
(c) Towards the end of the project execution.
(d) After completion of the project execution.
[ESE Prelims : 2017]

2.3 **Statement-I:** The concept of Just-in-Time is operationalized when the exact number of units required are bought at each successive stage of production at the appropriate time.
Statement-II: Just-in-Time concept has been expanded to mean a manufacturing philosophy of eliminating waste.
(a) Both statement I and statement II are individually true and statement II is correct explanation of statement I.
(b) Both statement I and statement II are individually true but statement II is not the correct explanation of statement I.
(c) Only statement I is true.
(d) Only statement II is true.
[ESE Prelims : 2018]

2.4 **Statement-I:** Quality is essential for survival and growth of an organization in present era of tough competition.
Statement-II: The concept of quality is confined only to construction and manufacturing organization.
(a) Both Statement (I) and Statement (II) are individually true and Statement (II) is the correct explanation of Statement (I)
(b) Both Statement (I) and Statement (II) are individually true, but Statement (II) is not the correct explanation of Statement (I)
(c) Statement (I) is true, but Statement (II) is false
(d) Statement (I) is false, but Statement (II) is true
[ESE Prelims : 2018]

2.5 Consider the following components:
1. Knowledge of psychology
2. Knowledge of the theory of variation
3. Knowledge of process
4. Knowledge of the system and the theory of optimization

Which of the above components comprise the basis of Deming's Systems of Profound Knowledge?
(a) 1, 2 and 3 only (b) 1, 3 and 4 only
(c) 1, 2 and 4 only (d) 2, 3 and 4 only
[ESE Prelims : 2019]

2.6 Which of the following steps come under 14-step plan for quality improvement?
1. Ad hoc committee for the zero defects program

2. Cost of quality evaluation
3. Quantity measurements
4. Supervisor training

(a) 1, 2 and 3 (b) 1, 2 and 4
(c) 1, 3 and 4 (d) 2, 3 and 4

[ESE Prelims : 2020]

2.7 Most of Deming's deadly diseases involve

(a) immobility of management
(b) a long term orientation
(c) a lack of understanding of variation
(d) high degree of constancy of purpose

[ESE Prelims : 2021]

2.8 **Statement (I):** The machine shop produces parts machined from stock material and finishes castings, forgings, etc., requiring machined surfaces.

Statement (II): In machine shops, machining operations remove metal, either to make a smoother and more accurate surface, as by planning, facing, milling, etc. or to produce a surface previously existing, as by drilling, punching, etc.

(a) Both Statement (I) and Statement (II) are individually true and Statement (II) is the correct explanation of Statement (I).
(b) Both Statement (I) and Statement (II) are individually true but Statement (II) is **not** the correct explanation of Statement (I).
(c) Statement (I) is true but Statement (II) is false.
(d) Statement (I) is false but Statement (II) is true.

[ESE Prelims : 2023]

3. QUALITY CONTROL TOOLS OR, STATISTICAL TOOLS & TECHNIQUES IN TQM PRACTICE

3.1 Which of the following departments ensure the quality of the product?

1. Product design and development
2. Marketing and product planning
3. Packaging and shipping
4. Sales

(a) 1, 2 and 3 (b) 1, 2 and 4
(c) 1, 3 and 4 (d) 2, 3 and 4

[ESE Prelims : 2020]

3.2 Which of the following are the general subareas of quality control?

1. Off-line quality control
2. Sales/market share
3. Statistical process control
4. Acceptance sampling plans

(a) 1, 2 and 3 (b) 1, 3 and 4
(c) 1, 2 and 4 (d) 2, 3 and 4

[ESE Prelims : 2020]

3.3 The standard deviation σ for sampling in the case of concrete for construction engineering and management is

(a) $\sqrt{\dfrac{\Sigma(X+\bar{X})^2}{N+1}}$ (b) $\sqrt{\dfrac{\Sigma(X-\bar{X})^2}{N+1}}$

(c) $\sqrt{\dfrac{\Sigma(\bar{X}+X)^2}{N-1}}$ (d) $\sqrt{\dfrac{\Sigma(X-\bar{X})^2}{N-1}}$

where

$X_1, X_2, X_3, X_4, .., X_n$ = Compressive strengths of the individual cubes

N = Number of cubes tested

$\bar{X}$ = Average of series of compressive strengths values

[ESE Prelims : 2020]

3.4 Which of the following variations are true in piece part production?

1. Piece-to-piece variation
2. Time-to-time variation
3. Within-piece variation
4. Process-to-process variation

(a) 1, 2 and 4 (b) 1, 3 and 4
(c) 2, 3 and 4 (d) 1, 2 and 3

[ESE Prelims : 2020]

3.5 Which one of the following is the objective of attribute charts in production process control?

(a) To determine the acceptance criteria of a product before production
(b) To evaluate the quality performance of operating and management personnel
(c) To keep a periodic record of a particular characteristic
(d) To determine the highest quality level

[ESE Prelims : 2020]

3.6 Which one of the following rules is NOT used for identifying an out-of-control process?

(a) A process is assumed to be out-of-control if a single point plots outside the control limits
(b) A process is assumed to be out-of-control if there is a run of six or more consecutive points steadily increasing or decreasing

(c) A process is assumed to be out-of-control if nine or more consecutive points fall to one side of the center line

(d) A process is assumed to be out-of-control if two or more consecutive points fall beyond the 1 σ limit on the same side of the center line

[ESE Prelims : 2021]

3.7 Rearrange the following steps involved in construction of pareto diagram in the proper order.

1. Determine how relative importance is to be judged.
2. Decide on the data categorization system.
3. Rank the categories from most important to least important.
4. Plot a bar graph.
5. Compute the cumulative frequency of the data categories in their chosen order.

Select the correct answer using the code given below:

(a) 1, 3, 2, 4, 5 (b) 2, 3, 1, 5, 4
(c) 2, 1, 3, 5, 4 (d) 1, 2, 3, 4, 5

[ESE Prelims : 2021]

3.8 Consider the following advantages of p-charting method:

1. Requires only go-no-go data, intuitive.
2. No requirement for pre-tested "standard" units.
3. Accounts for all errors including systematic errors.

Which of the above advantages is/are correct?

(a) 1 only (b) 2 only
(c) 1 and 2 (d) 2 and 3

[ESE Prelims : 2024]

3.9 Which one of the following is an example of discrete random variables?

(a) Triangular Distribution
(b) Normal Distribution
(c) Central Limit Theorem
(d) Negative Binomial Distribution

[ESE Prelims : 2024]

3.10 Almost all the quality control problems can be solved if the following conditions for manufacturing the product are met:

1. The quality characteristics are within the appropriate specification tolerance limits determined based on customers' requirements.
2. The variability of the quality characteristics is minimized as much as possible.
3. The mean of each quality characteristic is as close as possible to the target value of the characteristic.

Which of the above conditions are correct?

(a) 1 and 2 only (b) 2 and 3 only
(c) 1 and 3 only (d) 1, 2 and 3

[ESE Prelims : 2024]

3.11 The manufacturing cost of the components will decrease as a result of using the probabilistic relationship because

(a) manufacturing cost decreases as the tolerance on the quality characteristic decreases
(b) manufacturing cost decreases as the tolerance on the quality characteristic increases
(c) manufacturing cost remains constant as the tolerance on the quality characteristic increases
(d) manufacturing cost increases as the tolerance on the quality characteristic increases

[ESE Prelims : 2024]

4. SAMPLING

4.1 The maximum percentage defects that a consumer may find definitely acceptable is called:

(a) AOQL (b) LTPD
(c) AQL (d) AOQ

[ESE Prelims : 2017]

4.2 A unit produces packing boxes. Out of hourly production of 4,000 boxes, 20 were found to be non-conforming. If the inspector randomly chooses a box from an hour's production, the probability of it being non-conforming is

(a) 0.02 (b) 0.10
(c) 0.005 (d) 0.05

[ESE Prelims : 2019]

4.3 An Average Outgoing Quality (AOQ) is

(a) $\frac{P_a p(N)}{N-n}$ (b) $\frac{P_a p(N+n)}{N}$

(c) $\frac{P_a p(N-n)}{n-N}$ (d) $\frac{P_a p(N-n)}{N}$

where

P_a = Probability of accepting the lot

p = Incoming lot quality

N = Lot size

n = Sample size [ESE Prelims : 2020]

4.4 Consider the following statements regarding evaluating sampling plans:

1. If rectifying inspection is conducted for lots rejected by the sampling plan is the average total inspection.
2. The average number of items inspected for a series of lots in order to make a decision is the average sample number.
3. The average quality level of a series of batches that leave the inspection station after coming in for inspection at a certain quality level is the average outgoing quality limit.

Which of the above statements are correct?

(a) 1 and 2 only (b) 2 and 3 only
(c) 1 and 3 only (d) 1, 2 and 3

[ESE Prelims : 2022]

5. TOTAL QUALITY MANAGEMENT

5.1 One of the fundamental attribute of TQM is

(a) Having team meetings
(b) Top management direct involvement
(c) Drawing control chart
(d) All of the above

[ESE Prelims : 2017]

5.2 Which one of the following unique characteristics of the construction process makes TQM difficult to implement?

(a) The construction process is relatively short in duration
(b) A low percentage of the labour at a construction project only work for the construction firm for a short time period
(c) Project owners take a long term view to control projects
(d) Construction projects are multiple, each project being somewhat same

[ESE Prelims : 2021]

5.3 Which of the following are the elements of TQM?

1. Teamwork and Employee empowerment
2. Feedback mechanisms
3. Strong division of labour
4. Result oriented management

Select the correct answer using the code given below:

(a) 1 and 2 only
(b) 3 and 4 only
(c) 2, 3 and 4 only
(d) 1, 2, 3 and 4

[ESE Prelims : 2022]

6. ISO STANDARDS

6.1 For which of the following sectors do comprehensive safety and health statutes, for regulating Occupational Safety and Health (OSH) at work places, exist at present in India?

(a) Mining, Factories, Ports and Agriculture
(b) Factories, Mining Agriculture and Construction
(c) Mining, Factories, Ports and Construction
(d) Factories, Ports, Agriculture and Construction

[ESE Prelims : 2017]

6.2 Internal audits are used to verify whether:

1. Products conform to technical specifications.
2. Quality management system are effectively implemented.

Which of the above statements is/are correct?

(a) 1 only (b) 2 only
(c) Both 1 and 2 (d) Neither 1 nor 2

[ESE Prelims : 2017]

7. SIX SIGMA

7.1 Consider the following statements with reference to six-sigma.

1. It is a set of techniques and tool for process improvement.
2. It postulates that any process must not produce more than 3.4 defects per one million opportunities.
3. It is an initiative of Motorola.

Which of the above statements are correct?

(a) Only 1 and 2 (b) Only 1 and 3
(c) Only 2 and 3 (d) Only 1, 2 and 3

[ESE Prelims : 2018]

7.2 Consider the following statements with reference to six-sigma:

1. It is a methodology for structured, process oriented and systematic quality improvement.
2. It provides a systematic approach for quality and process in improvement, rather than being just a collection of tools.
3. It is a rigorous, data-driven, decision-making approach to analyse the root causes of problems.

Which of the above statements are correct?

(a) 1 and 3 only (b) 2 and 3 only
(c) 1 and 2 only (d) 1, 2 and 3

[ESE Prelims : 2023]

7.3 Six sigma gives a precision of

(a) 99.9997% (b) 98.4599%
(c) 97.7333% (d) 96.2799%

[ESE Prelims : 2023]

7.4 As sigma level increases,

(a) cost of poor quality and customer satisfaction both go up
(b) cost of poor quality goes up and customer satisfaction goes down
(c) cost of poor quality goes down and customer satisfaction goes up
(d) cost of poor quality and customer satisfaction both go down

[ESE Prelims : 2023]

7.5 Consider the following statements regarding the design for six-sigma:

1. The concept of six-sigma originated at Motorola.
2. The goal is to arrive at 3.4 defects per million opportunities.
3. Sigma is used to compare expected outcomes versus failures in population.

Which of the above statements are correct?

(a) 1 and 2 only (b) 2 and 3 only
(c) 1 and 3 only (d) 1, 2 and 3

[ESE Prelims : 2023]

7.6 Which one of the following refers to efforts to tailor thousands of items. Such as cars or hamburgers to specific customers' needs?

(a) Miniaturization (b) Mass customization
(c) Reactive mode (d) Fire-fighting

[ESE Prelims : 2024]

7.7 Which one of the following is associated with developing a qualitative and/or quantitative evaluation of how changes to system inputs affect system outputs ?

(a) Define (b) Measure
(c) Analyze (d) Control

[ESE Prelims : 2024]

8. INVENTORY

8.1 Factors which govern the operating cost of an equipment are:

1. Purchase price of the equipment.
2. Depreciation due to regular use.
3. Cost of operation, maintenance and repairs.

Which of the above statements are correct?

(a) 1 and 2 only (b) 1 and 3 only
(c) 2 and 3 only (d) 1, 2 and 3 only

[ESE Prelims : 2017]

8.2 In which one of the following types of industrial activities does the problem of loading and scheduling become more difficult?

(a) Single product continuous
(b) Multi-product continuous
(c) Batch production
(d) Continuous process production

[ESE Prelims : 2017]

8.3 In a company assembly line, assembling gear boxes, five workers are assigned tasks who take 10, 8, 6, 9 and 10 minutes respectively. The balance delay for this assembly line is

(a) 13.3% (b) 14.0%
(c) 16.0% (d) 43.3%

[ESE Prelims : 2017]

8.4 If the EOQ is 360 units, order cost is Rs. 5 order and carrying cost is Rs. 0.20 per unit, what is the usage?

(a) 2654 units (b) 2592 units
(c) 1872 units (d) 1574 units

[ESE Prelims : 2018]

8.5 Which one of the following is the responsiveness to business issues in commercial performances?

(a) Frequency of overshipments
(b) Quotations
(c) Timely reconciliation of cumulative shipments
(d) Timely supplier response to problems

[ESE Prelims : 2021]

9. QUALITY IN CONSTRUCTION

9.1 The process of removing irregular portions of stones and facilitating their easy transportation is known as

(a) Quarrying (b) Reticulating
(c) Dressing (d) Pointing

[ESE Prelims : 2018]

9.2 Which one of the following tests can be resorted to in order to check the structural soundness conformance to the specified standards, when all other tests fail?

(a) Destructive (b) Non-destructive
(c) Full scale load (d) Masonry

[ESE Prelims : 2019]

9.3 Which of the following are constraints to the use of TQM in construction process?

1. A transient labour force.
2. The construction process is relatively short in duration.
3. Hierarchical and vertical organization structure.
4. The construction process has not focused on the detailed needs of the customer.

Select the correct answer using the codes given below:

(a) 1 and 4 only (b) 2 and 3 only
(c) 1 and 2 only (d) 3 and 4 only

[ESE Prelims : 2019]

9.4 Which of the following are the sources of variation in quality control process in construction?

1. Material 2. Operator
3. Inspection activity

Select the correct answer using the codes given below:

(a) 1, 2 and 3 (b) 1 and 2 only
(c) 1 and 3 only (d) 2 and 3 only

[ESE Prelims : 2019]

9.5 Which of the following factors affect the quality of conformance in construction?

1. Site construction methods
2. Technical specifications
3. Engineering and design process
4. Supervision and control

(a) 1 and 3 (b) 2 and 3
(c) 2 and 4 (d) 1 and 4

[ESE Prelims : 2020]

9.6 Which of the following considerations are important during inspection of the work in prestressed concrete works?

1. Sheathings are threaded properly including correct threading of couplers and taping of joints
2. Only approved sheathings, HTS strands, anchor heads and wedges are to be used
3. Splicing of reinforcement to be provided at the joints

(a) 1, 2 and 3 (b) 1 and 2 only
(c) 1 and 3 only (d) 2 and 3 only

[ESE Prelims : 2020]

9.7 Which one of the following is NOT a major quality control method?

(a) Inspection (b) Testing
(c) Loading (d) Sampling

[ESE Prelims : 2021]

10. QUALITY PRACTICES IN SERVICES

10.1 Which of the following are relevant factors regarding quality in service sector?

1. Timeliness of service
2. Customer participation
3. Company personnel motivation
4. Company culture

Select the correct answer using the codes given below:

(a) 1, 3 and 4 only (b) 1, 2 and 3 only
(c) 1, 2, 3 and 4 (d) 2, 3 and 4 only

[ESE Prelims : 2019]

10.2 Which one of the following is the measure of service quality that correlates with the human factors and behavioural characteristics of service quality?

(a) Number of complementary responses based on human traits in delivery of service
(b) Proportion of income tax returns prepared by an agency that have errors
(c) Shabby appearance of a receptionist in a bank or hotel
(d) Inadequate temperature control in a convention meeting room

[ESE Prelims : 2020]

11. RELIABILITY AND MAINTENANCE

11.1 Consider the following statements with reference to maintenance and service of the product after delivery to the customer.

1. Reliability analysis uses statistics of failures to estimate understand and improve the performance of equipment and its maintenance.
2. Reliability analysis records supports or guides the maintenance engineer in improving the performance of product under service.

Which of the above statement is correct:

(a) 1 only (b) 2 only
(c) Both 1 and 2 (d) Neither 1 nor 2

[ESE Prelims : 2017]

11.2 **Statement-I:** Total productive maintenance (TPM) is productive maintenance involving total participation as a group activity.

Statement-II: Under the aegis of TPM, individual operators generally take care of minor maintenance aspects.

(a) Both Statement (I) and Statement (II) are individually true and Statement (II) is the correct explanation of Statement (I)
(b) Both Statement (I) and Statement (II) are individually true, but Statement (II) is not the correct explanation of Statement (I)
(c) Statement (I) is true, but Statement (II) is false
(d) Statement (I) is false, but Statement (II) is true

[ESE Prelims : 2018]

11.3 Which one of the following is NOT a component of total variability of measured observations?

(a) Variation between operators
(b) Variability due to operators
(c) Variability between parts dimensions
(d) Variation due to interaction between operators and parts

[ESE Prelims : 2021]

11.4 Consider the following non-reliability performance measures of automobile industry related objects:

1. Fuel efficiency (km/l)
2. Economic efficiency (cost/km/kg)
3. Quality of ride
4. Emissions (ppm)

Which of the above performance measures are correct?

(a) 1, 3 and 4 only (b) 1, 2 and 3 only
(c) 2, 3 and 4 only (d) 1, 2, 3 and 4

[ESE Prelims : 2023]

11.5 Match the following:

List-I (Severity of failure)

P. Catastrophic
Q. Critical
R. Marginal
S. Negligible

List-II (Impact of failure)

1. Less than minor injury or system damage
2. Minor injury or minor system damage
3. Result in death or total system loss
4. Result in severe injury or major system damage

Select the correct pair using the code given below:

	P	Q	R	S
(a)	3	4	2	1
(b)	4	3	1	2
(c)	2	1	3	4
(d)	1	2	4	3

[ESE Prelims : 2023]

11.6 Consider the following statements for the multi-state characterization (infinite number of states) with $K = \infty$.

1. $X(t)$ is non-decreasing.
2. $X(t)$ is continuous-time stochastic process.
3. Higher value of $X(t)$ implies greater degradation and the item failure time.

Which of the above statements are correct?

(a) 1 and 2 only (b) 2 and 3 only
(c) 1 and 3 only (d) 1, 2 and 3

[ESE Prelims : 2023]

12. MISCELLANEOUS

12.1 If the efficiencies of the boiler turbine and generator are 85%, 45% and 95% respectively, then the efficiency of power plant:

(a) 75% (b) 45%
(c) 36.3% (d) 28.7%

[ESE Prelims : 2017]

12.2 What is the maximum value of z, if $z = 10x + 6y$ subject to the constraints $3x + y \le 12$, $2x + 5y \le 34$, $x \ge 0$, $y \ge 0$?

(a) 56 (b) 52
(c) 50 (d) 40

[ESE Prelims : 2018]

12.3 The objective function $z = 3x_1 + 5x_2$ is to be maximized subject to constraints:

$$x_1 + 2x_2 \le 200, \quad x_1 + x_2 \le 150, \quad x_1, x_2 \ge 0$$

The value of x_1 and x_2 in this context are, respectively

(a) 100 and 75 (b) 125 and 75
(c) 100 and 50 (d) 125 and 50

[ESE Prelims : 2018]

12.4 Which one of the following is NOT a source of variation present in every process of construction?

(a) The equipment (b) The material
(c) The environment (d) The specifications

[ESE Prelims : 2022]

■■■■

Standards and Quality Practices in Production : Answer Key

1.1 (a)	**1.2** (c)	**1.3** (b)	**1.4** (d)	**1.5** (b)	**1.6** (d)	**1.7** (b)	**1.8** (b)
2.1 (b)	**2.2** (b)	**2.3** (b)	**2.4** (c)	**2.5** (c)	**2.6** (b)	**2.7** (c)	**2.8.** (a)
3.1 (a)	**3.2** (b)	**3.3** (d)	**3.4** (d)	**3.5** (b)	**3.6** (d)	**3.7** (c)	**3.8** (c)
3.9 (d)	**3.10** (d)	**3.11** (b)	**4.1** (c)	**4.2** (c)	**4.3** (d)	**4.4** (a)	**5.1** (b)
5.2 (a)	**5.3** (a)	**6.1** (c)	**6.2** (c)	**7.1** (d)	**7.2** (d)	**7.3** (a)	**7.4** (c)
7.5 (a)	**7.6** (b)	**7.7** (b)	**8.1** (c)	**8.2** (c)	**8.3** (b)	**8.4** (b)	**8.5** (b)
9.1 (c)	**9.2** (c)	**9.3** (c)	**9.4** (a)	**9.5** (d)	**9.6** (b)	**9.7** (c)	**10.1** (c)
10.2 (a)	**11.1** (c)	**11.2** (b)	**11.3** (c)	**11.4** (d)	**11.5** (a)	**11.6** (d)	**12.1** (c)
12.2 (a)	**12.3** (c)	**12.4** (d)					

EXPLANATIONS

1.1 (a)

Quality Circle: It is small group of people or employ in **same work area** or doing similar type of job, who volunteer meet and discuss the way of improving the product and process. It is about self and mutual development of grass root level employ (lower level employ). It is problem solving forum. It is human resource development technique.

ADDITIONAL INFORMATION

- Quality circles are effective in improving quality control in the work place and are easy to set up.
- It is useful for forward planning and analyses of current and future production quality.
- Kaouru Ishikawa (Japan) Father of "Quality Circle".

1.2 (c)

Preventive Cost: It is cost related to attempt made to prevent the failure. It increases with the introduction of quality system and may be a significant proportion of the total quality cost. For example, cost related to attempt made to prevent defect (or) failure and cost for education and training.

ADDITIONAL INFORMATION

- Cost of Quality
  - Cost of confirmance
    - Preventive cost
    - Appraisal cost
  - Cost of Non-confirmance
    - Cost of failure
      - Internal failure cost
      - External failure cost

1.3 (b)

Total Quality: Consists of continuous improvement activities involving everyone in the organization (managers and workers) in a totally integrated effort toward improving performance at every level.

ADDITIONAL INFORMATION

- Total quality management (TQM) is a process-oriented management system and consider the complete organisation as integrated, is customer focussed, demands continuous improvement and fact based decision making to meet the quality standards as planned and desired during the planning stage and provides an opportunity to improve and it helps in profitable survival in reaction to the competitive threat.
- Feigenbaum gave the term TQM.

1.4 (d)

The best approach to the prevention of product liabilities are

- Quality control.
- Comprehensive testing procedure.
- Good engineering in analysis and design.

ADDITIONAL INFORMATION

Product liabilities: This concept states that the manufacturer of an article is liable for any damage or harm that results because of a defects in product. There are three types of product defects that can result in product liability cases: Design defects, manufacturing defects and marketing defects.

1.5 (b)

Quality objectives: are specific, measurable goals that a company sets to improve it's quality management system. They are typically derived from the organization's quality policy and are aligned with it's strategic objectives.

ADDITIONAL INFORMATION

While setting quality objective, it's important to use the **SMART** goal-setting approach, which stands for specific, measurable, attainable, relevant and time-bound.

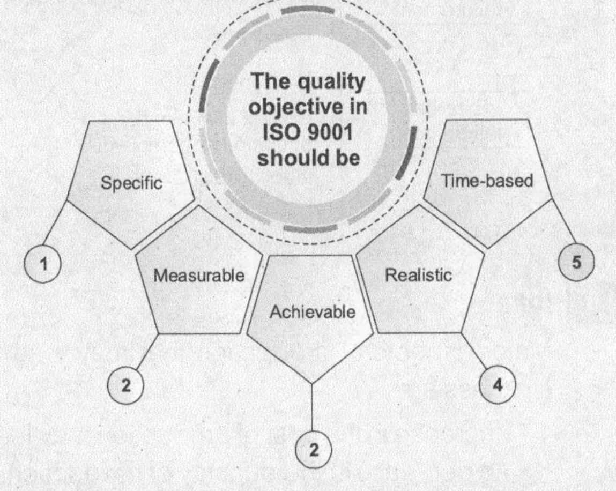

1.6 (d)

Quality Characteristics : One or more element define the intended quality level of product or services.

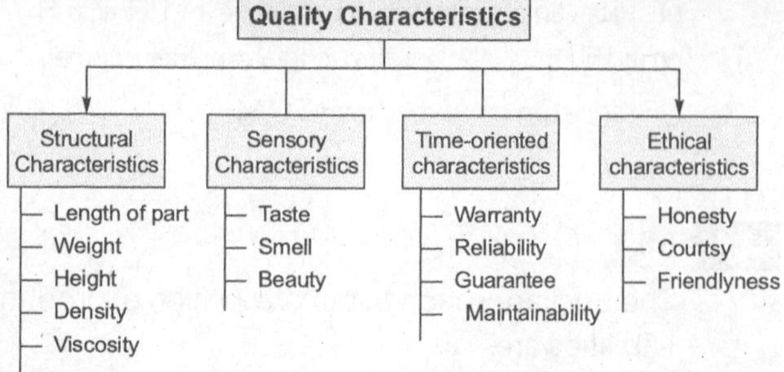

ADDITIONAL INFORMATION

Quality characteristics are the properties of a product or a service that are required to satisfy the customers. They can relate to the characteristics that customers expect or demand from the service.

1.7 (b)

Major categories for quality costs:

1. Preventive cost
2. Appraisal cost
3. Internal failure cost
4. External failure cost

Production cost : It is direct cost, does not come under quality cost.

ADDITIONAL INFORMATION

- Preventive cost
  - Education cost
  - Training cost
- Appraisal cost
  - Inspection cost
  - Auditing cost
  - Checking cost
- Internal failure cost
  - Some corrective action
  - Some changes
  - Some rework
- External failure cost
  - Warranty cost
  - Cost of investigation of customer complaint
  - Liability cost

1.8 (b)

Major aspects of production that may lead to sickness are

1. Increase in the cost of production may lead to decrease in the quantity of production.
2. Producing more quantity than can be sold leading to accumulation of stock that would increase inventory carrying cost.
3. Quality of product not meeting the standards/customer expectation would increase external failure cost.

ADDITIONAL INFORMATION

Bad Production Policies : It is one of major reason for industrial sickness can be related to the production decisions viz. Choosing inefficient site for conducting production activities, poor maintenance of plants and machineries, absence of quality control measures, absence of standard research and development technique.

2.1 (b)

Objective of quality circle activities includes:

For Individual :

1. To give opportunity to employees to lear new techniques for recognition.
2. To give opportunity to employees to identify and solve the problems in their work area.

For Organization :

1. To improve quality and productivity of products.
2. To promote quality, safety, cost and work consciousness.

ADDITIONAL INFORMATION

- Quality control circles originated in Japan, First company in Japan to introduce it was NIPPON WIRLESS AND TELEGRAPH COMPANY in 1962.
- Originally described by Edwards Demining in the 1950, praised Toyota.
- The ideas was later formalized across Japan in 1962 and expanded by others such as K. Ishikawa.

2.2 (b)

Concurrent Engineering : It is also known as simultaneous engineering, is a method of designing and developing products, in which the different stage run simultaneous rather than consecutively. It implements long before the project execution.

ADDITIONAL INFORMATION

It decreases product development time and also the time to market, leading to improve the productivity and reduced costs.

2.3 **(b)**

Both the statement given are correct for just-in-time concept.

Just-in-time(JIT): It is a management philosophy and a technique used to improve profit and it's associated cost like delivery cost, holding cost etc.

ADDITIONAL INFORMATION

- The core practice of JIT is use of KANBAN system.
- JIT leads to zero inventory, zero lead time, zero failure, zero wastage.
- It is pull type manufacturing system.
- It reduces over production.

2.4 **(c)**

The concept of quality is not confined only to construction and manufacturing organisation. It is used in other organisation also like automobile, medical, research and development, education, service sector etc.

ADDITIONAL INFORMATION

Quality : As it applies to an object (product, service, process), is defined as the "degree to which a set of inherent characteristics (attributes) of the object satisfies a set of requirements". Therefore, the quality of an object is determined by comparing a predetermined set of characteristics against a set of requirements. If those characteristics conforms to the requirements, high quality is achieved but if those characteristics to not conform, a low or poor level of quality is achieved.

2.5 **(c)**

Deming system of profound knowledge.

1. Knowledge of system and theory of optimization.
2. 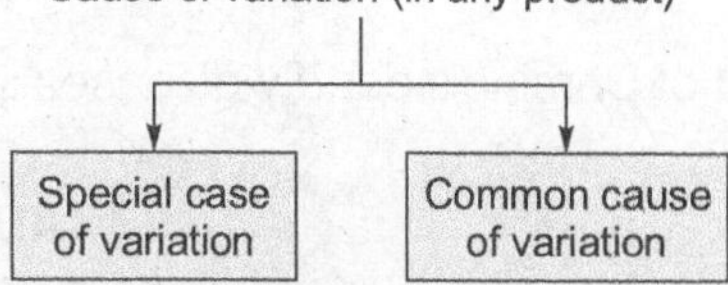

3. Exposure to the theory of knowledge.
4. Knowledge of psychology.

ADDITIONAL INFORMATION

About Dr. W. Edward Deming (U.S.)

- Father of quality control.
- Father of Japanese post war industrial revival.
- Deming prize → Highest quality award of Japan.

Contribution

- Deming cycle.
- 14 point's of management.
- 7 deadly disease.
- System of profound knowledge.

2.6 **(b)**

14-Step plan for quality improvement:

1. Management commitment
2. Quality improvement team
3. Cost of quality evaluation
4. Quality awareness
5. Quality measurement
6. Corrective action
7. Plan for zero defects program
8. Supervisor training
9. Zero defects day
10. Goal setting
11. Error cause removal
12. Recognition
13. Quality council
14. Do it over gain

ADDITIONAL INFORMATION

About Philip. B. Crosby

- Concept of doing it right first time.
- Quality is conformance to requirement.

Contribution

- Zero defect.
- Quality is free.
- Quality without tears.
- 14 points of quality improvement.

2.7 (c)

Most of Deming's deadly diseases involve a lack of understanding of variation.

ADDITIONAL INFORMATION

Deming 7 deadly disease:

1. Looking for short term profit.
2. Lack of constancy of purpose.
3. Mobility of management.
4. Excessive cost of warranty.
5. Excessive medical cost.
6. Running a company on visible figure only.
7. Evaluation by performance merit rating or annual review of performance.

2.8 (a)

Both the statement are correct and statement II is correct explanation of statement I.

ADDITIONAL INFORMATION

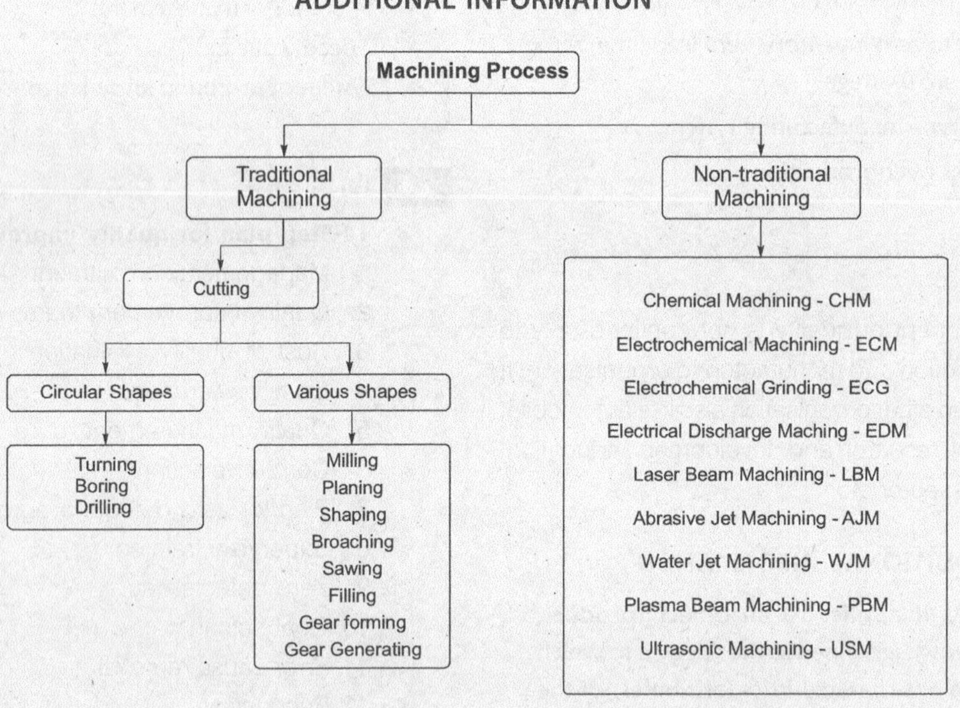

3.1 (a)

The department responsible for quality are product design engineering, procurement, process design, production, inspection and testing, packaging and shipping, after-sale service, and logistical support.

ADDITIONAL INFORMATION

Three types of Quality :

1. **Quality of Design:** It means the intention of designer to include or exclude feature in product or services.
2. **Quality of Conformance:** It means how well the product are services meet the specification determined by the designer.
3. **Quality of Performance:** It is associated with reliability of the product or services performing it's intended function under the prescribed set of condition.

3.2 (b)

Quality control can be divided into three main sub-areas :

1. Off-line quality control.
2. Statistical process control.
3. Acceptance sampling plan.

ADDITIONAL INFORMATION

The different types of acceptance sampling plane are:

- Single sampling plan.
- Double sampling plan.
- Multiple sampling plan.
- Sequential sampling plan.
- Skip lot sampling plan.

3.3 (d)

The formula for the calculation of the standard deviation for concrete is given as

$$\sqrt{\frac{\sum(X-\bar{X})^2}{N-1}}$$

The minimum number of samples tested should be 30 with three specimens from each samples.

ADDITIONAL INFORMATION

- The standard deviation (σ) is a statistic quantity that measures the dispersion of a set of data relative to it's mean.
- Variance (σ^2) in statistics is a measurement of the spread between numbers in a data set. It is calculated as the square of the standard deviation.

3.4 (d)

There are three categories of variation in piece part production:

1. Within-piece variation.
2. Piece-to-piece variation.
3. Time-to-time variation.

ADDITIONAL INFORMATION

Variation is present in every process due to a combination of the equipment, materials, environment, and operator

- First source of variation is the equipment.
- Second source of variation is the material.
- Third source of variation is the environment.
- Fourth source of variation is the operator.

3.5 (b)

The objectives of attribute charts are to

1. Determine the average quality level and the acceptance criteria of a product.
2. Bring to the attention of management any changes in the average quality of the product.
3. Improve the product quality by motivating the operating and management personnel to initiate ideas for quality improvement.

ADDITIONAL INFORMATION

There are two types of data:

1. **Attributes :** An attribute data is qualitative and discrete in nature.
2. **Variables :** A variable data is continuous in nature.

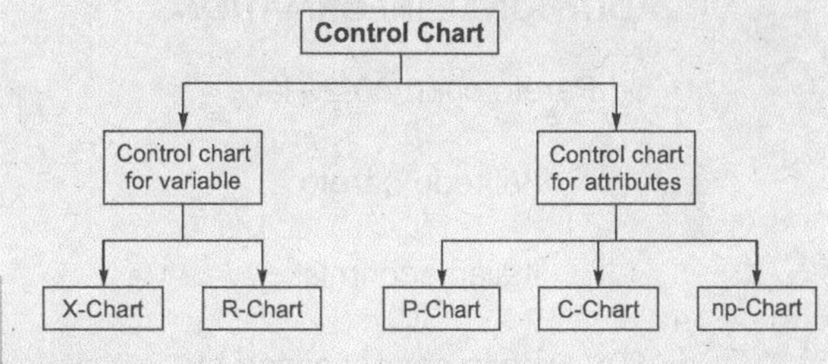

3.6 (d)

Rules for identifying an out of control process:

1. A process is assume to be out of control if a single point plot outside the control limit.
2. A process is assumed to be out of control if 2 out of 3 consecutive points fall outside the 2σ warning limit on same side of the central line.
3. A process is assumed to be out of control if 4 out of 5 consecutive point' fall beyond 1 σ limit on same side of centre line.
4. A process is assumed to be out of control if 9 or more consecutive points fall one side of central point.
5. A process is assumed to be out of control if there is a run of 6 or more points steadily decreasing or increasing.

ADDITIONAL INFORMATION

An out-of-control process indicates the presence of non-random variation. Non-random variation is caused by definite, specific causes that are called assignable causes. These assignable causes make the process go out of control or become statistically unstable.

3.7 (c)

Steps for constructing pareto chart:

1. Decide on data categorisation system
2. Determine how relative important is to be judge.
3. Rank the category from most important to least important.
4. Compute the commulative frequency of the data in their chosen order.
5. Plot a bar graph that showing relative importance.

ADDITIONAL INFORMATION

Pareto chart/80-20 law
↓
Vilfredo pareto
↓
Italian econmist
↓
80% wealth are possessed by 20% of population

Approximately 80% of the problem are created due to 20% of the cause.

4.1 (c)

Acceptance Quality Level (AQL): It represents the maximum proportion of defectives that the customer finds definitely acceptable. At AQL, the probability of acceptance is 100% under ideal condition.

ADDITIONAL INFORMATION

Examples of AQL :

(i) No electricity for 15 minutes in a day.

(ii) News paper not delivered 5 times in year.

Rejectable Quality Level (RQL) : The upper limit on the % of the lot size at which consumer is willing to accept the lot.

4.2 (c)

Hourly production = 4000 boxes

Non-conforming = 20 boxes

Probability of being non-conforming

$$= \frac{20}{4000} = 0.005$$

ADDITIONAL INFORMATION

Probability of event A

$$= \frac{\text{Number of outcomes favourable to event } A}{\text{Total number of possible outcomes}}$$

4.3 (d)

Average Outgoing Quality (AOQ):

- It is another method to evaluate the goodness of sampling plan.
- It represent average % defective in the outgoing product after the inspection including all accepting and all rejected.

$$\text{AOQ} = \frac{P_a\, p(N-n)}{N}$$

P_a = Probability of accepting the lot

p = Incoming lot quality, N = lot size

n = Sample size

ADDITIONAL INFORMATION

Average Outgoing Quality Limit (AOQL): Represents the maximum percentage defective in the outgoing product. The average outgoing quality depends on the incoming quality, the probability that the lot will be accepted.

4.4 (a)

Average outgoing quality limit, represents the maximum level of average percentage defective (defects) in the outgoing product after the inspection including all accepted and all rejected lot.

ADDITIONAL INFORMATION

Sampling: It is a technique in which a random sample is drawn and on basis of this random sample the decision of acceptance and rejection are taken.

- 100% Inspection — Generally very expensive.
- Handling by inspector causes defect.
- Can not use when product destroy while inspection.

5.1 (b)

TQM involves the integrated approach where a bottom level employee can interact with top management and a have involvement in quality-related parameters.

ADDITIONAL INFORMATION

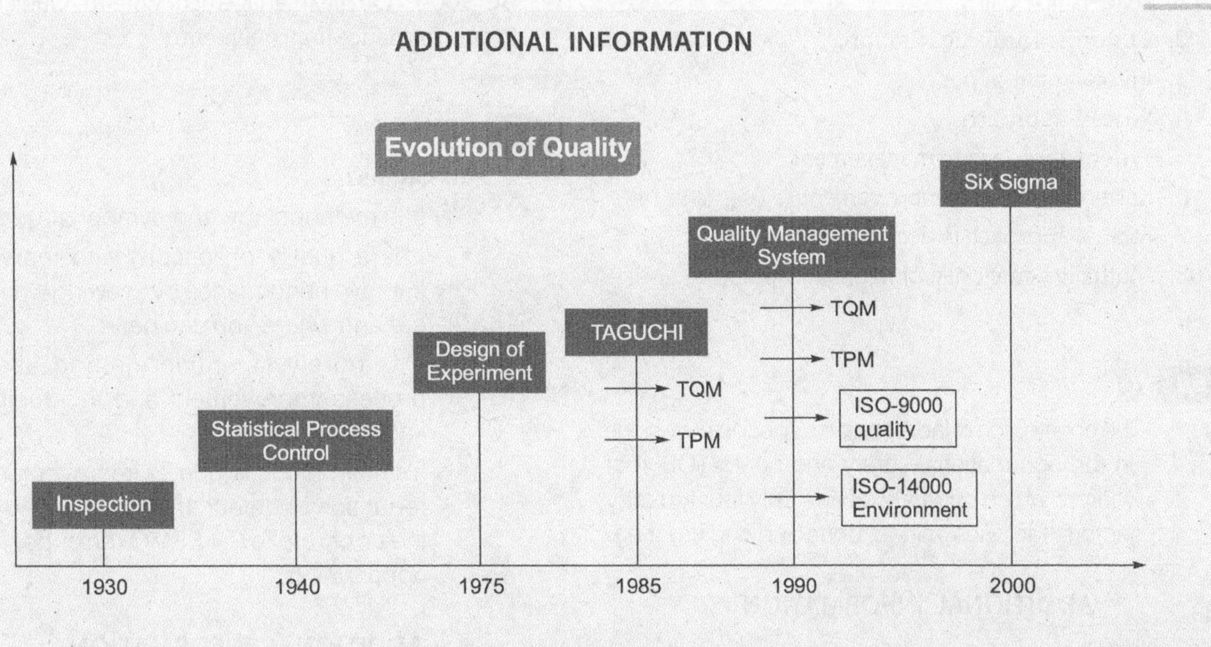

5.2 (a)

Construction process are new and relatively short in duration, that makes TQM difficult to implement.

ADDITIONAL INFORMATION

Taguchi has important influence specially in design area i.e. product design and process design for the improvement and developed in two ways:

1. Robust design. 2. Taguchi loss function.

- Robust design means, a process continue to produce good product with minimum deviation from target.
- According to Taguchi, there is always loss even if the product is made with in the specific limit, it means there is loss as long as product deviate from target.

$$L = k[y - y_o]^2$$

y = Measured value, y_o = Target value

L = Loss ; K = Taguchi constant

5.3 (a)

According to TQM

- People at all levels are the essence of an organisation and their involvement enables their abilities to be used for organisational benefits.
- The desired result is achieved more efficiently when related resources and activity are managed as a process.

Hence, option 3 and 4 are incorrect.

ADDITIONAL INFORMATION

Principle of TQM :

1. Leadership quality.
2. Customer focus organisation.
3. Involvement of people.
4. Process approach.
5. System approach to management.
6. Kaizen/continuous improvement.
7. Factual approach in decision making.
8. Mutually beneficial supplier relationship.

6.1 (c)

The constitution of India contains specific provisions on the occupational safety and health (OSH) of workers which currently covers only four industry sectors: factories, mines, ports and construction.

ADDITIONAL INFORMATION

- OHSAS (Occupational Health and Safely Management Standards) are international standard which provide a framework to identify, control and decrease the risks associated with health and safety within company.
- **OHSAS 18001:** It is worldwide known as occupational, health and safety management system series standard. It is based on the similar ISO 90001 and ISO 14001 Plan-Do-Check-Act structure.

6.2 (c)

As the system is being installed, it's effectiveness should be checked by regular internal quality audits. Internal quality audits are conducted to verify that the installed quality management system; Conform to the planned arrangements, to the requirements of the standard (ISO 9001 : 2000) and to the quality management system requirements establish by your organisation, and is effectively implemented and maintained.

ADDITIONAL INFORMATION

ISO standards are the sets of international standards on quality development by the international organisation for standardisation (ISO) based on Geneva, Switzerland. At present, three main standards of this family are :

ISO 9000 – Quality management system – Fundamentals and vocabulary

ISO 9001 – Quality management system – Requirement

ISO 9004 – Quality management system – Guidance for performance improvement.

7.1 (d)

Six-Sigma :

- It is methodology and symbol of quality.
- It is a quality philosophy and the way of improve performance by knowing where you are and where you can be.
- It is purely scientific method used for process improvement and this method is known as DMAIC.
- It's main focus is quality improvement and reduction of defect. It target to bring down the error rate to 3.4 EPMO (error per million opportunity).

ADDITIONAL INFORMATION

- Father of Six Sigma – Mr. Bill Smith.
- In 1987 motorola officially launch six sigma program and due to this Motorola known worldwide as quality leader and profit leader.
- In 1988 Motorola was awarded by Malcolm Baldrige National Quality Award (National award of USA).
- In 1991 Motoral certified it's first black belt.

7.2 (d)

All three statement with reference to six-sigma are correct.

ADDITIONAL INFORMATION

The six sigma methodology is widely used in many top corporations in the united states and around the world. It is normally defined as a set of practices that improve efficiency and eliminate defects.

Six sigma follows two project methodologies:

1. DMAIC 2. DMADV

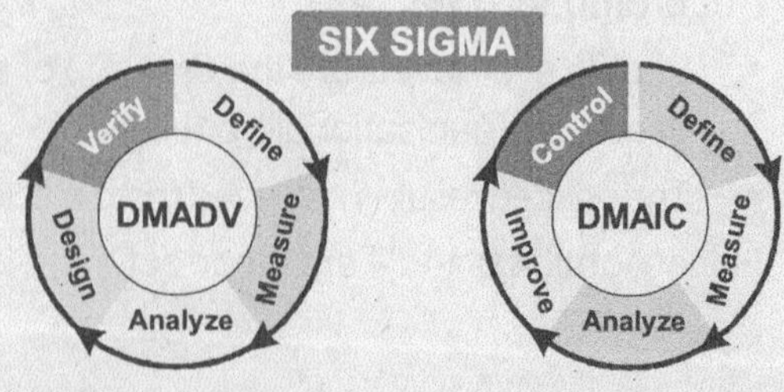

7.3 (a)

Six sigma gives a precision of 99.997%.

ADDITIONAL INFORMATION

Sigma Level	% Good
6σ	99.9997
5σ	99.98
4σ	99.78
3σ	93.37
2σ	69
1σ	31

7.4 (c)

As sigma level increases defects per Million opportunity decreases and cost of poor quality goes down and customer satisfaction goes up.

ADDITIONAL INFORMATION

σ level ↑	DPMO / EPMO ↑
6σ	3.4
4σ	233
3σ	6210
2σ	66810
1σ	308537

DPMO (Defect Per Million Opportunity)

$$= \frac{\text{No. of defect}}{\text{No. of unit produced}} \times 10^6$$

EPMO (Error per million opportunity)

$$= \frac{\text{No. of error}}{\text{No. of service provided}} \times 10^6$$

7.5 (a)

Sigma is only Greek word, which used to measure deviation or dispersion.

ADDITIONAL INFORMATION

Different level in six sigma :

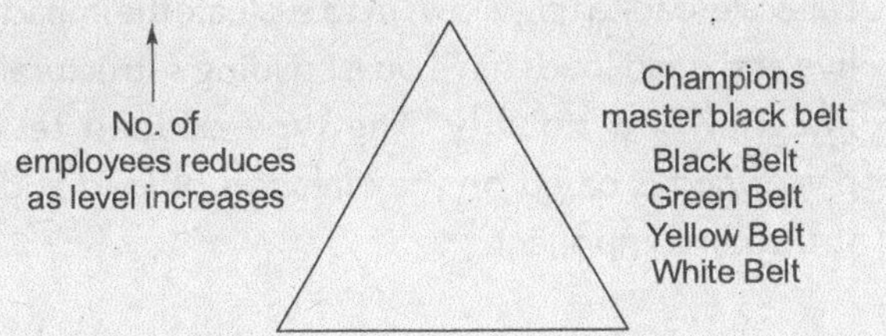

8.1 (c)

Operating cost incurred when the equipment is operated. The operating cost of equipment is influenced by various parameter namely number of operating hours, location of job site, operating condition, category of equipment etc.

The operating cost consist of following :

(a) Repair and maintenance cost

(b) Fuel cost

(c) Cost of lubricating oil

(d) Tire cost

(e) Equipment operator wages

(f) Cost of replacing higher wear item

ADDITIONAL INFORMATION

Classification of inventories

1. Direct Inventories :

(a) Production inventory

(b) Work-in-process inventory

(c) Finished goods inventory

(d) MRO inventory

(e) Miscellaneous inventory

2. Indirect Inventories :

(a) Transit or pipeline inventories

(b) Buffer inventories

(c) Decoupling inventories

(d) Seasonal inventories

(e) Lot size inventories

(f) Anticipation inventories

8.2 (c)

Batch Production : It is used for medium volume lot and for moderate product variety.
Batch production system has the most difficult loading and scheduling.

ADDITIONAL INFORMATION

Job Shop Production : This type of production involves very high variety and less lot size.

Continuous Production : It is a one-time activity and hence loading and scheduling is easier.

8.3 (b)

Five boxes:

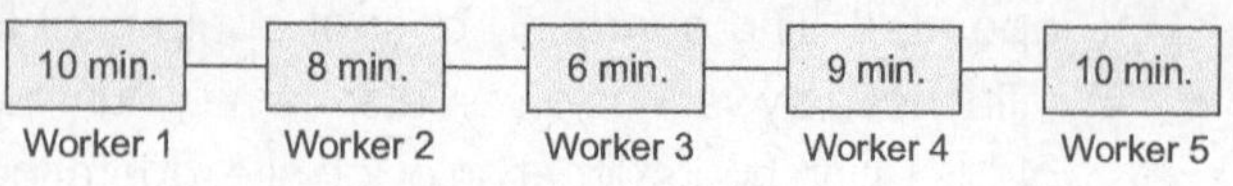

Cycle time = Max{10, 8, 6, 9, 10}

Cycle time = 10 minute

Now, Balance delay

$$= 1 - \frac{\text{Total work time}}{\text{No. of worker} \times \text{Cycle time}} = 1 - \frac{(10+8+6+9+10)}{5 \times 10}$$

$$\text{B.D.} = 0.14 = 14\%$$

ADDITIONAL INFORMATION

Balance Delay : It is the amount of idle time on production assembly lines caused by the uneven division of work among operators or stations. It is related to the extent and way the total task is subdivided.

Balance delay = 1 – Line efficiency

8.4 (b)

$$\text{EOQ} = \sqrt{\frac{2DC_o}{C_c}}$$

$$\text{or } D(\text{Demand}) = \frac{(EOQ)^2 \times C_C}{2C_o}$$

$$D = \frac{(360)^2 \times 0.2}{2 \times 5}$$

$$= 2592 \text{ units}$$

ADDITIONAL INFORMATION

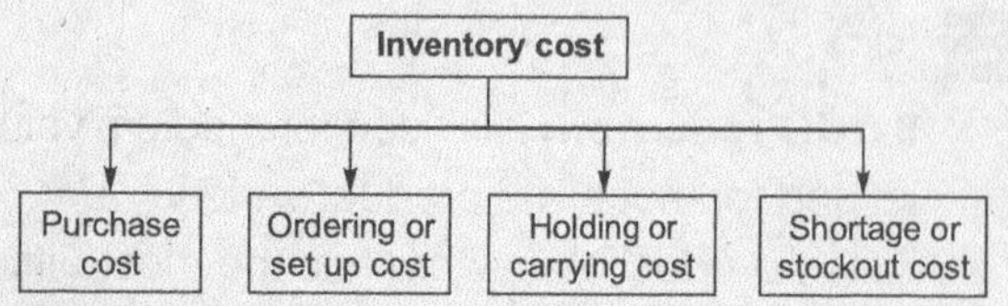

Purchase Cost (PC) = Annual Demand × Cost/Unit

Ordering Cost (OC) = Numbers of orders × Cost/Order

Holding or Carrying Cost (HC) = Average Inventory × Cost of holding one unit

Shortage (or) Stock out Cost (SC) = Number of units short × Shortage cost/unit

8.5 (b)

A quotation is a document that a seller provides to a buyer to offer goods or services at a stated price, under specified conditions. Thus, we can increase customer responsiveness by reducing time-to-quote.

ADDITIONAL INFORMATION

- Customer responsiveness refers to your business's ability to respond to service inquiries and fulfill them in a timely manner.
- To achieve high performance, managers try to improve their customer responsiveness, product quality and organizational efficiency.

9.1 (c)

The process of giving a proper size, shape and finish to the roughly broken stones as obtained from the quarry, is known as dressing of stone.

ADDITIONAL INFORMATION

- The process of extraction of suitable stones from their natural rock beds or layers is commonly called Quarrying of Stones.
- Dressing of stones is done immediately after quarrying and before seasoning to achieve less weight for transportation.

9.2 (c)

If all destructive and non-destructive test fail to check the structural soundness conformance to the specified standards then full scale test is used.

ADDITIONAL INFORMATION

Full-scale load testing allows us to evaluate the impact live and dead loads have on a building's structural integrity and stability. The type of load test performed is based on the direction of load and structural component.

9.3 (c)

Constrain to use TQM in construction company:

(i) Construction process is relatively shorter in duration.
(ii) A transient labour force.
(iii) Construction project are unique.
(iv) Penalties for non-performance.

ADDITIONAL INFORMATION

Importance of Quality in Construction:

- In construction project proper inspection and quality control become very important because ones the project is completed, it is very difficult and costly to later on modify.
- In many construction project, a good and safe design can be spoiled by careless execution, therefore, inspection and quality control are as important as investigation and design.

9.4 (a)

Variation is present in every process due to combination of

1. Equipment
2. Material
3. Environment
4. Operator

ADDITIONAL INFORMATION

Constraints to use TQM in construction company :

(i) Construction process is relatively shorter in duration.
(ii) A transient labour force.
(iii) Construction project are unique.
(iv) Penalties for non-performance.

9.5 (d)

Quality of conformance in construction is affected by the following :

1. Construction methods adopted, skill of workers, quality of material used.
2. Supervision to conform to the plans and specification and control to act immediately if anything goes out of the way to avoid any wastage.
3. Inspection through NDT and other sampling methods to assess the quality and plan control measures.

ADDITIONAL INFORMATION

Basic element of quality control in construction:

1. Quality characteristics.
2. Quality of design.
3. Quality of conformance.

9.6 (b)

There is no splicing of reinforcement be provided at the joints of prestressing steels in a prestressed concrete member. Hence, no such inspection is required.

ADDITIONAL INFORMATION

Prestressed concrete is a system devised to provide sufficient pre-compression in the concrete beam by tensioned steel wires, cables or rods that under working conditions, the concrete has no tensile stresses are so low that no visible cracking occurs.

9.7 (c)

Quality control method in construction:

1. Inspection
2. Testing
3. Sampling

ADDITIONAL INFORMATION

Quality control involves testing units and determining if they are within the specifications for the final product. The purpose of the testing is to determine any needs for corrective action in manufacturing process. Good quality control helps companies meet consumer demands for better products.

10.1 (c)

5-dimensions of service quality:

1. **Reliability:** The ability to complete the service on time and error free every time.

2. **Tangible :** The surrounding in which the service is provided.
3. **Responsiveness :** It is time taken by service provider to respond to the customer complaints.
4. **Assurance :** Gain the trust of customer.
5. **Empathy :** Individual attention to customer.

ADDITIONAL INFORMATION

Service Quality Gap Model :

- The aim of this model to identify the gap between customer expectation and actual service provided.
- Gap model is an important measurement tool for improving the service delivery.
- This is developed by Parasuraman, Zeithal & Bery in 1990 era.

10.2 (a)

Human Factor and Behavioural Characteristic :

- Number of complained based on the behavioural factor of person in service process.
- Number of complimentary response based on human traits in delivery of service.

ADDITIONAL INFORMATION

Service quality characteristics and their measurement :

1. Human factor and behavioural characteristic.
2. Timeliness characteristics
3. Service non-conformity characteristics.
4. Facility related characteristics.

11.1 (c)

The reliability analysis is recorded and the maintenance engineer works out on the basis of statistical data present about the poor performance of equipment and helps him identify the damage zone. This way reliability analysis records support the maintenance engineer in improving the performance of the product under service.

ADDITIONAL INFORMATION

- The reliability of product or system can be defined as the probability that the product will perform its required function under specific condition for certain period of time.
- Reliability is the measure of quality of the product over a long run.
- Reliability of system depend on reliability of individual component.

11.2 (b)

- Total Productive Maintenance (TPM) is well-defined and time-tested concept for maintaining plants and equipment. TPM increases production while, at the same time, increasing employee morale and job satisfaction.
- In TPM operators working at the equipment are trained and given a sense of ownership and hence they generally take care of minor maintenance aspects.

ADDITIONAL INFORMATION

Types of Maintenance :

1. **Reactive or Breakdown Maintenance :** It is basically run till breakdown. No action or effort taken to maintain equipment.
2. **Preventive Maintenance :** Routine servicing is planned to prevent failure.
3. **Predictive Maintenance :** Timely replacement of the component before the cause any breakdown.

11.3 (c)

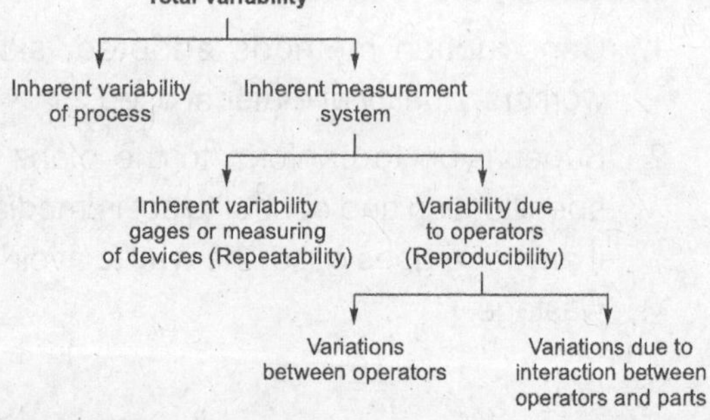

ADDITIONAL INFORMATION

There are four distinct causes of variation in production processes :

- **Common Cause (Noise) Variation :** Inherent in a process over time and affecting all outcomes.
- **Special Cause (Signal) Variations :** Unusual circumstances not inherent in a process.
- **Tampering :** When attempts to reduce variation increase variation instead.
- **Structural Variation :** Due to the fact that no two items are ever exactly the same.

11.4 (d)

The non-reliability measures include technical, operational, economic, environmental impact and so on. They are specific to the engineered object. Table below lists a few of non-reliability performance measures.

Engineered object	Type	Performance measures
Automobile	Consumer product	Fuel efficiency (km/*l*) Economic efficiency (cost/km/kg) Quality of ride Emissions (PPM)
Photocopier	Commercial product	Quality of image Throughput (copies/min)
Oil refinery	Plant	Efficiency Downtimes Emissions
Rail network	Infrastructure	Train delays (min/wk) Ride quality (noise. vibration, etc.)

ADDITIONAL INFORMATION

Reliability Performance Measures :

(a) **Interval Reliability :** The probability of no failure over a specified interval.

(b) **Interval Availability :** The fraction of the time in which the product or system is in the operational (non-failed) state over a specified.

(c) The number of failures over a specified interval.

11.5 (a)

The severity of a failure mode signifies the impact of the failure mode on the system as a whole and on the outside environment. A severity ranking classification scheme (MIL-STD-82D, 2000) is as follows :

- **Catastrophic :** Failure that result in death or total system loss.
- **Critical :** Failure that result in severe injury or major system damage.
- **Marginal :** Failure that result in minor injury or minor system damage.
- **Negligible :** Failures that result in less than minor injury or system damage.

ADDITIONAL INFORMATION

Failure causes may be classified (in relation to the life cycle of the object), as indicated below:

- **Design failure :** Due to inadequate design.
- Weakness failure : Due to weakness (inherent or induced) in the system so that the system cannot stand the stress it encounters in its normal environment.
- **Manufacturing failure :** Due to non-conformity during manufacturing.
- **Aging failure :** Due to the effects of age and/ or usage.
- **Misuse failure :** Due to misuse of the system (operating in environments for which it was not designed).
- **Mishandling failure :** Due to incorrect handling and/or lack of care and maintenance.

11.6 (d)

Multi-static characterization (infinite number of states) with $k = \infty$, $X(t)$ is a non-decreasing, continuous-time stochastic process as shown in figure below. Here, a higher value of $X(t)$ implies greater degradation and the item failure time is given by $T = \inf\{t = X(t) = x^*\}$

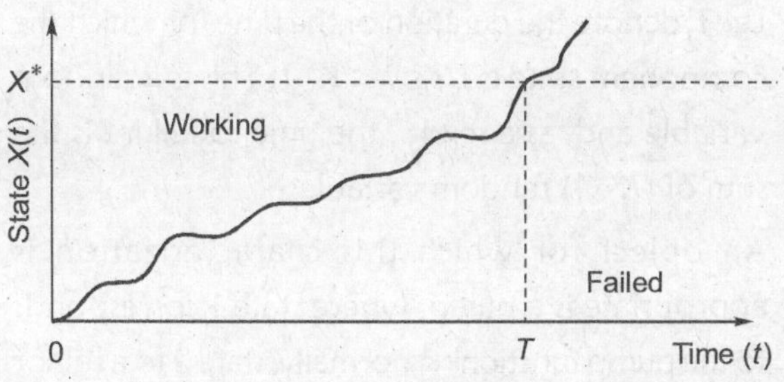

ADDITIONAL INFORMATION

Multi-State Characterization (Finite Number of States)

Here, $X(t)$ can assume values from the set (1, 2, ... , K) with:

- $X(t) = 1$ corresponding to item performance being fully acceptable (item is in a good working state).
- $X(t) = 1, 1 < i < K$, corresponding to item performance being partially acceptable (item is in a working state, with a higher value of i implying a higher level of degradation).
- $X(t) = K$ corresponding to item performance being unacceptable (item is in a failed state).

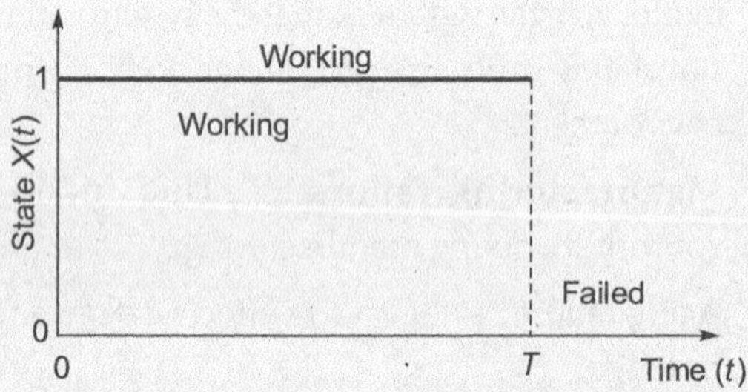

Time to failure (binary-state characterization)

The time to failure of the item is given by $T = \inf(t : X(t) = K)$, as shown in below figure.

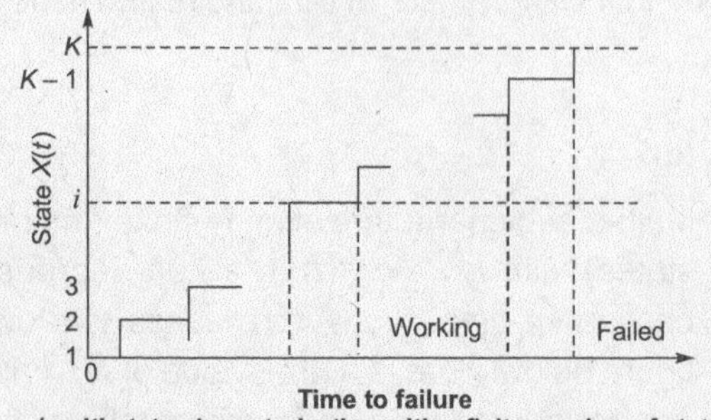

Time to failure (multi-state characterization with a finite number of state)

Let T_i denote the duration of the time for which the component state is i, $1 \le i < K-1$. This is a random variable and, as a result., the time to failure is the sum of $(K-1)$ random variables.

An object for which this characterization is appropriate is a pump, where state 1 corresponds to the pump functioning normally, state 2 to a minor leak, and state 3 to a major leak.

12.2 (a)

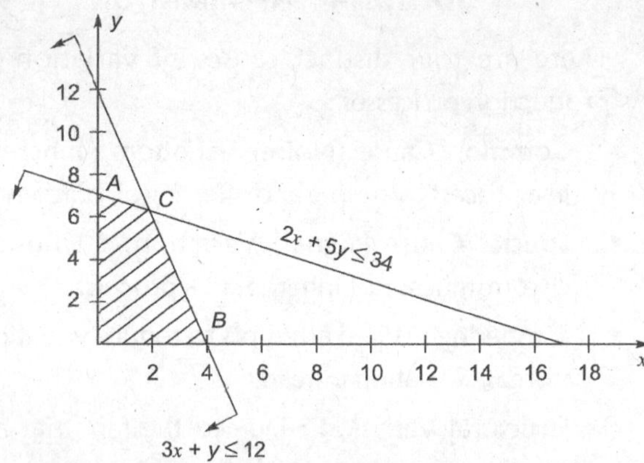

Maximize, $2 = 10x + 6y$

Subject to, $3x + y \le 12$

$$2x + 5y \le 34$$

$$x \ge 0, y \ge 0$$

Graphical Method : From feasible region, corner points are $A(0, 6.8)$, $B(0, 4)$, $C(2, 6)$.

To maximize, Z (objective function)

$$Z_A = 10 \times 0 + 6 \times 6.8 = 40.8$$

$$Z_B = 10 \times 0 + 6 \times 4 = 24$$

$$Z_C = 10 \times 2 + 6 \times 6 = 56 \text{ (max.)}$$

So, maximum value is 56 at $C(2, 6)$.

ADDITIONAL INFORMATION

The Linear Programming Problems (LPP) is a problem that is concerned with finding the optimal value of the given linear function. The optimal value can be either maximum value or minimum value. LPP applications may include production scheduling, inventory policies, investment portfolio, allocation of advertising budget, construction of warehouses, etc.

12.3 (c)

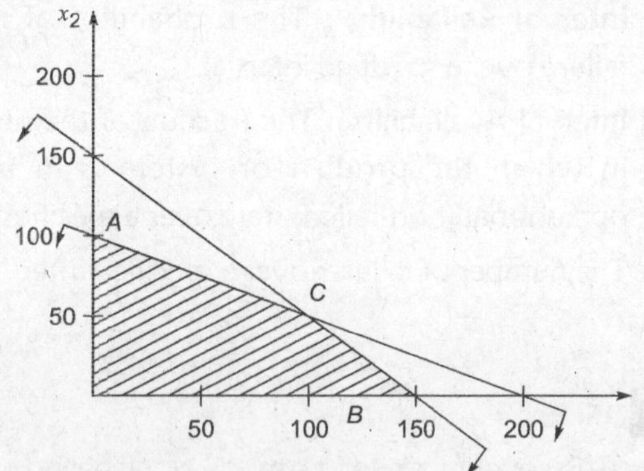

Maximize, $Z = 3x_1 + 5x_2$

Subjected to constraint,

$$x_1 + 2x_2 \le 200$$

$$x_1 + x_2 \le 150$$
$$x_1, x_2 \ge 0$$

Graphical Method : Corner points of feasible region are :

$A(0, 100)$, $B(150, 0)$, $C(100, 50)$

Objective function value of corner points :

$$Z_A = 3 \times 0 + 5 \times 100 = 500$$
$$Z_B = 3 \times 150 + 5 \times 0 = 450$$
$$Z_C = 3 \times 100 + 5 \times 50 = 550 \text{ (max.)}$$

So, maximum value of objective function occurs at $x_1 = 100$, $x_2 = 50$.

ADDITIONAL INFORMATION

The Linear Programming Problems (LPP) is a problem that is concerned with finding the optimal value of the given linear function. The optimal value can be either maximum value or minimum value. LPP applications may include production scheduling, inventory policies, investment portfolio, allocation of advertising budget, construction of warehouses, etc.

ADDITIONAL INFORMATION

Variation is basically defined as anything which is deviate from the standard/target values. Variation can be divided into two parts:

1. **Common/Random/Chance variation :** These variations are difficult to trace and difficult to control even under the best condition of operation. These type of variation are always present. These variations are of lower magnitude, always within limit and defective parts are not produce due to them . They don't whose a particular pattern.
2. **Assignable variation :** These variations are of higher magnitude which can be easily traced and detected. Defective parts are produced due to them and these may be due to some particular reasons like machine setting, defective raw material, tool wear, improper training of operator.

12.4 (d)

As for variability in construction process, several studies investigated the reasons and remedies of variability. Wambeke et. al. (2011) classified the causes of variations into eight main categories, which are pre-requisite work, labour force, detailed design and working method, material and components, tools and equipments, management, supervision, information flow, work and job site conditions and weather or external conditions.

■■■■

Basics of Energy and Environment

CONTENTS

Basics of Energy and Environment

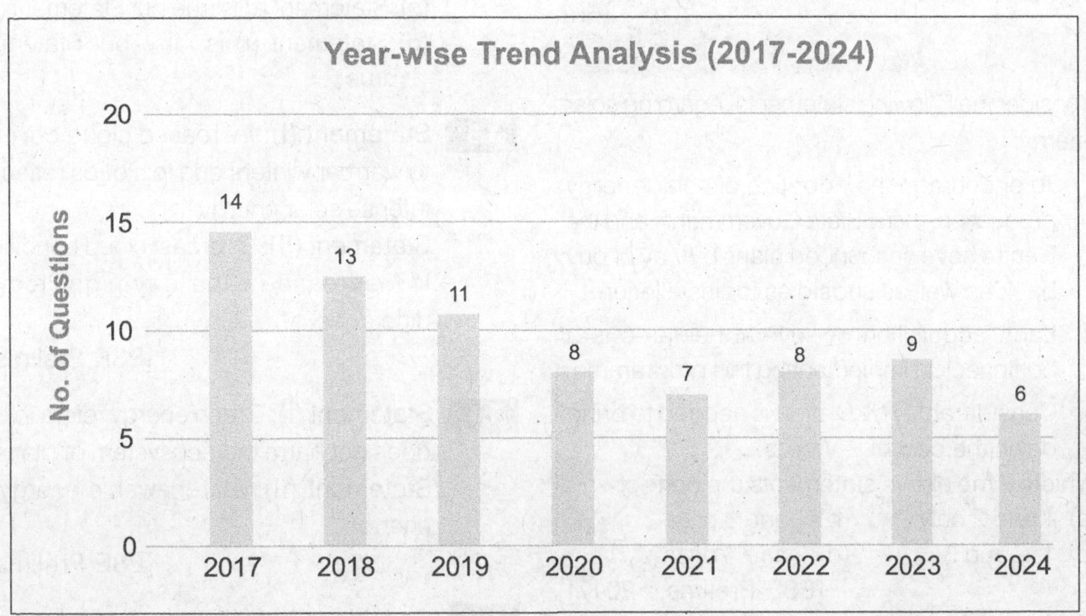

TOPIC-WISE TREND ANALYSIS (2017-2024)		
Sl. No.	**Topics**	**Total Number of Questions**
1.	Energy Resources : Conservation & Utilisation	18
2.	Basics of Environment, Ecology & Ecosystem	7
3.	Biodiversity & its Conservation	4
4.	Environmental Degradation & Pollution	23
5.	Global Warming & Climate Change	9
6.	Environmental Impact Assessment (EIA)	4
7.	Sustainable Development and Habitat Norms	5
8.	Disasters & their Management	6

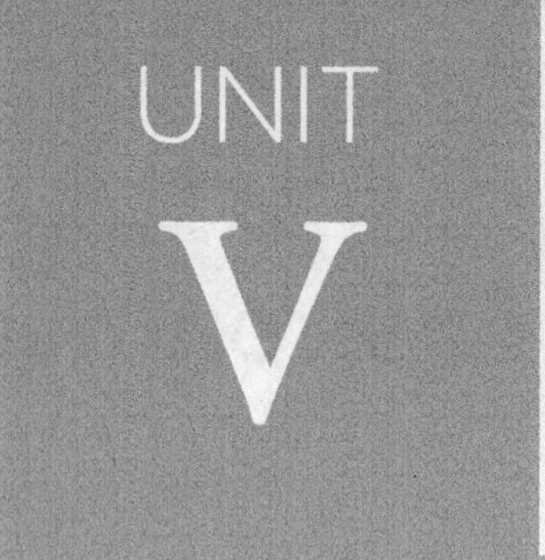

UNIT V

Basics of Energy and Environment

1. ENERGY RESOURCES : CONSERVATION & UTILISATION

1.1 Consider the following statements regarding solar energy:

1. To encourage the adoption of solar energy production, many State Governments and the Centre have announced plans by way of buy back as well as subsidies for installation.
2. Land acquisition of several hectares is a bottleneck in implementing this programme.
3. Considerable *R* & *D* effort is needed to bring down the cost of P-V cells.

Which of the above statements are correct?

(a) 1 and 2 only (b) 1 and 3 only
(c) 1, 2 and 3 (d) 2 and 3 only

[ESE Prelims : 2017]

1.2 Which one of the following statements is correct?

(a) The end product of fossil fuels is in the form of electrical energy.
(b) Watershed protection increases the rate of surface runoff of water.
(c) If timber is overharvested, the ecological functions of the forest are improved.
(d) Rivers change their course during floods and lots of fertile soils are lost to the sea.

[ESE Prelims : 2017]

Directions: Each of the next items consists of two statements, one labelled as '**Statement (I)**' and the other as '**Statement (II)**'. You are to examine these two statements carefully and select the answers to these items using the code given below:

Code:

(a) Both Statement (I) and Statement (II) are individually true and Statement (II) is the correct explanation of Statement (I)
(b) Both Statement (I) and Statement (II) are individually true but Statement (II) is not the correct explanation of Statement (I)
(c) Statement (I) is true but Statement (II) is false
(d) Statement (I) is false but Statement (II) is true

1.3 **Statement (I):** Increased cloud cover will lead to warmer winters due to clouds reflecting more intense solar energy.

Statement (II): Overcast cloud conditions result in decrease in the day-night temperature difference.

[ESE Prelims : 2017]

1.4 **Statement (I):** Green energy refers to one which does not harm the ecosystem of planet Earth.

Statement (II): All renewable energy is green energy.

[ESE Prelims : 2018]

1.5 Consider the following statements regarding Insolation:

1. It is the solar radiation that reaches the Earth's surface.
2. It is measured by the amount of solar energy received per square centimeter per minute.
3. It is the amount of solar energy absorbed by the stratosphere.

Which of the above statements are correct?

(a) 1, 2 and 3 (b) 1 and 2 only
(c) 1 and 3 only (d) 2 and 3 only

[ESE Prelims : 2018]

1.6 What are the limitations of solar energy?

1. Collecting solar energy over large areas and converting it to other forms that can be conveniently transported, stored and used in existing equipment is not economical.
2. Low density of solar energy as compared to coal, oil and gas.
3. Its major applications are photothermal conversion, solar water heating, green housing technology and photo voltaic conversion.

Select the correct answer using the codes given below:

(a) 1, 2 and 3 (b) 1 and 2 only
(c) 1 and 3 only (d) 2 and 3 only

[ESE Prelims : 2019]

1.7 What are the advantages of Biomass as a source of energy?

1. Its storage and transportation is possible
2. It is ecologically safe and is inoffensive
3. Can be developed with present man and material abilities
4. Low capital input required

Select the correct answer using the codes given below:

(a) 1, 2, 3 and 4 (b) 1, 2 and 3 only
(c) 1, 3 and 4 only (d) 2, 3 and 4 only

[ESE Prelims : 2019]

1.8 Energy used by man does NOT originate from which one of the following sources ?

(a) Radiant energy (b) Geothermal power
(c) Frictional energy (d) Gravitational energy

[ESE Prelims : 2022]

1.9 The largest tidal range in the world is

(a) Bay of Fundy
(b) Ungava Bay
(c) Bristol Channel
(d) Turnagain Arm of Cook Inlet

[ESE Prelims : 2022]

1.10 Which one of the following scientists called entropy time arrow?

(a) Thomas Young (b) Arthur Eddington
(c) Max Planck (d) Thompson

[ESE Prelims : 2023]

1.11 Consider the following statements:
Energy balances are fundamental for energy planning, since they allow analysing aspects such as:

1. distribution of final energy consumption per end-use sector.
2. storage and refinement of each fuel or group of energies in the matrix.
3. self-sufficiency in energy, foreign dependence and foreign trade.
4. efficiency in processes for transforming primary energy into secondary.

Which of the above statements is/are NOT correct?

(a) 2 only (b) 1 and 2 only
(c) 1, 2 and 3 only (d) 4 only

[ESE Prelims : 2023]

1.12 Which one of the following is NOT an advantage of energy efficiency?

(a) The cost of energy economy is usually smaller than that of its generation
(b) Security of supply increases and resources which are finite are saved
(c) There are micro and macro-economic gains associated with an increase in productivity and in industrial competitivity
(d) The access to energy services is decreased

[ESE Prelims : 2023]

1.13 Consider the following factors determining the evolution of energy intensity:

1. dematerialization. 2. fuel use intensity
3. recycling

Which of the above factors is/are correct?

(a) 1, 2 and 3 (b) 2 and 3 only
(c) 1 and 3 only (d) 3 only

[ESE Prelims : 2023]

1.14 Surface rocks on Earth are cool, but below the surface the temperature increases with depth. This is called

(a) the geothermal gradient
(b) the homogeneous accretion
(c) the pangaea
(d) the mesocrates [ESE Prelims : 2023]

1.15 Which one of the following is NOT correct?

(a) The formation of a mountain chain by the compression of crustal rocks is known as an orogeny
(b) Rock between the two extremes is called mesocratic
(c) Sediments are deposited in horizontal layers called clay plates
(d) Particles deposited as sediments are changed into rock by the pressure of later deposits at low temperature is called diagenesis

[ESE Prelims : 2023]

1.16 Which one of the following statements is NOT correct regarding thermal mass?
(a) In solar buildings, it reduces temperature variations between day and night.
(b) It is useful in ordinary buildings as it serves as a reservoir or sink for both heating and cooling.
(c) It provides a means of storing the solar energy that enters through the windows.
(d) The heavier a material is available, then the less thermal mass is available.
[ESE Prelims : 2024]

1.17 Consider the following statements about ethanol:
1. Ethanol is primarily produced from corn and sugarcane.
2. Ethanol provides a major part of the liquid fuel requirement in Brazil.
3. The production of ethanol accounts for around 90% of the production of biofuels in the world.

Which of the above statements is/are correct?
(a) 1, 2 and 3 (b) 2 and 3 only
(c) 3 only (d) 1, and 2 only
[ESE Prelims : 2024]

1.18 The expression of the ability of surfaces to reflect sunlight is known as
(a) the albedo effect
(b) the greenhouse effect
(c) the genshin effect
(d) the permafrost [ESE Prelims : 2024]

2. BASICS OF ENVIRONMENT, ECOLOGY & ECOSYSTEM

2.1 Which one of the following is the major characteristic of deciduous trees?
(a) They do not lose their leaves.
(b) They shed their leaves annually.
(c) They synthesize their own food.
(d) They depend on other factors for their food.
[ESE Prelims : 2018]

2.2 An association of two organisms of different species for mutual benefit, and where the individuals may not be able to survive separately, is called
(a) Commensalism (b) Parasitic
(c) Non-symbiotic (d) Symbiotic
[ESE Prelims : 2018]

2.3 Which one of the following is a terrestrial type of ecosystem?
(a) Limnetic (b) Estuary
(c) Prairie (d) Reefs
[ESE Prelims : 2019]

2.4 Which one of the following is NOT a principle of CERES?
(a) Controlled production
(b) Energy conservation
(c) Informing the public
(d) Protection of the biosphere
[ESE Prelims : 2022]

2.5 **Statement (I):** Alterations in both physico-chemical (abiotic) and biological (biotic) components of the biosphere by mankind resulted in environmental degradation world over.
Statement (II): Major environmental problems are in fact the manifestations of the degraded environments at global level.
(a) Both Statement (I) and Statement (II) are individually true and Statement (II) is the correct explanation of Statement (I).
(b) Both Statement (I) and Statement (II) are individually true but Statement (II) is **not** the correct explanation of Statement (I).
(c) Statement (I) is true but Statement (II) is false.
(d) Statement (I) is false but Statement (II) is true.
[ESE Prelims : 2023]

2.6 Match the following lists:
List-I
P. Lamarack
Q. Lyell
R. Malthus
S. Wallace
List-II
1. Evolutionary theory
2. Gradual geological processes have gradually shaped Earth's shaped surface.
3. Human population grow faster than the resources they depend on
4. Inheritance of acquired characteristics

Select the correct answer using the code given below:

	P	Q	R	S
(a)	2	1	4	3
(b)	4	2	3	1
(c)	1	4	3	2
(d)	4	3	1	2

[ESE Prelims : 2024]

2.7 What is Chaparral?

(a) Chaparrals are plants that grow on other plants
(b) Chaparral is a shrub forest biome dominated by densely-growing evergreen shrubs or small trees, such as scrub oak
(c) Chaparrals are temperate biomes that consist mainly of grasses
(d) Chaparrals are aquatic organisms that live on the surface below a body of water

[ESE Prelims : 2024]

3. BIODIVERSITY & ITS CONSERVATION

3.1 Besides species diversity and ecosystem diversity, which of the following is included in the term 'biodiversity'?

(a) Genetic diversity (b) Climatic diversity
(c) Cultural diversity (d) Lingual diversity

[ESE Prelims : 2017]

3.2 Which one of the following regions is characterized by sensitive ecosystems, enhanced occurrences of extreme weather events and natural catastrophes?

(a) Mountain region
(b) Evergreen forest region
(c) Tropical region
(d) Tundra region

[ESE Prelims : 2020]

3.3 Which one of the following is NOT correctly matched pair regarding the regional biodiversity?

(a) Point richness: The number of species that can be found at a single point in a given space.
(b) Alpha richness: The number of species found in a small heterogeneous area.
(c) Beta richness: The rate of change in species composition across different habitats.
(d) Gamma richness: The rate of change across large landscape gradients.

[ESE Prelims : 2021]

3.4 The Thirteenth Meeting of the Conference of the Parties to the Convention on the Conservation of Migratory Species of Wild Animals (CMS COP 13) in 2020 was held in

(a) Brazil (b) India
(c) Indonesia (d) Canada

[ESE Prelims : 2021]

4. ENVIRONMENTAL DEGRADATION & POLLUTION

4.1 Consider the following tertiary treatment methods for treatment of waste water:

1. Ion-exchange method
2. Reverse osmosis
3. Chemical oxidation method
4. Activated sludge process

Which of the above methods are correct?

(a) 1, 2 and 4 (b) 1, 3 and 4
(c) 2, 3 and 4 (d) 1, 2 and 3

[ESE Prelims : 2017]

4.2 Which industrial waste is commonly used in construction industry?

(a) Fly ash (b) Slag
(c) Sludge (d) Red oxide

[ESE Prelims : 2017]

4.3 Consider the following statements concerning environmental pollution:

1. Nuclear explosions cause radioactive radiation.
2. Earthquakes do not cause Tsunamis.
3. Acid rain is not a major environmental issue.
4. Air pollution has some impact on meteorology.

Which of the above statements are correct?

(a) 1 and 2 (b) 2 and 3
(c) 3 and 4 (d) 1 and 4

[ESE Prelims : 2017]

4.4 Consider the following statements regarding SMOG:

1. SMOG was coined during the 1950's to describe a mixture of smoke and fog experienced in London.
2. The principal pollutants in London SMOG are particulates and sulphur compounds.
3. The London SMOG occurs generally early in the morning on cold wet winter days.

Which of the above statements are correct?

(a) 1, 2 and 3 (b) 1 and 2 only
(c) 1 and 3 only (d) 2 and 3 only

[ESE Prelims : 2017]

4.5 The Rotterdam Convention deals with
(a) Reducing nuclear weapon stockpiles
(b) Limiting the use of toxic chemicals
(c) Protecting the oceans
(d) Banning of human clone experiment

[ESE Prelims : 2017]

4.6 Sanitary/municipal fills and waste heaps are unavoidably hazardous due to
1. Leachates
2. Emanating gases
3. Rodents and wandering animals
4. Automobile workshops that seems to have an affinity for such neighbourhoods.

Which of the above are correct?
(a) 1 and 4 only (b) 1 and 2 only
(c) 2 and 3 only (d) 3 and 4 only

[ESE Prelims : 2018]

4.7 Increased biological oxygen demand is an indication of
1. Low microbial contamination.
2. Absence of microbial pollution.
3. High level of microbial contamination.

Which of the above statements is/are correct?
(a) 1 only (b) 2 only
(c) 3 only (d) 1, 2 and 3

[ESE Prelims : 2018]

Directions: Each of the next items consists of two statements, one labelled as '**Statement (I)**' and the other as '**Statement (II)**'. You are to examine these two statements carefully and select the answers to these items using the code given below:

Code:
(a) Both Statement (I) and Statement (II) are individually true and Statement (II) is the correct explanation of Statement (I)
(b) Both Statement (I) and Statement (II) are individually true but Statement (II) is not the correct explanation of Statement (I)
(c) Statement (I) is true but Statement (II) is false
(d) Statement (I) is false but Statement (II) is true

4.8 **Statement (I):** Volcanic eruption is often accompanied by earthquakes.
Statement (II): Volcanoes erupt dust particles in the atmosphere.

[ESE Prelims : 2018]

4.9 **Statement (I):** Normally carbon dioxide is not considered an air pollutant.
Statement (II): Carbon dioxide is a constituent of atmospheric air.

[ESE Prelims : 2018]

4.10 Venturi scrubber, a device used to remove particulate matter from the atmosphere, works on the principle of
(a) Settling by gravitational force
(b) Removal by centrifugal force
(c) Removal by electrically charged particles
(d) Removal by atomized water vapour

[ESE Prelims : 2019]

4.11 Acid rain results when gaseous emissions of Sulfur oxides (SO_X) and nitrogen oxides (NO_X) interact with water vapour and
(a) Moonlight, and are chemically converted to strong acidic compounds such as sulfuric acid (H_2SO_4) and nitric acid (HNO_3)
(b) Sunlight, and are chemically converted to strong acidic compounds such as sulfuric acid (H_2SO_4) and nitric acid (HNO_3)
(c) Moonlight, and are chemically converted to weak acidic compounds such as sulfuric acid (H_2SO_4) and nitric acid (HNO_3)
(d) Sunlight, and are chemically converted to weak acidic compounds such as sulfuric acid (H_2SO_4) and nitric acid (HNO_3)

[ESE Prelims : 2019]

4.12 The '*Minamata Tragedy*' was caused by the eating of fish growing in the Minamata Bay contaminated with
(a) Peroxy alylnitrate
(b) Methyl isocyanate
(c) Potassium cyanide
(d) Methylmercury [ESE Prelims : 2019]

4.13 BOD of a waste water sample is estimated to be 180mg/*l*. Assuming 4 mg/*l* BOD can be consumed in the BOD bottle, the volume of undiluted sample to be added to a 300 m*l* bottle is nearly

(a) 6.7 ml (b) 5.6 ml
(c) 4.4 ml (d) 3.3 ml
[ESE Prelims : 2019]

4.14 Which one of the following is a hygienic way of disposing solid waste and is more suitable if the waste contains more hazardous material and organic content?
(a) Composting (b) Incineration
(c) Oxidation (d) Subgrading
[ESE Prelims : 2020]

4.15 Which one of the following gases is colourless with strong odour, irritates mucous membranes at common levels, can cause cough, fatigue and interference with lung functions at higher concentration?
(a) Carbon monoxide (b) Hydrogen
(c) Ozone (d) Nitrogen
[ESE Prelims : 2020]

4.16 Which of the following are the suggested ways of reducing NO_X emissions from stationary sources?
1. By reducing the peak temperature.
2. By increasing the availability of N_2 for reaction with O_2.
3. By minimizing the availability of O_2 for reaction with N_2.

(a) 1 and 2 only (b) 1 and 3 only
(c) 2 and 3 only (d) 1, 2 and 3
[ESE Prelims : 2020]

4.17 Basel Convention provides
(a) Indian standards for pollution measurement and prevention
(b) International guidelines to control the transboundary movements of hazardous wastes between different countries
(c) Indian standards for the disposal of municipal and industrial wastes
(d) International standards to categorize pollution in air and wastewater
[ESE Prelims : 2020]

4.18 According to UNEP, which of the following is/are the major component/s of air pollution?
1. SO_2 2. O_3
3. CO 4. NO_2

Select the correct answer using the code given below:
(a) 2 and 3 only (b) 2 only
(c) 2, 3 and 4 only (d) 1, 2, 3 and 4
[ESE Prelims : 2021]

4.19 Which one of the following is a 'soft coal'?
(a) anthracite (b) bituminous
(c) lignite (d) magnetite
[ESE Prelims : 2021]

4.20 Match the following:

I
A. Oligotrophic lakes
B. Dystrophic lakes
C. Meromictic lakes
D. Impoundments

II
1. created due to construction of dams
2. low pH and high humic acid content
3. low nutrient concentration
4. rich in salts and permanently stratified

Select the correct matching using the code given below:

	A	B	C	D
(a)	4	1	3	2
(b)	4	1	2	3
(c)	3	2	1	4
(d)	3	2	4	1

[ESE Prelims : 2021]

4.21 **Statement (I)** : Environmental pollution has become global problem.
Statement (II) : The rapidly growing human population, rapid urbanization, intensive agriculture and industrialization together with human activities resulted in the environmental pollution.
(a) Both Statement (I) and Statement (II) are individually true and Statement (II) is the correct explanation of Statement (I)
(b) Both Statement (I) and Statement (II) are individually true, but Statement (II) is not the correct explanation of Statement (I)
(c) Statement (I) is true, but Statement (II) is false
(d) Statement (I) is false, but Statement (II) is true
[ESE Prelims : 2022]

4.22 **Statement (I):** Ozone depletions are mostly harmful to biological systems in a variety of ways.

Statement (II): Ozone depletion in stratosphere leads to the loss of filtering ability of UV light.

(a) Both Statement (I) and Statement (II) are individually true and Statement (II) is the correct explanation of Statement (I)
(b) Both Statement (I) and Statement (II) are individually true but Statement (II) is not the correct explanation of Statement (I)
(c) Statement (I) is true but Statement (II) is false
(d) Statement (I) is false but Statement (II) is true

[ESE Prelims : 2023]

4.23 Which one of the following statements is NOT correct regarding carbon dioxide?

(a) Carbon dioxide is given off when dead organisms and other organic materials decompose.
(b) When volcanoes erupt, they give off carbon dioxide that is stored in the mantle.
(c) Ocean water releases dissolved carbon dioxide into the atmosphere when water temperature rises.
(d) A good amount of carbon in the atmosphere is present as methane gas.

[ESE Prelims : 2024]

5. GLOBAL WARMING & CLIMATE CHANGE

5.1 Consider the following statements regarding ozone:

1. It is continuously being produced and destroyed.
2. It helps to sustain life on earth.
3. It is formed in the atmosphere through photochemical reaction.

Which of the above statements are correct?

(a) 1 and 2 only (b) 1 and 3 only
(c) 2 and 3 only (d) 1, 2 and 3

[ESE Prelims : 2017]

5.2 **Statement (I):** Global warming is the cause for climate change.

Statement (II): Ozone depletion will cause global warming.

(a) Both Statement (I) and Statement (II) are individually true and Statement (II) is the correct explanation of Statement (I)
(b) Both Statement (I) and Statement (II) are individually true but Statement (II) is not the correct explanation of Statement (I)
(c) Statement (I) is true but Statement (II) is false
(d) Statement (I) is false but Statement (II) is true

[ESE Prelims : 2017]

5.3 The meaning of 'Carbon Footprint' is described by the amount of

(a) Carbon dioxide released into the atmosphere as a result of the activities of a particular individual, organization or community.
(b) Greenhouse gases emitted by industries contributing to global warming.
(c) Carbon emission released by the burning of jet fuel.
(d) Increase in the carbon content of the atmosphere due to the felling of trees.

[ESE Prelims : 2018]

5.4 Consider the following statements regarding depletion of the ozone layer:

1. Excessive release of chlorine and bromine in the environment from man-made compounds, such as chlorofluorocarbons.
2. Occurrence of certain natural phenomena such as sunspots, and stratospheric winds.
3. Degradation of materials by ultra-violet radiation.
4. Major volcanic eruptions.

Which of the above can be categorized as causing ozone depletion?

(a) 1, 2 and 3 only (b) 1, 3 and 4 only
(c) 1, 2 and 4 only (d) 2, 3 and 4 only

[ESE Prelims : 2018]

5.5 **Statement (I):** A greenhouse gas is any gas in the atmosphere which absorbs and re-emits heat and thereby keeps the plant's atmosphere warmer than it otherwise would be.

Statement (II): In the Earth's atmosphere, water vapour is one of the main greenhouse gases.

(a) Both Statement (I) and Statement (II) are individually true and Statement (II) is the correct explanation of Statement (I)
(b) Both Statement (I) and Statement (II) are individually true but Statement (II) is not the correct explanation of Statement (I)
(c) Statement (I) is true but Statement (II) is false
(d) Statement (I) is false but Statement (II) is true

[ESE Prelims : 2018]

5.6 Consider the following statements with regard to atmospheric humidity:

1. Absolute humidity is the amount of water vapour per unit volume.
2. Hygrometer is used to measure relative humidity.
3. Dew point is the temperature at which the relative humidity is 75%.

Which is the above statements are correct?

(a) 1 and 2 only (b) 1 and 3 only
(c) 2 and 3 only (d) 1, 2 and 3

[ESE Prelims : 2018]

5.7 Ozone layer present in the atmosphere protects life on earth by not permitting harmful radiations present in the sunlight to penetrate through it. Ozone layer is formed by the reaction of

(a) Chlorofluorocarbons (CFCs) on oxygen (O_2)
(b) Chlorine (Cl) on oxygen (O_2)
(c) Solar Ultraviolet rays on oxygen (O_2)
(d) Chlorine nitrate ($ClNO_3$) on oxygen (O_2)

[ESE Prelims : 2019]

5.8 Kyoto Protocol operationalizes the UN framework convention on

(a) sustainable development
(b) renewable energy
(c) climate change
(d) soil erosion **[ESE Prelims : 2022]**

5.9 The British economist Nicholas Stern gave the most impressive analysis in the year 2006 on

(a) Ozone layer depletion
(b) Renewable energy sources
(c) Climate change
(d) Deforestation **[ESE Prelims : 2023]**

6. ENVIRONMENTAL IMPACT ASSESSMENT (EIA)

6.1 Environmental Impact Assessment (EIA) is aimed to help

(a) Estimate future needs of the society
(b) Smooth implementation of a project
(c) Cope with rapid increase in population
(d) Resource conservation

[ESE Prelims : 2019]

Directions: Each of the next items consists of two statements, one labelled as '**Statement (I)**' and the other as '**Statement (II)**'. You are to examine these two statements carefully and select the answers to these items using the code given below:

Code:

(a) Both Statement (I) and Statement (II) are individually true and Statement (II) is the correct explanation of Statement (I)
(b) Both Statement (I) and Statement (II) are individually true but Statement (II) is not the correct explanation of Statement (I)
(c) Statement (I) is true but Statement (II) is false
(d) Statement (I) is false but Statement (II) is true

6.2 **Statement (I)** : Training should be conducted among the line and low management for ensuring the importance of environmental protection plan.

Statement (II) : Environmental science is a developing subject and the people implementing environment strategies should remain up to date with the environmental control processes.

[ESE Prelims : 2019]

6.3 **Statement (I)** : An EIA is a study of the probable changes in socio-economic and biophysical characteristics of the environment that may result from a proposed action.

Statement (II) : The purposes of an EIA is to help design projects, which do not disturb the quality of the environment by examining alternatives.

[ESE Prelims : 2019]

6.4 NEPA stands for

(a) National Ecological Physical Area
(b) Natural Environmental Policy Act
(c) National Environmental Policy Act
(d) Natural Ecological Primary Area

[ESE Prelims : 2020]

7. SUSTAINABLE DEVELOPMENT AND HABITAT NORMS

7.1 The LEED and GRIHA:

1. Are green building rating systems.
2. Issue energy compliance certificate.
3. Refer to Global standards.
4. Are India standards under finalization.

Select the correct answer using the codes given below:

(a) 1 and 2 (b) 2 and 3
(c) 3 and 4 (d) 1 and 4
[ESE Prelims : 2017]

7.2 What is the full form of GPP?
(a) Green Public Policy
(b) Green Private Procurement
(c) Green Public Procurement
(d) Green Private Policy [ESE Prelims : 2017]

7.3 **Statement (I):** Environmental considerations are not necessarily the same as, or congruent with, Ecological considerations.
Statement (II): Environmental considerations address more towards maintaining, whereas Ecological considerations address more towards sustaining.
(a) Both Statement (I) and Statement (II) are individually true and Statement (II) is the correct explanation of Statement (I)
(b) Both Statement (I) and Statement (II) are individually true but Statement (II) is not the correct explanation of Statement (I)
(c) Statement (I) is true but Statement (II) is false
(d) Statement (I) is false but Statement (II) is true
[ESE Prelims : 2017]

7.4 Which one of the following is an iterative and evolutionary process for achieving sustainable development?
(a) Flood Control Management (FCM)
(b) Solid Waste Management (SWM)
(c) Integrated Coastal Zone Management (ICZM)
(d) Natural Ecosystem Zone (NEZ)
[ESE Prelims : 2020]

7.5 Energy services for sustainable development are directly linked to:
1. Poverty 2. Lifestyles
3. Women 4. Deforestation
Select the correct answer using the codes given below:
(a) 1, 2 and 3 only (b) 1, 2 and 4 only
(c) 2, 3 and 4 only (d) 1 and 3 only
[ESE Prelims : 2021]

8. DISASTERS & THEIR MANAGEMENT

8.1 **Statement (I):** To practise terraced cultivation in hill slopes, it can be admissible to have the vertical face of the terraced boundary run perpendicular to the ground trace of the fault line, if any, in the underlying land.
Statement (II): Fault lines are susceptible to slips and should be guarded against in land use.
(a) Both Statement (I) and Statement (II) are individually true and Statement (II) is the correct explanation of Statement (I)
(b) Both Statement (I) and Statement (II) are individually true but Statement (II) is not the correct explanation of Statement (I)
(c) Statement (I) is true but Statement (II) is false
(d) Statement (I) is false but Statement (II) is true
[ESE Prelims : 2018]

8.2 Which of the following practices are adopted for river basin water resources management?
1. Soil conservation in catchments of river valley projects and flood-prone rivers
2. Soil and land use surveys
3. Control of shifting cultivation
4. Deforestation
(a) 1, 2 and 4 (b) 1, 3 and 4
(c) 2, 3 and 4 (d) 1, 2 and 3
[ESE Prelims : 2020]

8.3 Which one of the following is NOT included in the 27 principles issued at the Rio-92 UN Conference on the Environment and Development?
(a) The right to development that meets the needs of present and future generations
(b) Right to safety from natural disasters
(c) Protection to the environment in times of armed conflict
(d) Youth mobilization for a global partnership
[ESE Prelims : 2021]

8.4 According to Carson, which one of the following approaches argues that nature has intrinsic value and we should protect it because of this value?

(a) Instrumental approach
(b) Axiological approach
(c) Eco-critical approach
(d) Anthropological approach
[ESE Prelims : 2022]

8.5 The Gaia hypothesis, which suggested that the earth should be seen as a single organism, was devised by
(a) James Lovelock
(b) Francoise d' Eaubonne
(c) Earnest Haeckel
(d) Paul Ehrlich [ESE Prelims : 2022]

8.6 The international dispute about modern environmental movement began with the publication of Silent Spring by
(a) Mary Daly
(b) Rachel Carson
(c) Carolyn Merchant
(d) Maria Mies [ESE Prelims : 2022]

■■■■

Basics of Energy and Environment : Answer Key

1.1	(c)	**1.2**	(d)	**1.3**	(d)	**1.4**	(c)	**1.5**	(b)	**1.6**	(b)	**1.7**	(a)	**1.8**	(c)
1.9	(a)	**1.10**	(b)	**1.11**	(c)	**1.12**	(d)	**1.13**	(a)	**1.14**	(a)	**1.15**	(c)	**1.16**	(d)
1.17	(d)	**1.18**	(a)	**2.1**	(b)	**2.2**	(d)	**2.3**	(c)	**2.4**	(a)	**2.5**	(a)	**2.6**	(b)
2.7	(b)	**3.1**	(a)	**3.2**	(a)	**3.3**	(b)	**3.4**	(b)	**4.1**	(d)	**4.2**	(a)	**4.3**	(d)
4.4	(a)	**4.5**	(b)	**4.6**	(b)	**4.7**	(c)	**4.8**	(b)	**4.9**	(b)	**4.10**	(d)	**4.11**	(b)
4.12	(d)	**4.13**	(a)	**4.14**	(b)	**4.15**	(c)	**4.16**	(b)	**4.17**	(b)	**4.18**	(d)	**4.19**	(c, b)
4.20	(d)	**4.21**	(a)	**4.22**	(b)	**4.23**	(d)	**5.1**	(d)	**5.2**	(b)	**5.3**	(a)	**5.4**	(c)
5.5	(b)	**5.6**	(a)	**5.7**	(c)	**5.8**	(c)	**5.9**	(c)	**6.1**	(d)	**6.2**	(b)	**6.3**	(c)
6.4	(c)	**7.1**	(a)	**7.2**	(c)	**7.3**	(a)	**7.4**	(c)	**7.5**	(a, d)	**8.1**	(a)	**8.2**	(d)
8.3	(b)	**8.4**	(b)	**8.5**	(a)	**8.6**	(b)								

EXPLANATIONS

1.1 (c)

- Union Government encourages the adoption of solar energy production and provide subsidies for installation of solar units.
- Rule of thumb is to take 100 sq. ft. for every 1 kW of solar panels. Extrapolating this, a 1 MW solar PV power plant should require about 100000 sq. ft. (about 2.5 acres, or 1 hectare).
  So, we need a large area for solar energy production the land acquisition is necessarily for this purpose which is a major bottleneck.
- The capital cost of solar power plant is very high as the cost of solar panels is high, for reduction of cost of photovoltaic cells, the research and development need to be enhanced.

1.5 (b)

Insolation is defined as the thermal radiation received from the Sun per unit surface area of the Earth.

1.6 (b)

Third statement is advantage regarding solar energy rather than limitation. Therefore, third statement is wrong and 1 and 2 are correct.

Disadvantages of Solar Energy:

- The high initial costs of installing panels. The most commonly cited solar energy disadvantage, cost, is declining as the industry expands.
- Solar energy storage is expensive.
- Solar doesn't work for every roof type.
- Solar panels are dependent on sunlight.

1.11 (c)

- The energy balance is an assessment of the relation between the energy consumption of the product and the energy production.
- It helps us to understand how products are transformed into one another, highlight the various relationships among these products, and show how all energy types are ultimately used.
- The energy balance presents all the data in a common energy unit.

1.12 (d)

Energy efficiency is the use of less energy to perform the same task or produce the same result. Energy-efficient homes and buildings use less energy to heat, cool, and run appliances and electronics, and energy-efficient manufacturing facilities use less energy to produce goods.

1.16 (d)

In general, the denser a material is, the more thermal storage is available per mass. Hence, option (d) is not correct.

2.1 (b)

Deciduous trees are giant flowering plants. They include oaks, maples, and beeches, and they grow in many parts of the world. The word deciduous means to "fall off," and every fall these trees shed their leaves. Most deciduous trees are broad-leaved, with wide, flat leaves.

2.2 (d)

- Symbiosis is a relationship or interaction between two or more species that share a common habitat.
- Mutualism, Commensalism, and Parasitism are the three types of behaviors seen in symbiotic interactions.
- **Mutualism:** Mutualism occurs when both creatures involved benefit from each other.
- **Commensalism:** Only one organism benefits from commensalism, whereas the other is neither benefited nor injured.
- **Parasitism:** Parasitic relationships benefit one creature while harming the other.
- Example of a symbiotic relationship is Lichens.

2.3 (c)

Ecosystem is broadly terrestrial and aquatic. Aquatic includes pond, river, lake, wetlands, ocean (Coral reef, limnetic). While terrestrial ecosystems are forest, grassland (Temperate grassland-Prairies/Steppes).

3.3 (b)

Explanation of Alpha richness is not correct because in case of Alpha richness, the number of species found in a in this question small "homogeneous area" but in question it is given "heterogeneous area". Therefore, it is incorrect.

3.4 (b)

CMS COP-13 was held in Gandhinagar, Gujrat from 17th to 22nd February 2020. The theme of CMS COP-13 was " Migratory species connect the planet and we welcome them home"

4.10 (d)

Venturi scrubber can remove particulate matter as well as larger particles. This type of technology is a part of the group of air pollution control devices collectively referred as 'wet scrubbers'. It is designed to effectively use the energy from the inlet gas stream to atomize the liquid being used to scrub the gas stream.

ADDITIONAL INFORMATION

The term 'scrubber' is used for those pollution control devices that use liquid to wash unwanted pollutants from a gas stream.

4.11 (b)

- Acid rain has constituents of strong acid (H_2SO_4 and HNO_3).
- Acid rain contains a mixture of H_2SO_4 and HNO_3 which lower the pH of rainwater. Since NOx undergoes photodissociation before forming the HNO3, therefore, acid rain contains more H_2SO_4 than HNO_3.

4.12 (d)

- Minamata disease is caused by mercury presence (especially water pollution).

- Minamata Disease is a poisoning disease that nervous system, mainly central nervous system, is damaged by methylmercury. It is established that the disease differs from inorganic mercury poisoning which damages kidneys, etc., and also, it isn't confirmed that it damages organs other than nervous system.

4.18 (d)

According to UNEP, major components of air pollution are:

1. SO_2
2. O_3
3. CO
4. NO_2
5. Lead
6. Particulates

4.19 (c, b)

Based on Characteristics: Lignite
Based on Popular Nickname: Bituminous
Different Countries have different designations attributed to various types of coals. Generally, Anthracite coal is called as Hard Coal, whereas Bituminous coal is called as "soft coal and Lignite is often called as Brown Coal. But in many countries, both Bituminous and Lignite types of coal are referred to as "Soft Coal".
However, based on their characteristics, in comparison to Bituminous, Lignite is quite soft. Hence, Lignite shall be the more preferred option.

4.20 (d)

Oligotrophic lakes : Low nutrient concentration.
Dystrophic lakes : Low pH and high humic acid content.
Meromictic lakes : Rich in salts and permanently stratifies..
Impoundments : Created due to construction of dam.

5.1 (d)

Ozone (composed of three atoms of oxygen) is a gas that occurs both in the Earth's upper atmosphere and at ground level. Ozone can be "good" or "bad" for your health and the environment, depending on its location in the atmosphere.

5.3 (a)

A carbon footprint is the total amount of greenhouse gases (including carbon dioxide and methane) that are generated by our actions. The average carbon footprint for a person in the United States is 16 tons, one of the highest rates in the world. Globally, the average carbon footprint is closer to 4 tons.

5.7 (c)

UV (Ultra violet) rays once fall on the O2 molecule in stratosphere then O2 molecule is split into nascent oxygen (O2 = O + O). Nascent oxygen is combined with O2 molecule and forming ozone (O3) {(O + O2 = O3)}. Thus, ozone formation is natural phenomena in stratosphere unlike the troposphere.

5.8 (c)

The Kyoto Protocol operationalizes the United Nations Framework Convention on Climate Change by committing industrialized countries and economies in transition to limit and reduce greenhouse gases (GHG) emissions in accordance with agreed individual targets.

6.1 (d)

Environmental impact assessment (EIA) study is required for the new projects regarding future impacts on the environmental components. It aims to achieve sustainable development taking into care the resource conservation.

6.4 (c)

The National Environmental Policy Act is a United States environmental law that promotes the enhancement of the environment and established the President's Council on Environmental Quality. The law was enacted on January 1, 1970

7.1 (a)

- The GRIHA full form is Green Rating for Integrated Habitat Assessment. This tool of rating the habitats helps people to assess the overall performance of their dwelling places, precisely buildings. This takes into consideration the various benchmarks that are nationally accepted.
- Leadership in Energy and Environmental Design (LEED) is one of the most popular green building certification programs used worldwide. Developed by the non-profit U.S. Green Building Council (USGBC) it includes

a set of rating systems for the design, construction, operation, and maintenance of green buildings, homes, and neighbourhoods.

Its aim is to help building owners and operators be environmentally responsible and use resources efficiently.

7.2 (c)

Green Public Procurement (GPP) means that public authorities seek to purchase goods, services and works with a reduced environmental impact throughout their life-cycle compared to goods, services and works with the same primary function which would otherwise be procured.

7.5 (a, d)

Energy services are directly linked to poverty and women. Energy services also directly have the impacts on the lifestyles, living conditions, income and livelihood, especially of poor people. So the preferred option shall be (a).
However, sometimes we have seen UPSC people do have some interpretations of their own.

8.3 (b)

The Rio Declaration on Environment comprises a proclamation of 27 principles and submission of recommendations. Some of the important principles relating to environmental protection in the Rio declaration are as follows:-

1. Human beings are at the centre of concerns for sustainable development. They are entitled to a healthy and productive life in harmony with nature.
2. States have, in accordance with the Charter of the United Nations and the principles of international law, the sovereign right to exploit their own resources pursuant to their own environmental and developmental policies, and the responsibility to ensure that activities within their jurisdiction or control do not cause damage to the environment of other States or of areas beyond the limits of national jurisdiction.
3. The right to development must be fulfilled so as to equitably meet developmental and environmental needs of present and future generations.
4. In order to achieve sustainable development, environmental protection shall constitute an integral part of the development process and cannot be considered in isolation from it.
5. All States and all people shall co-operate in the essential task of eradicating poverty as an indispensable requirement for sustainable development, in order to decrease the disparities in standards of living and better meet the needs of the majority of the people of the world.
6. The special situation and needs of developing countries, particularly the least developed and those most environmentally vulnerable, shall be given special priority. International actions in the field of environment and development should also address the interests and needs of all countries.
7. States shall co-operate in a spirit of global partnership to conserve, protect and restore the health and integrity of the Earth's ecosystem. In view of the different contributions to global environmental degradation, States have common but differentiated responsibilities. The developed countries acknowledge the responsibility that they bear in the international pursuit of sustainable development in view of the pressures their societies place on the global environment and of the technologies and financial resources they command.
8. To achieve sustainable development and a higher quality of life for all people, States should reduce and eliminate unsustainable patterns of production and consumption and promote appropriate demographic policies.
9. States should co-operate to strengthen endogenous capacity-building for sustainable development by improving scientific understanding through exchanges of scientific and technological knowledge, and by enhancing the development, adaptation,

diffusion, and transfer of technologies, including new and innovative technologies.

10. Environmental issues are best handled with the participation of all concerned citizens, at the relevant level. At the national level, each individual shall have appropriate access to information concerning the environment that is held by public authorities, including information on hazardous materials and activities in their communities, and the opportunity to participate in decision-making processes. States shall facilitate and encourage public awareness and participation by making information widely available. Effective access to judicial and administrative proceedings, including redress and remedy, shall be provided.
11. States shall enact effective environmental legislation, Environmental standards, management objectives, and priorities should reflect the environmental and developmental context to which they apply. Standards applied by some countries may be inappropriate and of unwarranted economic and social cost to other countries, in particular developing countries.
12. States should co-operate to promote a supportive and open international economic system that would lead to economic growth and sustainable development in all countries, to better address the problems of environmental degradation. Trade policy measures for environmental purposes should not constitute a means of arbitrary or unjustifiable discrimination or a disguised restriction on international trade. Unilateral actions to deal with environmental challenges outside the jurisdiction of the importing country should be avoided. Environmental measures addressing transboundary or global environmental problems should, as far as possible, be based on an international consensus.
13. States shall develop national law regarding liability and compensation for the victims of pollution and other environmental damage. States shall also co-operate in an expeditious and more determined manner to develop further international law regarding liability and compensation for adverse effects of environmental damage caused by activities within their jurisdiction or control to areas beyond their jurisdiction.
14. States should effectively co-operate to discourage or prevent the relocation and transfer to other States of any activities and substances that cause severe environmental degradation or are found to be harmful to human health.
15. In order to protect the environment, the precautionary approach shall be widely applied by States according to their capabilities. Where there are threats of serious or irreversible damage, lack of full scientific certainty shall not be used as a reason for postponing cost-effective measures to prevent environmental degradation.
16. National authorities should endeavour to promote the internalization of environmental costs and the use of economic instruments, taking into account the approach that the polluter should, in principle, bear the cost of pollution, with due regard to the public interest and without distorting international trade and investment.
17. Environmental impact assessment, as a national instrument, shall be undertaken for proposed activities that are likely to have a significant adverse impact on the environment and are subject to a decision of a competent national authority.
18. States shall immediately notify other States of any natural disasters or other emergencies that are likely to produce sudden harmful effects on the environment of those States. Every effort shall be made by the international community to help States so afflicted.
19. States shall provide prior and timely notification and relevant information to potentially affected States on activities that may have a significant adverse transboundary environmental effect and shall consult with those States at an early stage and in good faith.

20. Women have a vital role in environmental management and development. Their full participation is therefore essential to achieve sustainable development.
21. The creativity, ideas, and courage of the youth of the world should be mobilized to forge a global partnership in order to achieve sustainable development and ensure a better future for all.
22. Indigenous people and their communities, and other local communities have a vital role in environmental management and development because of their knowledge and traditional practices. States should recognize and duly support their identity, culture, and interests and enable their effective participation in the achievement of sustainable development.
23. The environment and natural resources of people under oppression, domination, and occupation shall be protected.
24. Warfare is inherently destructive of sustainable development. States shall, therefore, respect international law providing protection for the environment in times of armed conflict and co-operate in its further development, as necessary.
25. Peace, development and environmental protection are interdependent and indivisible.
26. States shall resolve all their environmental disputes peacefully and by appropriate means in accordance with the Charter of the United Nations.
27. States and people shall co-operate in good faith and in a spirit of partnership in the fulfillment of the principles embodied in this Declaration and in the further development of international law in the field of sustainable development.

8.4 **(b)**

- Axiological is the philosophical study of values and value systems, it is a branch of philosophy which concerns with determining factors associated with values specially moral values.
- Options (a), (c) and (d) are incorrect as all indicate to the hypothesis that nature is a means to an end (instrument) for the objectives of human beings. Hence it is humans who can determine the value nature carries.

8.5 **(a)**

The Gaia Hypothesis proposed by James Lovelock (1972) suggests that living organisms on the planet interact with their surrounding inorganic environment to form a synergetic and self-regulating system that created, and now maintains, the climate and biochemical conditions that make life on Earth possible.

8.6 **(b)**

The overarching theme of Silent Spring is the powerful—and often negative—effect humans have on the natural world. Carson's main argument is that pesticides have detrimental effects on the environment;
She says these are more properly termed "biocides" because their effects are rarely limited to solely targeting pests.

■■■■

Basics of Project Management

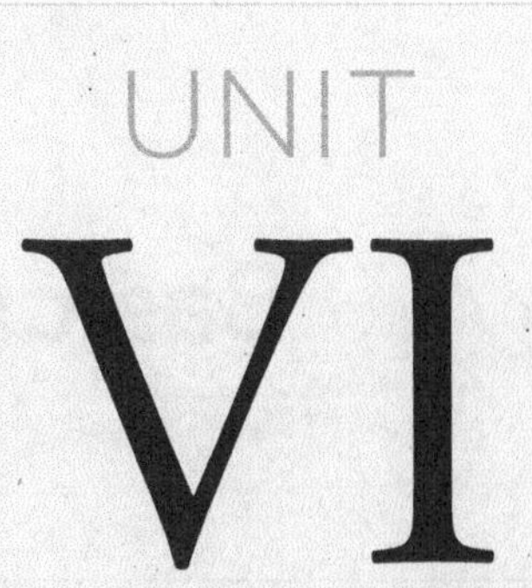

CONTENTS

Basics of Project Management

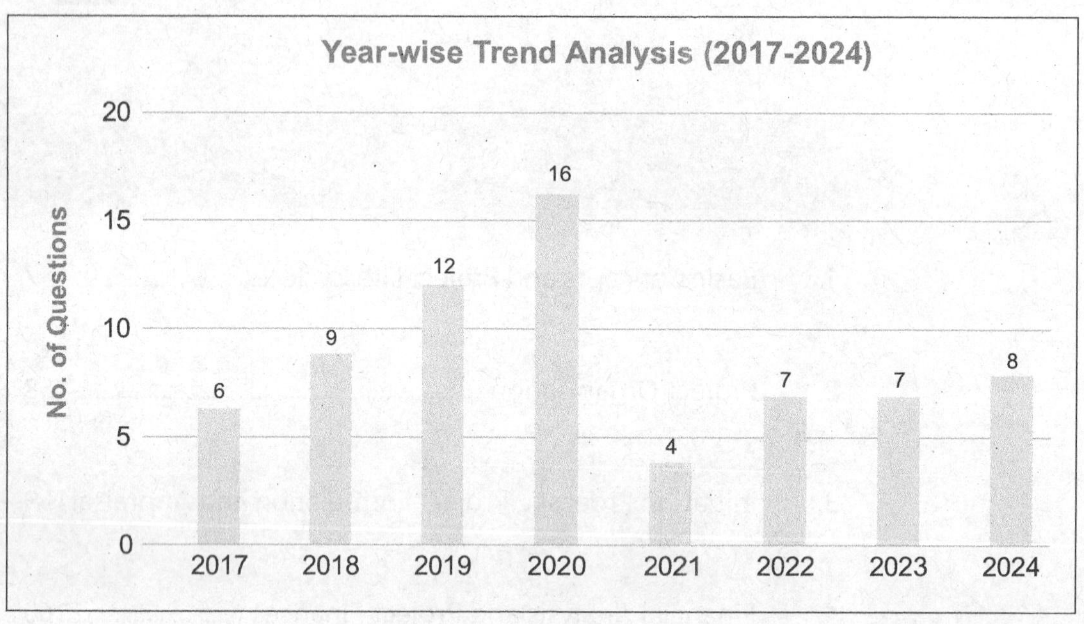

TOPIC-WISE TREND ANALYSIS (2017-2024)		
Sl. No.	**Topics**	**Total Number of Questions**
1.	Basic Concepts and Project Life Cycle	10
2.	Project Organisation	6
3.	Initiation Process : Project Formulation and Appraisal	11
4.	Financial Analysis and Project Finances	11
5.	Planning Process : Project Planning and Scheduling	22
6.	Execution Process: Project Monitoring and Control	4
7.	Tender and Contract	5

Basics of Project Management

1. BASIC CONCEPTS AND PROJECT LIFE CYCLE

1.1 Consider the following statements regarding project management:

1. It is the process of attaining project objectives in a stipulated time to produce quantified and qualified deliverables.
2. It is the art of bringing together the responsibilities, resources and people necessary to accomplish the business goals and objectives within the specified time limitations and within the financial grant.

Which of the above statements is/are correct?

(a) 1 only (b) 2 only
(c) Both 1 and 2 (d) Neither 1 nor 2

[ESE Prelims : 2017]

1.2 Consider the following phases of project management:

1. Identification 2. Formulation
3. Appraisal 4. Implementation

Which of the above phases are relevant, sequentially?

(a) 1, 2 and 3 only (b) 1, 2 and 4 only
(c) 3 and 4 only (d) 1, 2, 3 and 4

[ESE Prelims : 2017]

1.3 **Statement (I):** Project management is essentially the process to plan its implementation and to pre-determine the period-wise need of resources including funds and personnel, given the choice of total duration and quality standards.

Statement (II): Of the four dimensions (not denying that there can be some more) of a project. viz., scope, cost, time and quality, only any two can be pre-assigned; others have to abide by these two prescriptions.

(a) Both Statement (I) and Statement (II) are individually true and Statement (II) is the correct explanation of Statement (I).
(b) Both Statement (I) and Statement (II) are individually true but Statement (II) is **not** the correct explanation of Statement (I).
(c) Statement (I) is true but Statement (II) is false.
(d) Statement (I) is false but Statement (II) is true.

[ESE Prelims : 2018]

1.4 In progress of a project, the percentage of error will be less in

(a) Definitive cost estimate
(b) Detailed estimate
(c) Preliminary estimate
(d) Study estimate

[ESE Prelims : 2019]

1.5 In a project life cycle, the maximum percentage of effort is done in

(a) Concept phase
(b) Definition phase
(c) Planning and organizing phase
(d) Implementation phase

[ESE Prelims : 2019]

1.6 **Statement (I) :** All projects have constraints or limitations that inhibit their ability to reach goals and objectives.

Statement (II) : Time and money are universal constraints in projects.

(a) Both Statement (I) and Statement (II) are individually true; and Statement (II) is the correct explanation of Statement (I)
(b) Both Statement (I) and Statement (II) are individually true; but Statement (II) is NOT the correct explanation of Statement (I)
(c) Statement (I) is true; but Statement (II) is false
(d) Statement (I) is false; but Statement (II) is true

[ESE Prelims : 2019]

1.7 Which one of the following is a measure of sustainable income level that can be secured without decreasing the stock of natural assets?
(a) Natural Capital Stock
(b) Environmental Value
(c) Green Accounting
(d) Social Discount Rate

[ESE Prelims : 2020]

1.8 According to Black's Law Dictionary, the law treats the corporation itself as a person which can
(a) monitor (b) maintain
(c) sue and be sued (d) give response

[ESE Prelims : 2024]

1.9 Those who drive automobiles are familiar with blind spots. Applying this term to organizational and business arenas, Dennis Moberg draws an analogy between business blind spots and those we experience when driving. Blind Spot is one of the significant common impediments to responsibility. Which one of the following is NOT the method under Blind Spot?
(a) Self-deception (b) Willful blindness
(c) Inattentional blindness (d). Illusion of invulnerability of group

[ESE Prelims : 2024]

1.10 Which one of the following is NOT a factor for large scale diversification into unrelated areas by some of the industry Conglomerate in India?
(a) Restriction in growth in the existing line of business
(b) Policies with respect to imports, duties, pricing and reservations
(c) Opening up of newer areas of investments
(d) Desire not to avail tax incentives

[ESE Prelims : 2024]

2. PROJECT ORGANISATION

2.1 Which term refers to a single person having authority to oversee all aspects of a product's production scheduling, inventory, dislocation and sales?
(a) Project management
(b) Product management
(c) Commercial management
(d) Venture management

[ESE Prelims : 2019]

2.2 Who is responsible for establishing, documenting and maintaining procedures for post-production handling functions such as storage, packaging and delivery?
(a) Production Manager
(b) Marketing Manager
(c) Vendor
(d) Quality Supervisor **[ESE Prelims : 2019]**

2.3 Who is responsible for the following activities in a project?
1. Achieving a unity of control over project activities.
2. Having an authority to control project matters and disburse funds from the budget.
3. Having no actual line of authority over workers.

(a) Project Expeditor (b) Project Coordinator
(c) Matrix Manager (d) Project Manager

[ESE Prelims : 2020]

2.4 **Statement (I):** If the project influence is more in decision-making for the project, then the arrangement is considered a strong matrix.

Statement (II): If functional departments are seen to be influencing the decision-making more, the arrangement is considered a weak matrix.
(a) Both Statement (I) and Statement (II) are individually true and Statement (II) is the correct explanation of Statement (I)
(b) Both Statement (I) and Statement (II) are individually true but Statement (II) is not the correct explanation of Statement (I)
(c) Statement (I) is true but Statement (II) is false
(d) Statement (I) is false but Statement (II) is true

[ESE Prelims : 2020]

2.5 The Boston Consulting Group matrix classifies business in four categories as "STAR", "QUESTION MARK", "CASH COWS". Which one of the following is the fourth one?
(a) CATS (b) HORSES
(c) DOGS (d) HENS

[ESE Prelims : 2021]

2.6 Boston Consulting Group, the BCG matrix classifies the various businesses in a firm's portfolio on the basis of
(a) Relative Share and Relative Growth Rate
(b) Relative Market Share and Substantial Market Share
(c) Relative Market Share and Relative Market Growth Rate
(d) Substantial Growth Rate and Relative Market Growth Rate
[ESE Prelims : 2024]

3. INITIATION PROCESS : PROJECT FORMULATION AND APPRAISAL

3.1 Which of the following methods are considered as most favourable to production organizations liable for taxation?
1. Straight line method of depreciation.
2. Declining balance method of depreciation.
3. Sum of the years digits methods of depreciation.
4. Sinking fund method of depreciation.

Select the correct answer using the codes given below:
(a) 1 and 2 only (b) 2 and 3 only
(c) 3 and 4 only (d) 4 and 1 only
[ESE Prelims : 2017]

3.2 Factors which govern the operating cost of an equipment are:
1. Purchase price of the equipment.
2. Depreciation due to regular use.
3. Cost of operation, maintenance and repairs.

Which of the above statements are correct?
(a) 1 and 2 only (b) 1 and 3 only
(c) 2 and 3 only (d) 1, 2 and 3 only
[ESE Prelims : 2017]

3.3 Consider the following statements with reference to maintenance and service of the product after delivery to the customer:
1. Reliability analysis uses statistics of failures to estimate, understand and improve the performance of the equipment and its maintenance.
2. Reliability analysis records supports or guides the maintenance engineer in improving the performance of the product under service

Which of the above statements is/are correct?
(a) 1 only (b) 2 only
(c) Both 1 and 2 (d) Neither 1 nor 2
[ESE Prelims : 2017]

3.4 The 'Economic life' of a building is considered to be at an end
1. When the net income from the building fails to justify existence
2. When the building facilities become obsolete
3. When the capitalization rate becomes high

Which of the above statements are correct?
(a) 1 and 2 only (b) 1 and 3 only
(c) 2 and 3 only (d) 1, 2 and 3
[ESE Prelims : 2017]

3.5 Which of the following steps are involved in the product improvement cycle?
1. Sell it in the market
2. Determine quality of performance
3. Design the product based on customer needs
4. Test it in the laboratory

Select the correct answer using the codes given below:
(a) 1 and 3 only (b) 2 and 4 only
(c) 1, 3 and 4 only (d) 1, 2, 3 and 4
[ESE Prelims : 2019]

3.6 What is the break-even sale for the following products?

	Products		
	A	B	C
Sales (units)	5,000	6,000	4,000
Unit selling price (Rs.)	10	15	18
Unit variable price (Rs.)	6	4	13
Fixed cost of the product is (Rs. 20,000)			

(a) Rs. 90,000 (b) Rs. 80,000
(c) Rs. 60,000 (d) Rs. 40,000
[ESE Prelims : 2019]

3.7 Which one of the following is a viable alternative to term-loans and are instruments for raising debt finance by large publicly traded firms?
(a) Shares (b) Debentures
(c) Asset loans (d) Gold loans
[ESE Prelims : 2019]

3.8 Which of the following is a resource allocation as per Chenery's development process?
(a) Investment
(b) Structure of domestic demand
(c) Labour allocation
(d) Government revenue

[ESE Prelims : 2020]

3.9 Phillip Kotlar argues that the 4 Ps which represent the seller's thinking more than buyer's thinking can be translated into the 4 Cs. Match the following:

4 Ps of Marketing Planning
A. Product
B. Price
C. Place
D. Promotion

4 Cs of Marketing Planning
1. Customer communication
2. Customer value
3. Customer costs
4. Customer convenience

Select the correct matching using the code given below:

	A	B	C	D
(a)	2	3	4	1
(b)	2	3	1	4
(c)	1	4	3	2
(d)	1	4	2	3

[ESE Prelims : 2021]

3.10 Which one of the following policies is concerned with changing the supply of money stock and the rate of interest, for the purpose of stabilizing the economy at full potential output level?
(a) Commercial policy
(b) Fiscal policy
(c) Monetary policy
(d) Social policy

[ESE Prelims : 2022]

3.11 Consider the following statements :

The broad areas of corporate appraisal and the few important aspects to be considered under them are
1. Marketing and Distribution
2. Production and Operation
3. Research and Development
4. Project Rating

Which of the above statements are correct?
(a) 1, 2, 3 (b) 2, 3, 4
(c) 1, 2, 4 (d) 1, 3, 4

[ESE Prelims : 2024]

4. FINANCIAL ANALYSIS AND PROJECT FINANCES

4.1 During an assessment of economic viability of the project, the ratio of average annual earnings after tax to the average book investment after depreciation is called
(a) Benefit-Cost Ratio (BCR)
(b) Net Present Value (NPV)
(c) Pay-Back Period (PBP)
(d) Return on Investment (ROI)

[ESE Prelims : 2020]

4.2 The market price per share of a company is ₹125. The dividend per share (DPS) expected a year is ₹12 and DPS is expected to grow at a constant rate of 8% per annum. The cost of the equity capital to the company will be
(a) 17.6% (b) 15.4%
(c) 13.2% (d) 11.8%

[ESE Prelims : 2020]

4.3 An individual investor who invests in the e-project usually during an early stage is
(a) corporate strategic investor
(b) founder capital
(c) angel investor
(d) venture capital

[ESE Prelims : 2020]

4.4 If the nominal rate of interest is 12% and is compounded quarterly, the effective rate of interest per annum will be nearly
(a) 10.8% (b) 12.6%
(c) 14.4% (d) 16.2%

[ESE Prelims : 2020]

4.5 In a bank, deposits can be made for periods ranging from 6 months to 10 years. Every quarter, an interest will be added on to the principal. The rate of interest applied is 9% per annum for periods from 12 months to 23 months and 10% per annum for periods from 24 months to 120 months. An amount of ₹1,000 invested for 2 years to grow, will be nearly
(a) ₹1,218 (b) ₹1,334
(c) ₹1,414 (d) ₹1,538

[ESE Prelims : 2020]

4.6 A company has issued ₹20 million worth of non-convertible debentures, each at a face value of ₹100 at the rate of 12%. Each debenture is redeemable at a premium of 5%, after 10 years. If the net amount realized is ₹95 and tax rate is 40%, the cost per debenture will be

(a) 5.8% (b) 6.6%
(c) 7.4% (d) 8.2%

[ESE Prelims : 2020]

4.7 **Statement (I):** Raw materials are taken as traded items and their values at domestic and world prices are estimated.

Statement (II): Raw materials, which have a high value-to-volume ratio and involve proportionately high transport cost and are imported, are regarded as non-traded items.

(a) Both Statement (I) and Statement (II) are individually true and Statement (II) is the correct explanation of Statement (I)
(b) Both Statement (I) and Statement (II) are individually true but Statement (II) is not the correct explanation of Statement (I)
(c) Statement (I) is true but Statement (II) is false
(d) Statement (I) is false but Statement (II) is true

[ESE Prelims : 2020]

4.8 The creative technique applied when the available and required inputs as well as the desired outputs are listed, is

(a) Attribute listing (b) Direct dreaming
(c) Black box (d) Delphi

[ESE Prelims : 2020]

4.9 Under which one of the following circumstances is the project accepted as worthwhile, keeping the principal non-discounting criteria?

(a) The payback period (PBP) > target period
(b) The payback period (PBP) < target period
(c) The payback period (PBP) = target period
(d) The payback period (PBP) = 0

[ESE Prelims : 2021]

4.10 Consider the following statements regarding production:

The major aspects of production that may lead to sickness are

1. Increase in the cost of production.
2. Decrease in the quantity of production.
3. Quality of product not meeting the standards/customer expectation.
4. Producing more quantity than can be sold, leading to accumulation of stock.

Which of the above statements are correct?

(a) 1 and 2 only (b) 1, 2, 3 and 4
(c) 2 and 3 only (d) 1, 3 and 4 only

[ESE Prelims : 2022]

4.11 What are the factors that contribute to decline in unit cost with respect to the accumulated volume of production?

(a) Pioneering stage, Rapid growth stage and Economies of scale stage
(b) Learning effects, Technological improvements and Economies of scale
(c) Technological improvements stage, Maturity stage and Decline stage
(d) Pioneering stage, Rapid growth stage and Decline stage

[ESE Prelims : 2024]

5. PLANNING PROCESS : PROJECT PLANNING AND SCHEDULING

5.1 In a project there are 9 activities : *A*, *B*, *C*, *D* which are sequential; *E*, *F*, *G* which are sequential; *H*, *K* which are sequential. Also *E*, *F*, *G* run parallel to *B*, *C*, *D*, and *H*, *K* run parallel to *A*, *B*, *C*, *D*. Besides these activity dependencies. It is also needful that *B* be completed before taking up G; A and H be completed before taking up *D* and *K*. How many dummies are to be drawn on the activity network?

(a) 5 (b) 4
(c) 3 (d) 2

[ESE Prelims : 2018]

5.2 Consider the following statements:

1. In work breakdown structure, top-down approach is adopted.
2. Duration along critical path is the shortest duration permissible.
3. PERT is probabilistic in its approach.

Which of the above statements are correct?

(a) 1 and 2 only (b) 1 and 3 only
(c) 2 and 3 only (d) 1, 2 and 3

[ESE Prelims : 2018]

5.3 If L_j = the latest occurrence time for event j.
E_i = the earliest occurrence time for event i,
T_{ij} = the duration of activity $\vec{ij}$.

What is the float for activity $\vec{i_j}$?

(a) $E_i - L_j - T_{ij}$
(b) $L_j - L_i - T_{ij}$
(c) $T_{ij} - E_i - L_j$
(d) $T_{ij} - L_j - E_i$

[ESE Prelims : 2018]

5.4 **Statement (I):** Preparation of bar charts is merely a scheduling operation while the preparation and analysis of a network is a planning function.
Statement (II): A bar chart, prima facie, does not show the interrelationships between activities.

(a) Both Statement (I) and Statement (II) are individually true and Statement (II) is the correct explanation of Statement (I).
(b) Both Statement (I) and Statement (II) are individually true but Statement (II) is **not** the correct explanation of Statement (I).
(c) Statement (I) is true but Statement (II) is false.
(d) Statement (I) is false but Statement (II) is true.

[ESE Prelims : 2018]

5.5 CPM method of network analysis is

1. Ideally suited for linearly extending works.
2. Meant essentially for research and development activities.
3. Activity-oriented.
4. Used for planning, scheduling and controlling purposes.

Which of the above statements are correct?

(a) 1 and 2 only
(b) 2 and 3 only
(c) 3 and 4 only
(d) 1 and 4 only

[ESE Prelims : 2018]

5.6 A simple project comprises of two start-to-end parallel paths, each with three activities in series, with no interpath dependencies. The *a, m, b* data (in days) for each activity are shown in the diagram. Assuming that three activities in series are enough for further computations, what will be the total project duration and its standard deviation?

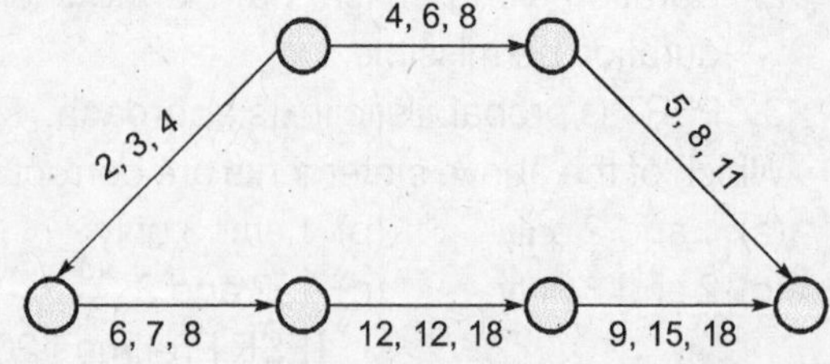

(a) $35\frac{1}{2}$ day and $\frac{14}{3}$ days
(b) $34\frac{1}{2}$ day and $\frac{5}{2}$ days
(c) $35\frac{1}{2}$ day and $\frac{13}{2}$ days
(d) $34\frac{1}{2}$ day and $\frac{11}{6}$ days

[ESE Prelims : 2018]

5.7 Which one of the following schedules shows the specific activities necessary to complete an activity or work package?

(a) Project schedule
(b) Master schedule
(c) Task schedule
(d) Internal schedule

[ESE Prelims : 2019]

5.8 In principle, the network should not be made complex. No control system, for that matter, can operate unless it is kept simple. This principle is called

(a) CPM
(b) PERT
(c) KISS
(d) GERT

[ESE Prelims : 2019]

5.9 Which one of the following methods of planning is an attempt to work out the implications of the development effort in terms of factor allocations and product yields so as to maximize income and employment?

(a) Perspective planning
(b) Physical planning
(c) Financial planning
(d) Indicative planning

[ESE Prelims : 2020]

5.10 Which one of the following risks can be reduced by investing in projects or acquiring other firms that have a negative correlation with the earnings of the firm?

(a) Investment risk
(b) Business risk
(c) Financial risk
(d) Portfolio risk

[ESE Prelims : 2020]

5.11 Which one of the following is NOT the purpose of the organization breakdown structure?

(a) To provide a framework to summarize organization unit work performance
(b) Do not tie the organizational unit to cost control accounts
(c) Identify organization units responsible for work packages

(d) How the firm has organized to discharge work responsibility

[ESE Prelims : 2021]

5.12 Which one of the following is NOT a condition for preferring Top-Down Time and Cost Estimates?

(a) Strategic decision making
(b) Cost and time important
(c) High uncertainty
(d) Internal, small project [ESE Prelims : 2022]

5.13 In Network Computation Process, which one of the following is correct for forward pass?

(a) It starts with the first project activity(ies) and traces each path (chain of sequential activities) through the network to the last project activity(ies).
(b) This is the longest path in the network, which will delay the project.
(c) It starts with the last project activity(ies) on the network.
(d) It starts with the last project activity(ies) and traces each path (chain of sequential activities) through the network to the first project activity(ies)

[ESE Prelims : 2022]

5.14 Which one of the following projects are those which are to be completed within a stipulated time, even at the cost of ending up with a higher project cost?

(a) Normal projects
(b) Business projects
(c) Crash projects
(d) Research projects [ESE Prelims : 2022]

5.15 Consider the following statements for hammock activities:

1. It derives its name because it spans over a segment of a project.
2. The hammock activity duration is determined after the network plan is not drawn.
3. The hammock activities are frequently used to identify the use of fixed resources or costs over a segment of the project.
4. The maximum amount of time an independent activity must be delayed to begin or end.

Which of the above statements are correct?

(a) 2 and 3 only (b) 1 and 3 only
(c) 1 and 4 only (d) 2 and 4 only

[ESE Prelims : 2023]

5.16 Consider the following strategies for mitigating risk under risk response development:

1. Reduce the likelihood that the event will occur.
2. Reduce the impact that the adverse event would have on the project.
3. Analyze the project to identify sources of risk.
4. Assess risks in terms of severity of impact.

Which of the above strategies are correct?

(a) 1 and 2 only (b) 3 and 4 only
(c) 1 and 4 only (d) 2 and 3 only

[ESE PRELIMS : 2023]

5.17 Consider the following statements:

The strategy is to assign extra time at critical moments in the project, buffers are added to:

1. activities with no risk.
2. merge activities that are prone to delays due to one or more preceding activities being late.
3. non-critical activities to reduce the likelihood that they will create another critical path.
4. activities that require scarce resources to ensure that the resources are available when needed.

Which of the above statements are correct?

(a) 1, 2 and 3 only (b) 1, 2 and 4 only
(c) 2, 3 and 4 only (d) 1, 3 and 4 only

[ESE Prelims : 2023]

5.18 Which one of the following does NOT always yield an optimal schedule, however it is capable of yielding a "good" schedule for very complex networks having many types of resources?

(a) Algorithm (b) Optimum
(c) Backhoes (d) Heuristics

[ESE Prelims : 2023]

5.19 According to CCPM, using 50/50 estimates will discourage Parkinson's law, the student syndrome, and self-protection from coming into play because there is less "free time" available. What does the abbreviation CCPM stand for?

(a) Control – Chain Project Management
(b) Creating – Chain Project Management
(c) Computer – Control Project Management
(d) Critical – Chain Project Management

[ESE Prelims : 2023]

5.20 According to project cost-duration graph, any reduction in project duration means a reduction in

(a) direct costs (b) indirect costs
(c) total costs (d) optimum costs

[ESE Prelims : 2023]

5.21 Which one of the following methods is an important qualitative method under demand forecasting?

(a) Jury of Executive Method
(b) Trend Projection Method
(c) Chain Ratio Method
(d) Bass Diffusion Method

[ESE Prelims : 2024]

5.22 Why Delphi method of demand forecasting appeals to many organizations?

(a) The biases underlying are subjective, and it seems to be more accurate and less expensive than the traditional face-to-face group meetings
(b) It is intelligible to users, it is a fancy name, and it seems to be more accurate and less expensive than the traditional face-to-face group meetings
(c) It has immense appeal, the biases underlying are subjective, and it seems to be more accurate and less expensive than the traditional face-to-face group meetings.
(d) It is an expedition's method, it has immense appeal, and it seems to be more accurate and less expensive than the traditional face-to-face group meetings.

[ESE Prelims : 2024]

6. EXECUTION PROCESS: PROJECT MONITORING AND CONTROL

6.1 Crashing is

(a) Abandoning the project
(b) Completing the project with all possible haste
(c) Reduction of duration for a few of the activities
(d) Reducing the cost of the project with all needful modifications

[ESE Prelims : 2018]

6.2 **Statement (I) :** Social involvement discourages additional government regulation and intervention.

Statement (II) : Social involvement can create a weakened international balance of payments situation.

(a) Both Statement (I) and Statement (II) are individually true and Statement (II) is the correct explanation of Statement (I)
(b) Both Statement (I) and Statement (II) are individually true, but Statement (II) is not the correct explanation of Statement (I)
(c) Statement (I) is true, but Statement (II) is false
(d) Statement (I) is false, but Statement (II) is true

[ESE Prelims : 2022]

6.3 Which of the following factors are included in product realization process?

1. Marketing functions to assess customer requirements
2. Documentation of the design
3. Legal requirements

Select the correct answer using the code given below:

(a) 1 and 2 only (b) 2 and 3 only
(c) 1 and 3 only (d) 1, 2 and 3

[ESE Prelims : 2022]

6.4 **Statement (I):** The machine shop produces parts machined from stock material and finishes castings, forgings, etc., requiring machined surfaces.

Statement (II): In machine shops, machining operations remove metal, either to make a smoother and more accurate surface, as by planning, facing, milling, etc. or to produce a surface previously existing, as by drilling, punching, etc.

(a) Both Statement (I) and Statement (II) are individually true and Statement (II) is the correct explanation of Statement (I)
(b) Both Statement (I) and Statement (II) are individually true, but Statement (II) is not the correct explanation of Statement (I)
(c) Statement (I) is true, but Statement (II) is false
(d) Statement (I) is false, but Statement (II) is true

[ESE Prelims : 2023]

7. TENDER AND CONTRACT

7.1 ABC analysis in materials management is a method of classifying the inventories based on the
(a) Economic order quantity
(b) Value of annual usage of the items
(c) Volume of material consumption
(d) Quantity of material used

[ESE Prelims : 2018]

7.2 It is expected to receive Rs. 5,000 annually for 3 years with each receipt occurring at the end of the year. With a discount rate of 10%, the present value of the annuity will be nearly
(a) Rs. 12,435 (b) Rs. 9,945
(c) Rs. 4,975 (d) Rs. 2,487

[ESE Prelims : 2019]

7.3 A machine is expected to generate cash saving (after-tax) of Rs. 50,000 per annum over a period of 5 years. Salvage value of machine is 40% of the original cost. If accounting rate of return is 20%, cost of two such machines will be
(a) Rs. 78,125 (b) Rs. 1,56,250
(c) Rs. 3,12,500 (d) Rs. 6,25,000

[ESE Prelims : 2019]

7.4 Which one of the following is a particular form of collusive price-fixing behaviour by which firms coordinate their bids on procurement or project contracts?
(a) Predatory pricing
(b) Horizontal price-fixing (collusion)
(c) Bid rigging
(d) Exclusive territory

[ESE Prelims : 2020]

7.5 Which one of the following is an example of horizontal practice of firm?
(a) Refusal to deal
(b) Retail price maintenance
(c) Predatory pricing
(d) Exclusive territory

[ESE Prelims : 2020]

■■■■

Basics of Project Management : Answer Key

1.1 (c)	**1.2** (d)	**1.3** (a)	**1.4** (a)	**1.5** (d)	**1.6** (a)	**1.7** (c)	**1.8** (c)
1.9 (a)	**1.10** (a)	**2.1** (b)	**2.2** (c)	**2.3** (b)	**2.4** (b)	**2.5** (c)	**2.6** (c)
3.1 (b)	**3.2** (c)	**3.3** (c)	**3.4** (a)	**3.5** (d)	**3.6** (d)	**3.7** (b)	**3.8** (b)
3.9 (a)	**3.10** (c)	**3.11** (a)	**4.1** (d)	**4.2** (a)	**4.3** (c)	**4.4** (b)	**4.5** (a)
4.6 (d)	**4.7** (c)	**4.8** (a)	**4.9** (b)	**4.10** (b)	**4.11** (a)	**5.1** (c)	**5.2** (d)
5.3 (b)	**5.4** (a)	**5.5** (c)	**5.6** (d)	**5.7** (c)	**5.8** (c)	**5.9** (b)	**5.10** (d)
5.11 (b)	**5.12** (b)	**5.13** (a)	**5.14** (c)	**5.15** (b)	**5.16** (a)	**5.17** (c)	**5.18** (d)
5.19 (d)	**5.20** (b)	**5.21** (a)	**5.22** (a)	**6.1** (c)	**6.2** (c)	**6.3** (d)	**6.4** (c)
7.1 (b)	**7.2** (a)	**7.3** (c)	**7.4** (c)	**7.5** (c)			

EXPLANATIONS

1.1 (c)

Project management is the practice of leading and organizing resources, tasks and people to achieve specific goals within a defined timeline and budget. It involves the planning, execution, monitoring, and control of all aspects of a project from initiation to completion.

The process involves defining project goals and objectives, identifying tasks and resources required to achieve those goals, creating a project plan and executing the plan while monitoring progress and making adjustments as necessary.

ADDITIONAL INFORMATION

The project management process typically involves various stages:

1. **Initiation:** During the initiation stage, the project manager defines the project scope, objectives, and deliverables, and identifies the resources needed to complete the project.
2. **Planning:** During the planning stage, the project manager creates a detailed project plan that outlines the tasks, timelines, and budgets for each phase of the project. This plan also identifies potential risks and outlines strategies for managing them.
3. **Execution:** During the execution stage, the project manager oversees the actual work of the project team, ensuring that tasks are completed on time and within budget.
4. **Monitoring and control:** The monitoring and control stage involves tracking project progress, identifying potential problems, and making necessary adjustments to keep the project on track.
5. **Closure:** Finally, during the closure stage, the project manager ensures that all deliverables have been completed and that the project is closed out properly. This involves obtaining feedback from stakeholders to evaluate the success of the project and identifying lessons learned for future projects.

1.2 (d)

Project Identification : The purpose of project identification is to develop a preliminary proposal for the most appropriate set of interventions and course of action, within specific time and budget frames, to address a specific development goal in a particular region or setting. It involves.

(a) a review of alternative approaches or options for addressing a set of development, problems and opportunities.

(b) Identifying major issues that must be tackled and questions to be addressed before a project based on concept can be implemented.

Project formulation : It refers to mechanics of formulating a project report. It comprises of various steps that are very essential to analyse the collected data and subsequently reaching at reasonable conclusions relevant to objectives. The process of formulation is carried out in following steps.

1. Selection of project
2. Collection of information
3. Analysis of data
4. Preparation of statement
5. Writing of project report
6. Text check and presentation of report.

Project appraisal : Project appraisal is a generic term that refers to process of assesing, in a structured way, the case for proceeding with a project. In short, project appraisal is effort of calculating a project's viability. It often involves comparing various options, using economic appraisal or some other decision analysis technique.

Project implementation : Project implementation is process of putting a project plan into action to produce deliverables, otherwise knowns as products or services, for clients or stakeholders. Implementation involves coordinating resources and measuring performance to ensure the project remains within its expected scope and budget.

ADDITIONAL INFORMATION

The scheduling of implementation is generally depicted through progressive listing of different activities through sophisticated techniques of CPM and PERT.

1.3 (a)

Project management is an application of knowledge, skills, tools and techniques to meet the project requirements. The objective is to use the optimum resources to accomplish a goal in minimum time.

With any Project, there are limitations and risks that need to be taken into account and addressed to ensure the project's ultimate success. The three primary constraints that project managers should be familiar with are time, scope and cost. These are frequently known as the triple constraints or the project management triangles. Each constraint is connected to the other two; so, for example, increasing the scope of the project will likely require more time and money while speeding up the timeline for the project may cut costs, but also diminish the scope.

In reality, the four dimensions of a project(scope, cost, time, and quality) are interdependent and cannot be fully separated from each other. Therefore, while it may be possible to prioritize or emphasize two dimensions over the others, all four dimensions must be considered and managed throughout the project lifecycle.

ADDITIONAL INFORMATION

There are four main parameters of a project management :

1. **Time:** Time is very crucial to any project. Alteration of Project completion time change the cost.
2. **Scope:** Scope tells what needs to be achieved and the work that must be done to deliver a project.
3. **Cost:** Cost is the monetary value of a Project. Cost is directly dependent on time and the quality to be produced.
4. **Quality:** This is the standard of something as measured against other things of a similar kind, the degree of excellence of something.

The sextuple constraint : Now a days apart from time, cost, scope and quality, risk and resources are also constraints for better understanding.

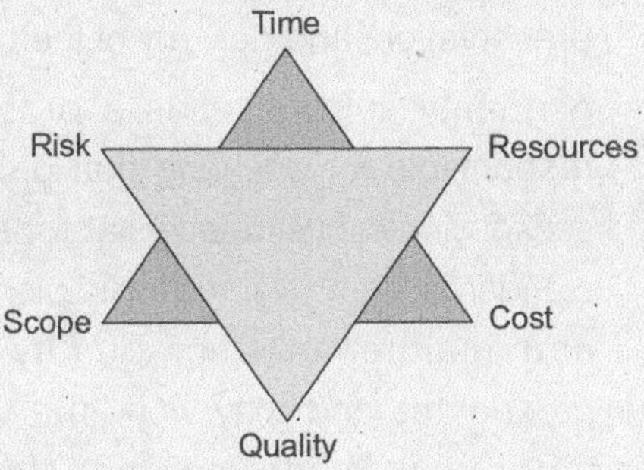

1.4 (a)

Definite estimate : A definitive estimate is a very specific estimate that project managers use to determine the amount of work that will be required to complete a given project. It is the most accurate estimate that can be made and ensures that project is completed on time and within budget. It is made on basis of current knowledge and past experiences. It is prepared during the closure stage of a project.

Note : UPSC has given answer of this question as option **(b)** i.e. detailed estimate.

ADDITIONAL INFORMATION

1. Detailed estimate is prepared during implementation stage of a project.
2. Preliminary estimate is prepared during 'planning' phase of a project.

1.5 (d)

The level of effort is a useful indicator for the project manager to quantify the amount of work to be performed and the amount of work completed within each phase.

The maximum percentage of effort is done in implementation or execution phase in a project cycle.

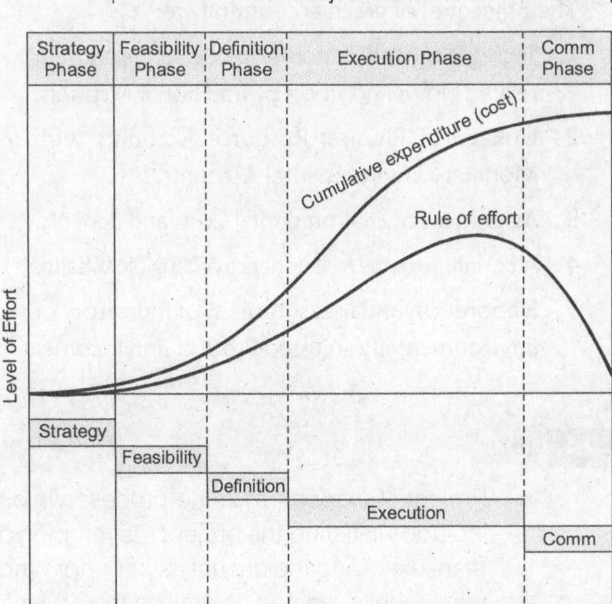

1.6 (a)

All projects have constraints or limitations that affect their ability to achieve their goals and objectives. These constraints can include limitations on resources, such as time, money, and personnel, as well as technical constraints, regulatory constraints, and environmental constraints.

Time and money are indeed universal constraints in projects. Time constraints refer to the deadline or timeline for completing the project, while budget constraints refer to the amount of money allocated for the project. These constraints can have a significant impact on the project's ability to achieve its goals and objectives.

In many cases, time and money constraints are interrelated, as extending the timeline of a project may result in increased costs, or reducing the budget may impact the project timeline or quality.

1.7 (c)

A new system of sustainable accounting, known as Green Accounting, has emerged. "It permits the computation of income for a nation by taking into account the economic damage and depletion in the natural resource base of an economy." It is a measure of sustainable income level that can be secured without decreasing the stock of natural assets. This requires adjustment of the System of National Accounts (SNA) in terms of stock of natural assets.

ADDITIONAL INFORMATION

The objectives of green accounting are:

1. Segregation and Elaboration of all Environment related Flows and Stocks of Traditional Accounts.
2. Linkage of Physical Resource Accounts with Monetary Environmental Accounts.
3. Assessment of Environmental Costs and Benefits.
4. Accounting for the Maintenance of Tangible Wealth.
5. Elaboration and Measurement of indicators of Environmentally adjusted Product and Income.

2.1 (b)

- Project Management is the process which includes initiating the project, developing a plan, executing the project as per approved plan, monitoring and controlling the project, handing over the output of the project to the client and finally closing the project.
- Product management is the process of overseeing all aspects of a product's life cycle, from its conception to its eventual retirement. It involves working closely with cross-functional teams, such as engineering, design, marketing, sales, and customer support, to ensure that the product meets the needs of both the business and its customers.
- Commercial Management is identification and development of business opportunities and profitable management of projects and contracts from in option to completion.
- Venture management is a discipline that focuses on being both innovative and challenging in the realm of introducing what could be a completely new product or entering a promising new emerging market.

2.2 (c)

- A production manager is a professional responsible for overseeing and managing the manufacturing process within a company. Their primary responsibilities include ensuring that production runs efficiently, safely, and within budget, while also maintaining high-quality standards.
- A marketing manager is responsible for developing and executing marketing strategies that help to promote a company's products or services to potential customers. The role involves a wide range of activities, including market research, product development, advertising, and sales promotion.
- As per ISO 9000-2 clause 4.15, the vendor shall establish and maintain documented procedures for handling storage, packaging, preservation, and delivery of the product.
- A quality supervisor is a professional responsible for ensuring that products or services meet or exceed established quality standards. They oversee the implementation and maintenance of quality control procedures and may work in a variety of industries, such as manufacturing, healthcare, and food service.

2.3 (b)

A project coordinator is a professional who supports project managers in planning, executing, and monitoring projects. The role of a project coordinator is to ensure that projects are completed on time, within budget, and to the satisfaction of stakeholders. The main responsibilities of a project coordinator may include:

1. Assisting project managers in developing project plans and schedules.
2. Coordinating project team meetings and ensuring that project tasks are assigned and completed on time.
3. Monitoring project progress and providing regular updates to the project manager.
4. Identifying potential issues or delays and working with the project manager to develop and implement solutions.
5. Tracking project expenses and ensuring that they are within budget.
6. Coordinating project documentation and ensuring that it is accurate and up to date.
7. Communicating project status to stakeholders and providing regular reports on project progress. A matrix manager is a professional who manages employees who work on multiple projects or tasks for different managers within the same organization.

A matrix manager is responsible for ensuring that the employees under their supervision have the resources and support needed to complete their tasks efficiently and effectively. A project expeditor usually carries the responsibility of staff assistant/ communication coordinator. A project expeditor is not given the authority (or very low authority) to make or enforce decisions.

2.4 (b)

Matrix structure can be strong, weak, or balanced. The strength of the matrix is based upon who has more influence over the daily performance of the workers: project manager or line managers. If the project manager has more influence over the worker, then the matrix structure functions as a strong matrix as seen by the project manager. If the line manager has more influence than does the project manager, then the organisation functions as a weak matrix as seen by project manager.

2.5 (c)

There are four quadrants in BCG matrix.

1. **Question mark :** In this quadrant, those products are placed that have high market growth but a low market share. These types of products are the most managerially intensive products and require extensive investment and resources to increase their market share.
2. **Stars :** Products placed in these projects have high market growth and a high market share. These products consume a significant amount of cash but also generate large cash flows. As the market matures and products remains successful, stars migrate to become cash cows.
3. **Dogs :** In this quadrant, products with low market growth and a low market share are placed. These type of products are typically able to sustain themselves and provide cash flows, but the products will never reach the 'star' quadrant.
4. **Cash Cows :** Products with low market growth but a high market share are placed in this quadrant. These types of products are considered as leaders in their respective markets.

ADDITIONAL INFORMATION

The Boston consulting group matrix (BCG Matrix), also referred to as product portfolio matrix, is a business planning tool used to evaluate the strategic position of a firm's brand portfolio. It is one of the most popular portfolio analysis methods.

3.2 (c)

The operating cost of equipment is the cost associated with running and maintaining the equipment over its useful life. The factors that govern the operating cost of equipment may vary depending on the type of equipment, its usage, and the operating environment. However, some common factors that can impact the operating cost of equipment include:

1. **Maintenance:** Regular maintenance is essential to keep equipment running efficiently and reliably. The cost of maintenance, including routine inspections,

repairs, and replacement of parts, can significantly impact the operating cost of equipment.

2. **Energy consumption:** Equipment that requires a lot of energy to operate, such as heavy machinery or industrial equipment, can have a high operating cost. Factors that can impact energy consumption include the type of fuel used, operating temperature, and operating time.
3. **Material costs:** The cost of materials, including lubricants, fluids, and filters, can add up over the life of an equipment, and impact its operating cost.
4. **Operating conditions:** Equipment operating in harsh or extreme conditions, such as high-temperature environments or dusty areas, may require more maintenance and repairs, increasing the operating cost.
5. **Equipment utilization:** The frequency and duration of equipment usage can impact its operating cost. Equipment that is used more frequently or for longer periods may require more maintenance, repairs, and replacement of parts, increasing its operating cost.
6. **Depreciation:** Equipment depreciates over time, and the depreciation cost can add up over the life of the equipment, impacting its operating cost.

Note: UPSC has given answer as option (d), but the correct answer is option (c) because purchase price of the equipment does not determine its operating cost as it comes as initial cost.

3.3 (c)

Reliability analysis is the process of evaluating the probability of a product or system performing its intended function without failure over a specific period. When it comes to maintenance and service of a product after delivery, reliability analysis plays a crucial role in ensuring that the product meets the expected reliability standards and requirements. Reliability analysis helps to identify potential areas of failure in a product, allowing for preventative maintenance and servicing to be performed before an actual failure occurs. This analysis can be done by monitoring product performance, analyzing historical data on product failures, and conducting testing to determine the product's reliability and failure rate. Reliability analysis can also help to identify areas for improvement in a product's design, manufacturing, or servicing processes. By analyzing data on product failures, companies can identify patterns and trends that can be used to improve the design or production process, leading to a more reliable and efficient product.

Note: UPSC has given option (d) as correct answer but, the correct answer is option (c).

3.4 (a)

The economic life of a building is considered to be at an end when the cost of maintaining and repairing the building exceeds its potential income or value. This means that the building has reached a point where it is no longer financially viable to continue using or investing in it, and the cost of demolition or redevelopment may be more appropriate.

Other factors that determines the economic life of a building are physical condition of the building, the cost of maintenance and repairs, the demand for the building's use, and the potential for future revenue generation.

3.5 (d)

The product improvement cycle refers to the process of continually refining and enhancing a product over time to meet the changing needs and preferences of customers and to stay ahead of competitors. The cycle typically involves several stages including:

1. **Idea Generation:** The first stage of the product improvement cycle involves generating new ideas about how to improve the product. This can be done through customer feedback, market research, or brainstorming sessions with the product development team.
2. **Product Design:** In this stage, the ideas are developed into concrete plans for how to modify the product. This may involve creating prototypes, conducting feasibility studies, and making decisions about which improvements to prioritize.

3. **Product Development:** Once the plans are finalized, the product development team works on implementing the changes to the product. This may involve updating the product's design, sourcing new materials or components, and making changes to the manufacturing process.
4. **Testing and Validation:** Before the new version of the product is released to the market, it undergoes testing and validation to ensure that it meets the desired performance standards and customer expectations.
5. **Launch:** Once the product has been validated, it is launched in the market. At this stage, it is important to monitor customer feedback and make any necessary adjustments to ensure that the product continues to meet customer needs and preferences.

3.6 (d)

Break even point

$$= \frac{\text{Total fixed price}}{\text{Average selling price} - \text{Average variable prices}}$$

Average selling price

$$= \frac{5000 \times 10 + 6000 \times 15 + 4000 \times 18}{15000} = 14.13$$

Average variable price

$$= \frac{5000 \times 6 + 6000 \times 4 + 4000 \times 13}{15000} = 7.067$$

$\therefore$ Break even point

$$= \frac{20000}{14.13 - 7.067} = 2831.65 \text{ units}$$

$\therefore$ Break even sales

$= \text{Rs.}2831.66 \times 14.13 = \text{Rs.}40011.36 \simeq 40000$

3.7 (b)

Shares refer to units of ownership in a company that are bought and sold on stock exchanges. When a company goes public, it issues shares of stock to the public in an initial public offering (IPO), allowing individuals and institutional investors to buy a stake in the company. By purchasing shares of a company's stock, investors become part owners of the company and can potentially benefit from its profits and growth.

A debenture is a type of debt instrument that companies can use to raise funds from investors. Essentially, a debenture is a loan that a company takes out from investors, with the promise to pay back the loan with interest over a specified period of time. Unlike other forms of debt, such as bank loans or lines of credit, debentures are not secured by specific assets of the company. Instead, they are backed only by the creditworthiness of the company issuing the debenture. This means that if the company defaults on the loan, debenture holders are considered unsecured creditors and are not guaranteed to recover their investment.

Asset loans are loans where borrowers use their valuable assets such as real estate, vehicles, or investments as collateral.

Gold loans are a specific type of asset loan where borrowers use their gold jewelry or coins as collateral.

3.8 (b)

Hollis B. Chenery defines the development pattern as "A systematic variation in any significant aspect of the economic or social structure associated with a rising level of income." But, he is primarily interested in those structural changes that are needed to achieve sustained increase in per capita income. He adopted policies like:

- Shift from agricultural to industrial production.
- Steady accumulation of physical and human capital.
- Growth of cities and urban areas.
- Change in composition of consumer demand i.e. Structure of domestic demand.

3.9 (a)

THE FOUR P'S OF MARKETING:

Product: A product refers to an item that satisfies the consumer's needs or wants.

Products may be tangible (goods) or intangible (services, ideas or experiences).

Price: Price refers to the amount a customer pays for a product. Price may also refer to the sacrifice consumers are prepared to make to acquire a product (e.g. time or effort), and includes considerations of customer perceived value.

Place: Place describes giving the customer access to your product, traditionally at a brick-and-mortar place or through the mail.

Promotion: Promotion refers to marketing communications (advertising, PR, or sales/ promotions).

3.10 (c)

Commercial policy is an umbrella term describing the regulations and policies that dictate how companies in different countries can conduct commerce with each other. Commercial policy includes tariffs, import quotas, export constraints, and restrictions against foreign-owned companies operating domestically.

Fiscal policy refers to the use of government spending and tax policies to influence economic conditions, especially macroeconomic conditions, including aggregate demand for goods and services, employment, inflation, and economic growth.

Monetary policy is a set of tools that a nation's central bank has available to promote sustainable economic growth by controlling the overall supply of money that is available to the nation's banks, its consumers, and its businesses.

Fiscal policy is often contrasted with monetary policy, which is enacted by central bankers and not elected government officials.

Social policy is concerned with the ways societies across the world meet human needs for security, education, work, health and wellbeing. Social policy addresses how states and societies respond to global challenges of social, demographic and economic change, and of poverty, migration and globalisation.

4.1 (d)

- Return on investment (RoI) is a performance measure used to evaluate the efficiency of an investment or compare the efficiency of a number of different investments.
- It calculate the profitability of the company by measuring the earnings related to the amount of capital invested.
- It is calculated by dividing the benefit or return of an investment to its cost. The result is expressed as a percentage or ratio.

4.2 (a)

Dividend per share = Rs.12, Market price per share = Rs.125, rate = 8%

Cost of equity

$$= \frac{\text{Dividend per share}}{\text{Market price per share}} + \text{rate}$$

$$= \frac{12}{125} + 0.08 = 0.176 \text{ or } 17.6\%$$

4.3 (c)

An angel investor is a high net worth individual who provides financial backing for small startups or entrepreneurs, in exchange for ownership equity in the company. The funds that the angel investors provide may be a one-time investment to help the business get off the ground or an ongoing injection to support and carry the company through its difficult early stages.

4.4 (b)

Effective rate of interest per annum,

$$R = \left(1 + \frac{r}{c}\right)^c - 1$$

where 'r' is the nominal rate of interest and 'c' is the number of times compounding is done annually.

Here, r = 12% = 0.12, c = 4 (compounding is done quarterly)

$$\Rightarrow \quad R = \left(1 + \frac{0.12}{4}\right)^4 - 1$$

$$= 0.1255 = 12.55\% \simeq 12.6\%$$

4.5 (a)

The investment is made for a period of 2 years.

So, rate of interest per annum = 10%

Compounding is done quarterly.

$$\text{Amount,} \quad R = P\left(1 + \frac{r}{c}\right)^{nc}$$

P is the principal amount

r is the rate of interest per annum

c is the number of times compounding is done annually

n is the number of years for which amount is invested

Here, P = Rs.1000, r = 10%, c = 4, n = 2 years

$$R = 1000\left(1 + \frac{0.1}{4}\right)^{2\times 4}$$

$$= 1218.4 \simeq \text{Rs.}1218$$

4.6 (d)

$$\text{Cost per debenture} = \frac{I(1-t) + \dfrac{RV - NP}{N}}{\dfrac{RV + NP}{2}} \times 100$$

where, t = Tax rate

I = Interest rate = 12%

RV = Redeemable value

= 100(1 + 0.05) = Rs.105

NP = Net proceeds from issue = Rs.95

⇒ Cost per debenture

$$= \frac{12(1-0.4) + \dfrac{105-95}{10}}{\dfrac{105+95}{2}} \times 100 = 8.2\%$$

∴ Cost per debenture = 8.2%

4.7 (c)

In general raw materials are taken as traded items and their value at domestic and world prices are estimated. However, raw materials which have a low value to volume ratio and involve disproportionately high transport cost and which are not imported are regarded as non-traded items.

4.8 (a)

Attribute listing is a means of getting you to focus on as many attributes of a product or problem as possible. In breaking down the elements of a problem or object, you can look at each in turn and generate new ideas. The technique is particularly useful for considering complex products or processes in that it allows you to consider each feature or stage and look at the associated attributes in detail.

4.9 (b)

Payback Period refers to the amount of time it takes to recover the cost of an investment. If it is less than the target period, investment is accepted as worthwhile. The lesser the Payback Period the more preferred is the investment.

4.10 (b)

The major causes that may lead to sickness in production are insufficient resources and improper production and marketing policies.

These causes lead to aspects such as

1. Increase in the cost of production.
2. Decrease in the quantity of production.
3. Quality of product not meeting the standards/ customer expectation.
4. Producing more quantity than can be sold, leading to accumulation of stock.

5.1 (c)

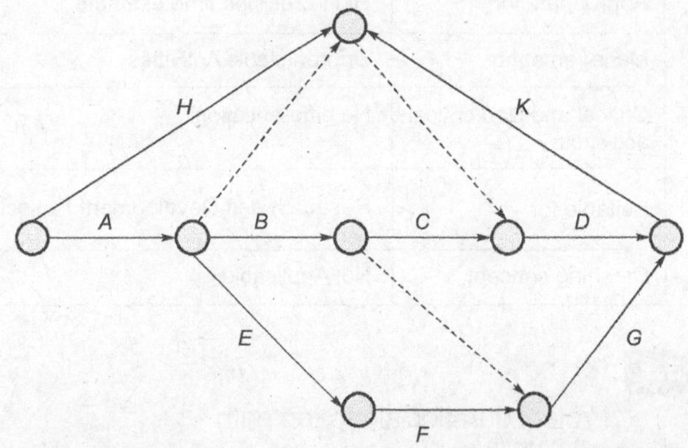

Number of dummies needed = 3

5.2 (d)

Project will be completed only when activities along critical path are completed. It means minimum project duration is duration of critical path it means shortest duration permissible of project is critical path duration.

5.3 (b)

$$F_T = T_{L_J} - T_E^i - t_{ij}$$
$$= L_j - E_i - T_{ij}$$

5.5 (c)

The Critical Path Method (CPM) is a project management technique used to identify the most critical tasks and activities required to complete a project. The critical path represents the sequence of tasks that must be completed on time in order to meet the project deadline.

CPM has been used for planning, scheduling and controlling in the construction project management.

Basis Comparison	PERT	CPM
Meaning	PERT is a project management technique, used to manage uncertain activities of a project	CPM is a statistical technique of project management that manages well defined activities of a project.
What is it?	A technique of planning and control of time.	A method to control cost and time.
Orientation	Event-oriented	Activity-oriented
Evolution	Evolved as Research & Development project	Evolved as Construction project
Model	Probabilistic Model	Deterministic Model
Estimates	Three time estimates	One time estimate
Appropriate for	High precision time estimate	Reasonable time estimate
Management of	Unpredictable Activities	Predictable activities
Critical and Non-critical activities	No differentiation	Differentiated
Suitable for	Research and Development Project	Non-research projects like civil construction, ship building etc.
Crashing concept	Not Applicable	Applicable

5.6 (d)

Project duration along top path

$$= \frac{2+4\times3+4}{6}+\frac{4+4\times6+8}{6}+\frac{5+4\times8+11}{6}$$

$= 3 + 6 + 8 = 17$ days

Project duration along bottom path

$$= \frac{6+4\times7+8}{6}+\frac{12+4\times12+18}{6}+\frac{9+4\times15+18}{6}$$

$$= 7 + 13 + 14.5 = 34.5 = 34\frac{1}{2}$$

Longest is $34\frac{1}{2}$, hence this path is critical.

$$\sigma = \sqrt{\left(\frac{8-6}{6}\right)^2+\left(\frac{18-12}{6}\right)^2+\left(\frac{18-9}{6}\right)^2} = \sqrt{\left(\frac{1}{3}\right)^2+1+\left(\frac{3}{2}\right)^2} = \sqrt{\frac{1}{9}+1+\frac{9}{4}}$$

$$= \sqrt{\frac{4+36+9\times9}{36}} = \sqrt{\frac{121}{36}} = \frac{11}{6}\text{days}$$

5.7 (c)

A **project schedule** is a document that outlines the timeline, tasks, and resources required to complete a project. It provides a roadmap for the project team to follow and helps to ensure that the project is completed on time and within budget.

A **task schedule** is a detailed plan that outlines the specific tasks and activities required to complete a project. It typically includes information such as the start and end dates of each task, the resources required, and any dependencies between tasks.

A **master schedule** is a project management document that outlines the overall plan and timeline for a project. It typically includes information about the start and end dates of the project, major milestones, and the sequence of tasks needed to complete the project.

ADDITIONAL INFORMATION

Work Breakdown Structure (WBS) is a project management technique that involves breaking down a complex project into smaller, more manageable components or tasks. The WBS helps to organize the project into smaller parts and assign responsibility for each task.

The steps to create a Work Breakdown Structure are:

Plan → Program → Project → Subproject → Activity → Work → Task

1. Define the project scope and deliverables: Identify the objectives, goals, and outcomes of the project. Determine what the final product or deliverables should look like.
2. Identify the major project phases: Divide the project into distinct phases, such as planning, design, development, testing, and implementation.
3. Break down each phase into smaller tasks: Break down each phase into smaller tasks, until they become manageable and easy to understand.
4. Assign each task a unique identifier: Assign a unique identifier to each task, such as a number or code, to help track progress and identify tasks easily.
5. Estimate the time and resources required: Estimate the time and resources required to complete each task, including the people, materials, and equipment needed.
6. Determine the dependencies: Identify the dependencies between tasks and determine which tasks need to be completed before others can start.
7. Review and refine: Review the WBS with stakeholders, project team members, and subject matter experts to ensure it is accurate, complete, and achievable. Refine the WBS as needed based on feedback and changes in the project scope.

5.8 (c)

KISS stands for "Keep It Simple, Stupid". It is a design principle that emphasizes the importance of simplicity and avoiding unnecessary complexity. The principle is commonly used in various fields, including software development, engineering, and design.

The KISS principle encourages designers to focus on the essentials and to avoid overcomplicating things. This can help improve the usability and efficiency of a design, as well as reduce costs and increase reliability. By keeping things simple, designers can make their products or services more accessible and user-friendly, which can lead to greater adoption and success.

5.9 (b)

Physical planning is an attempt to work out the implications of the development efforts in terms of factor allocation and product yields and maximize income and employment. It was quoted in the second five year plan(1956-1961).

Financial planning is the process of setting goals, creating a budget, and developing a plan to achieve those goals. It involves assessing your current financial situation, determining your short and long-term financial goals, and developing a strategy to achieve those goals.

Perspective planning is a long-term strategic planning process that involves developing a vision for the future and identifying the steps necessary to achieve that vision. It is often used by organizations, governments, and other entities to set goals and objectives that guide decision-making and resource allocation over a period of several years or even decades.

Indicative planning is a type of planning that involves setting broad goals and targets for economic and social development, without specifying in detail how those goals will be achieved. It is often used in situations where there is significant uncertainty about future trends and conditions, and where flexibility and adaptability are important.

5.10 (d)

Portfolio risk refers to the overall risk of a collection of investments held by an individual or an entity, such as a mutual fund or an exchange-traded fund (ETF). It is the risk of the entire portfolio declining in value due to market movements or other factors that affect the underlying investments.

Portfolio risk is a function of the risk of the individual investments in the portfolio and their correlations with one another. If the investments in the portfolio have a high positive correlation, then the portfolio is likely to be riskier than if the investments have a low correlation or are negatively correlated.

ADDITIONAL INFORMATION

Financial risk refers to the risk of loss that investors or businesses may face due to financial factors such as market conditions, creditworthiness, interest rates, liquidity, and currency fluctuations. Financial risk can arise from various sources such as investments, loans, bonds, and derivatives.

Business risk is the risk associated with a specific company's operations and its ability to generate profits or revenue. It is the risk of a company's earnings being impacted by internal or external factors, such as changes in the market, competition, economic conditions, management decisions, etc.

Investment risk is the potential for an investment to lose value or fail to meet its expected returns. All investments come with some degree of risk, and the level of risk varies depending on the type of investment, the market conditions, and other factors.

5.11 (b)

- A project Organisation Breakdown Structure (OBS) is a depiction of the project organization arranged to indicate the reporting relationships within the project context. The OBS reflects the way the project is functionally organized. It is a direct representation and description of the hierarchy and organizations that will provide resources to plan and perform work identified in the Work Breakdown Structure (WBS).
- The OBS helps management to focus on establishing the most efficient organization, by taking into consideration availability and capability of management and technical staff including subcontractors, to achieve project objectives.
- The OBS is depicted on the Responsibility Assignment Matrix (RAM), where it is used to identify the organization responsible and accountable for every element of the WBS and Scope of Work (SOW).

5.12 (b)

Top-down estimates is a technique used by upper-level management to estimate the total cost of a project by using information from a previous, similar project.

This approach is practical for the initial stage of strategic decision-making and in situations where the information required to develop accurate duration and costs estimates is not available in the initial phase of the project.

5.13 (a)

Forward pass scheduling is a technique used in project management to determine the earliest start and finish times for each activity in a project network diagram. It is a critical path method (CPM) technique used to calculate the project duration, identify the critical path, and determine the earliest completion time for a project.

The forward pass method involves the following steps:

1. Start by assigning an early start time of zero to the first activity in the project network diagram.
2. For each subsequent activity, calculate the early start time (EST) by adding the duration of the previous activity to its early finish time (EFT).
3. Calculate the early finish time (EFT) for each activity by adding its duration to its early start time (EST).
4. Continue this process until you reach the end of the project network diagram.
5. The earliest completion time for the project is the early finish time of the last activity in the project network diagram.

5.14 (c)

A normal project is a project that is completed using the normal time and resources required for each activity in the project. The project manager prepares a project plan with a timeline that includes the normal duration for each activity, and resources are allocated accordingly. Normal projects are typically completed within the planned schedule and budget.

A crash project, on the other hand, is a project that is completed in a shorter time than what is required in the normal project plan. This approach involves allocating additional resources, such as labor, equipment, or funds, to complete the project faster than the normal schedule. A crash project is used when there is a time constraint, such as a deadline that cannot be missed.

5.15 (b)

- A hammock or summary activity is used to gather together a number of sub-activities into one master activity.
- Hammock activities show the included sub-activities.
- Hammock activities are frequently used to identify the use of fixed resources or costs over a segment of project.

5.16 (a)

As per Krause, 2014, Mitigating risk involves two basic strategies : reducing the likelihood that the event will occur and/or reducing the impact that the adverse event would have on the project.

- The risk response plan development process considers risk transfer, reduction, avoidance and other mitigation tools to develop a risk response plan to ensure that the appropriate risk warning tools are in place, to handle risk efficiency.
- Proper risk management will attempt to reduce the probability of an event occurring and/or the magnitude of its impact as well as increase the probability of project success.

5.17 (c)

A buffer is extra time added to plan/schedule so that a task can be done and delays can be considered. The critical chain method has three different types of buffers, as:

1. **Resource Buffer :** It is inserted just before a critical chain activity where a critical resource is required. It is used to remind the project team that a resource is needed and to finish up prior activities.
2. **Feeding Buffer :** It is inserted as a safety margin in the non-critical chain of a network schedule. It is placed where the path feeds back into the critical chain path.
3. **Project Buffer :** It is the summation of all the internal buffers added to each project task or activity.

5.18 (d)

Heuristics do not always yield an optimal schedule, but they are very capable of yielding a "good" schedule for very complex networks with many types of resources. Since, each project is unique, it is good to test several set of heuristics on a network to determine the priority allocation rules that minimize project delay.

5.19 (d)

CCPM stands for critical-chain project management. The CCPM aims to eliminate project schedule delays due to uncertainties, overestimating task duration, etc.

5.20 (b)

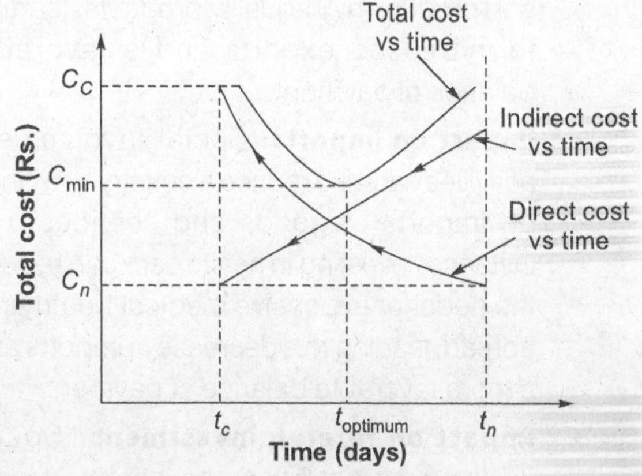

Total cost = Indirect cost + Direct cost

Any reduction in project duration, reduces the indirect cost but increases the direct cost.

6.1 (c)

Crashing is the method for shortening the project duration by reducing the time of one or more critical activities to less than their normal time.

ADDITIONAL INFORMATION

Schedule comparison refers to techniques used when a project manager wants to shorten the duration of the project without changing the scope of the project. It can be used when a project falls behind schedule and needs to catchup or to finish the project sooner than originally scheduled.

The two techniques that are used are fast tracking and crashing. Fast-tracking in project management refers to the practice of concurrently completing tasks that would normally be completed sequentially, while crashing is the technique to use when fast tracking has not saved enough time on the project schedule.

6.2 (c)

Social involvement refers to the participation of individuals, organizations, and governments in social and cultural activities such as education, health, and environmental protection. These activities can affect a country's international balance of payment in the following ways:

1. **Impact on exports:** Social involvement activities can improve the quality of goods and services produced in a country, making them more competitive in international markets. For example, investing in education can help develop a skilled workforce that can produce high-quality products, leading to increased exports and a favorable balance of payment.
2. **Impact on imports:** Social involvement activities can also reduce a country's reliance on imports of goods and services. For instance, investing in health care can reduce the need for expensive medical treatments abroad, leading to a decrease in imports and a more favorable balance of payment.
3. **Impact on foreign investment:** Social involvement activities can attract foreign investment to a country. For example, investing in environmental protection can make a country more attractive to foreign investors who are interested in socially responsible investments, leading to an increase in foreign investment and a more favorable balance of payment.

In summary, social involvement activities can have a positive impact

6.3 (d)

Product realisation combines market requirements, technological capabilities, and resources to define new product designs and the requisite manufacturing and field support processes. The relevance and viability of specific elements of the product realization process are determined by consideration related to the roles of

1. Customers, including channel and suppliers,
2. Technological feasibility, including information requirements and
3. Organization, including people, management, and the incentives and measures that affect productivity.

6.4 (c)

A machine shop or engineering workshop is a room, building, or company where machining, a form of subtroctive manufacturing, is done. In a machine shop, machinists use machine tools and cutting tools to make parts, usually of metal or plastic. The production consists of cutting, shaping drilling, finishing and other processes, frequently those related to metal working.

Machine shops often also contains the raw materials required for manufacturing the specific part, such as bar stocks. It is also stores an inventory of the parts that have been finished.

7.1 (b)

ABC analysis is based on principle of 'vital few and trivial many'. *ABC* analysis segregates the material items that are consumed in a production process/projects into three groups i.e. *A*-items, *B*-items and *C*-items.

***A*-items :** If an analysis is made on amount of money spent on different materials in a production process, then it can be seen that 70% to 75% of total money is spent on a few items of materials that accounts for only about 5% to 10% of total items of materials consumed. Such materials that form part of only 5% to 10% of total items of materials consumed, but that account for a major chunk of money spent are called *A*-items.

***B*-items :** These are those items that normally form part of about 10% to 15% of total number of items and accounts for about 10% of total expenditure on materials.

C-items : These are those items that normally form about 70% to 80% of total number of items but account for only 5% to 10% of total money spent.

For instance, in construction of an industrial shed,

(i) Materials like cement, steel angle sections are *A*-items.

(ii) Materials like steel rods and asbestos sheets etc. will be *B*-items.

(iii) Materials like sand, bricks, paint, wood etc. will be *C*-items.

ADDITIONAL INFORMATION

Economic order quantity (EOQ) is the ideal quantity of units a company should purchase to meet demand while minimising inventory costs such as holding costs, shortage costs, and order costs. This production-scheduling model was developed in 1913 by Ford W. Harries and has been refined over time. It is to be noted that economic order quantity formula assumes that demand, ordering and holding costs all remain constant.

7.2 (a)

Given,

Amount of each annuity payment,

P = Rs.5000

Discount rate, r = 10%

Number of years in which payment will be made,

n = 3 years

Now, $$\text{P.V.} = P\times\left(\frac{1-\left(\frac{1}{(1+r)^n}\right)}{r}\right) \quad \text{...(i)}$$

where, P.V. is present value of

Putting values in (i), we get,

$$\text{P.V.} = 5000\times\left(\frac{1-\left(\frac{1}{1+0.1}\right)^3}{0.1}\right) = 12434.26$$

So, correct answer is (a).

ADDITIONAL INFORMATION

An annuity is a financial product that provides a stream of payments to an individual over a period of time, typically in form of regular installments. Present value of annuity is an important concept for annuities because it allows individuals to compare the value of receiving a series of payments in future to value of receiving a lump sum payment today.

7.3 (c)

Let, the initial cost of 1 machine = x

Now, salvage value, S.V. = 40% of $x = 0.4x$

So, depreciation per annum = $\frac{x-0.4x}{5} = \frac{0.6x}{5}$

Now, accounting rate of return (%)

$$= \left(\frac{\text{Average annual profit after taxes}}{\text{Average investment over the whole life}\times 1000}\right) \quad \text{...(i)}$$

Now, average annual profit after taxes

$$= 50000 - \frac{0.6x}{5} \quad \text{...(ii)}$$

Also, average investment over whole life = x ...(iii)

Putting in equation (ii) and (iii), we get

$$0.2 = \frac{50000-\frac{0.6x}{5}}{x}$$

$$\Rightarrow \quad 0.2x + \frac{0.6x}{5} = 50000$$

$$\Rightarrow \quad 1.6x = 250000$$

$$\Rightarrow \quad x = 156250$$

So, total cost of 2 machines = 2 × 1562500

= Rs.312500

ADDITIONAL INFORMATION

The average rate or accounting rate of return is the average annual amount expected from an investment. The ARR formula divides an asset's average revenue by company's initial investment to derive the ratio or return that one may expect over the lifetime of an asset or project.

Key points:

1. ARR is helpful in determining the annual percentage rate of return of a project.
2. It is calculated as average annual profit/initial investment.
3. One of limitation of ARR is that it does not differentiate between investments that yield different cash flows over lifetime of a project.

7.4 (c)

Bid rigging : It is an illegal practice in which completing parties collude to determine winner of a bidding process. Bid rigging is a form of anti-competitive collusion and is an act of market manipulation. It undermines the bidding process and can result in a rigged price that is higher than what might have resulted from market competitive bidding.

Some key points :

1. Bid rigging is an illegal practice in which completing parties collued to determine the winner of a bidding process.
2. When bidders coordinate, it undermines the bidding process and can result in a rigged price that is higher than what might have resulted from a free market with competitive bidding process.
3. Some forms of bid rigging can be categorised as follows:
   (i) Predatory pricing is a method of pricing in which a seller sets a price which is so low that other sellers or suppliers cannot compete with him and they are forced to exit the market.
   (ii) Horizontal price fixing is an agreement between business, either explicitely or implicitely, to set the selling prices for a product or service. In this case, an agreement occurs between companies under same value chain level, for example, between producers, between wholesalers or between retailers. In other words, the agreement occurred between players who are actually in direct competition with each other.

7.5 (c)

Horizontal practice inside a firm is a practice where one company takes over another company or removes the competitors who are working at the same level in the industry. Predatory pricing is one of the examples of this strategy. It is also known as underrating. It is a strategy in which product or services are setup at very low prices. It is done to aim new customers through losses and remove competitors from the market. It also cut down the entry of new competitors in the market.

ADDITIONAL INFORMATION

Vertical practice inside a firm is a strategy where a company owns or have control of its supplier, retail location and distributors to control its supply chain. It allows them to control the process reduce the prices and increase efficiencies. However, it requires huge capital investment.

■■■■

Basics of Material Science and Engineering

CONTENTS

Basics of Material Science and Engineering

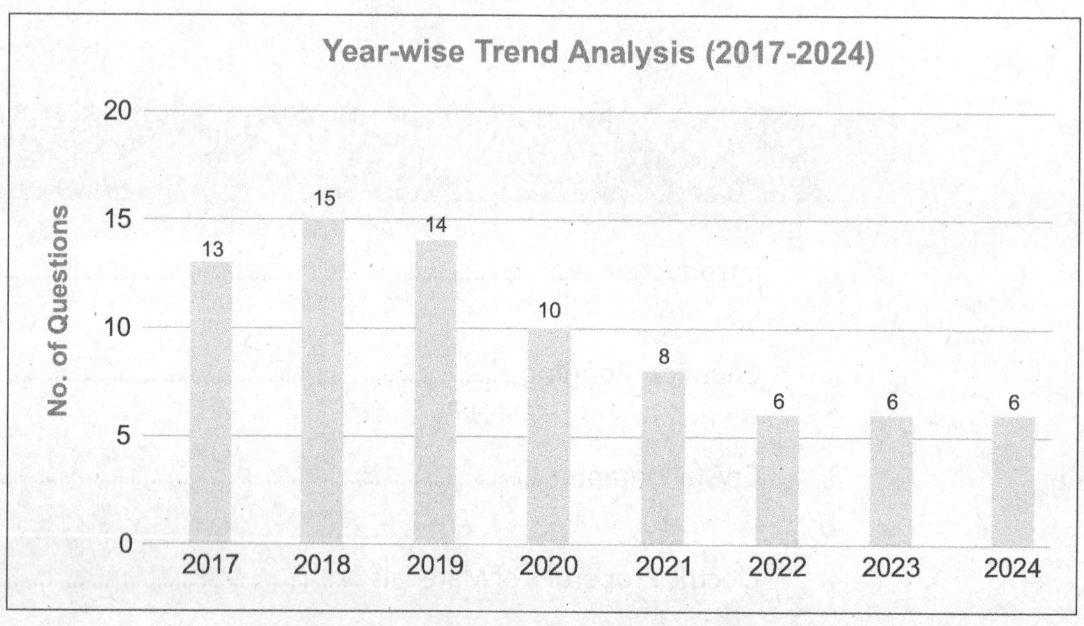

TOPIC-WISE TREND ANALYSIS (2017-2024)		
Sl. No.	Topics	Total Number of Questions
1.	Introduction	1
2.	Chemical Bonding	9
3.	Crystallography	9
4.	Electric Properties of Materials	9
5.	Magnetic Properties of Materials	4
6.	Ceramics	6
7.	Polymers	11
8.	Composites	3
9.	Mechanical Properties of Materials	10
10.	Ferrous Metals	6
11.	Non-Ferrous Metals and Alloys	6
12.	Basics of Corrosion, Creep and Heat Treatment	4

Basics of Material Science and Engineering

1. INTRODUCTION

1.1 Consider the following statements:
1. Material science deals with the strength and stiffness behaviour of components (buildings/machines/vehicle facilities) based on their response to imposed stresses (forces, moments, torque etc.).
2. Material properties are dependent on their micro-structure and response to force fields and surface interaction.

Which of the above statements is/are correct?
(a) 1 only (b) 2 only
(c) Both 1 and 2 (d) Neither 1 nor 2

[ESE Prelims : 2018]

2. CHEMICAL BONDING

2.1 In the case of ionic bonding, the molecule is stable as long as the number of bonding electrons is
(a) Equal to the number of antibonding electrons
(b) Less than the number of antibonding electrons
(c) Greater than the number of antibonding electrons
(d) Equal to the number of antibonding neutrons

[ESE Prelims : 2017]

2.2 Consider the following characteristics with respect to Alpha particles:
1. They have large specific ionization values.
2. They dissipate their energy rather slowly.
3. They can penetrate the outer layer of human skin.
4. Their emitters are heavy elements.

Which of the above statements are correct?
(a) 1 and 4 only (b) 1 and 3 only
(c) 2 and 4 only (d) 2 and 3 only

[ESE Prelims : 2018]

2.3 **Statement (I) :** Atoms can neither be created nor destroyed.

Statement (II) : Under similar conditions of temperature and pressure, equal volumes of gases do not contain an equal number of atoms.
(a) Both Statement (I) and Statement (II) are individually true and Statement (II) is the correct explanation of Statement (I)
(b) Both Statement (I) and Statement (II) are individually true, but Statement (II) is not the correct explanation of Statement (I)
(c) Statement (I) is true, but Statement (II) is false
(d) Statement (I) is false, but Statement (II) is true

[ESE Prelims : 2018]

2.4 In which one of the following types of bonds, the bond formation is by free moving electrons in an array of positive ions?
(a) Homopolar bond (b) Electrostatic bond
(c) Metallic bond (d) Covalent bond

[ESE Prelims : 2020]

2.5 Which of the following are the characteristics of covalent compounds?
1. They are mostly gases and liquids.
2. They are usually electric insulators.
3. They are directional in nature.
4. They are insoluble in polar solvents like water but are soluble in non-polar solvents.

(a) 1, 2 and 3 only (b) 1, 2 and 4 only
(c) 1, 3 and 4 only (d) 1, 2, 3 and 4

[ESE Prelims : 2020]

2.6 Match the following:

I

A. Thompson
B. James P. Joule
C. Max Planck
D. Albert Einstein

II
1. The concept of converting mechanical work into heat
2. The theory of relativity
3. The energy characteristics of light
4. The energy equivalence between heat, work and electric power

Select the correct matching using the code given below:

	A	B	C	D
(a)	3	4	1	2
(b)	1	4	3	2
(c)	3	2	1	4
(d)	1	2	3	4

[ESE Prelims : 2021]

2.7 Stereoisomerism denotes the situation in which atoms are linked together

(a) in the different order and also differ in their spatial arrangement.
(b) in the different order but same in their spatial arrangement.
(c) in the same order (head-to-tail) but differ in their spatial arrangement.
(d) in the same order (head-to-tail) and also same in their spatial arrangement.

[ESE Prelims : 2022]

2.8 Consider the following statements regarding metallic bonding :

1. The metallic sharing changes with time and the bonding electrons resonate between different atoms.
2. The metallic state can be visualized as an array of positive ions, with a common pool of electrons to which all the metal atoms have contributed their outer electrons.
3. These electrons have freedom to move anywhere within the crystal and act like an all-pervasive, mobile glue holding the ion cores together.

Which of the above statements are correct?

(a) 1 and 2 only (b) 1 and 3 only
(c) 2 and 3 only (d) 1, 2 and 3

[ESE Prelims : 2024]

2.9 Consider the following statements regarding the characteristics of covalent compounds and covalent solids:

1. Covalent compounds are soluble in paraffins.
2. Covalent solids do not form closed-packet structures because the covalent bonds are very strong and rigid.
3. The simplest covalent structure is that of diamond which is fairly open and empty and far from close-packed.

Which of the above statements are correct?

(a) 1 and 2 only (b) 1 and 3 only
(c) 2 and 3 only (d) 1, 2 and 3

[ESE Prelims : 2024]

3. CRYSTALLOGRAPHY

3.1 What is the volume of an FCC unit cell in terms of its atomic radius R?

(a) $\sqrt{3}R^3$ (b) $16R^3\sqrt{2}$
(c) $16R^3\sqrt{3}$ (d) $\sqrt{2}R^3$

[ESE Prelims : 2018]

3.2 The number of atoms per unit length whose centres lie on the direction vector for a specific crystallographic direction is called

(a) Linear density (b) Theoretical density
(c) Atomic density (d) Avogadro number

[ESE Prelims : 2019]

3.3 A state for ionic compounds wherein there is the exact ratio of cations to anions as predicted by the chemical formula is

(a) Electroneutrality (b) Stoichiometry
(c) Equiliometry (d) Frankel defect

[ESE Prelims : 2019]

3.4 In a simple cubic structure, atomic packing factor is nearly

(a) 0.9 (b) 0.7
(c) 0.5 (d) 0.3

[ESE Prelims : 2019]

3.5 If a pair of one cation and one anion is missing in an ionic crystal such that those pairs of ions are equal to maintain electrical neutrality, then that pair of vacant sites is called

(a) Schottky imperfection
(b) Pair of vacancies
(c) Frenkel defect
(d) Point imperfection

[ESE Prelims : 2020]

3.6 Which one of the following factors does NOT characterize the formation of non-crystalline structure?
(a) Presence of primary bonds in the directions
(b) Non-formation of three-dimensional primary bond
(c) Weak secondary bond
(d) Open network of the atomic packing
[ESE Prelims : 2021]

3.7 Which one of the following is NOT a classification of microscopic diffusion?
(a) Inter-diffusion (b) Vacancy diffusion
(c) Surface diffusion (d) Lattice diffusion
[ESE Prelims : 2022]

3.8 When a pair of one cation and one anion are absent from an ionic crystal, then the defect is called
(a) Schottky's defect (b) Frenkel's defect
(c) Cross-slip defect (d) Stacking defect
[ESE Prelims : 2023]

3.9 The density of $\alpha-$Fe is 7.87×10^3 kg/m^3. Atomic weight of Fe is 55.8. If $\alpha-$Fe crystallises in BCC space lattice, what is the lattice constant approximately? (Take Avogadro's number (N) = 6.02×10^{26}/kg/mole and number of atoms per unit cell is 2)
(a) 0.666 Å (b) 1.766 Å
(c) 2.866 Å (d) 3.966 Å
[ESE Prelims : 2023]

4. ELECTRIC PROPERTIES OF MATERIALS

4.1 A resistor measures 4 Ω at 40°C and 6Ω at 80°C. At 0°C the resistor will measure
(a) 1.5 Ω (b) 2 Ω
(c) 3 Ω (d) 4 Ω
[ESE Prelims : 2017]

4.2 The Hall effect may be used to
1. Determine whether the semiconductor is *p*-type or *n*-type.
2. Determine the carrier concentration.
3. Calculate the mobility.
Which of the above statements are correct?
(a) 1 and 2 only (b) 1 and 3 only
(c) 2 and 3 only (d) 1, 2 and 3
[ESE Prelims : 2018]

4.3 The photoelectric current depends on which of the following factors?
1. The frequency of the incident light.
2. The intensity of the incident light.
3. The potential difference between the electrodes.
4. The photosensitivity of the non-metal.
(a) 1, 2 and 4 (b) 1, 2 and 3
(c) 1, 3 and 4 (d) 2, 3 and 4
[ESE Prelims : 2020]

4.4 An FET is a semiconductor device with the output current controlled by an electric field and its current is carried predominantly by one type of carriers. It is known as
(a) junction transistor (b) unipolar transistor
(c) MOSFET (d) IGBT
[ESE Prelims : 2020]

4.5 For a semiconductor to be called as p-type semiconductor, which one of the following element impurities are added to a pure semiconductor?
(a) Phosphorus (b) Arsenic
(c) Antimony (d) Boron
[ESE Prelims : 2021]

4.6 Impure semiconductor
(a) has more conductivity in contrast to pure semiconductor
(b) has less conductivity in contrast to pure semiconductor
(c) has electrons and holes in equal number
(d) has a fermi level which is in the centre of conduction and valence bands
[ESE Prelims : 2021]

4.7 Which one of the following is the disadvantages of ion-implantation over diffusion doping?
(a) It is a low temperature process
(b) Point imperfections are not produced
(c) Shallow doping is possible
(d) Gettering is possible
[ESE Prelims : 2021]

4.8 Which one of the following is correct in n-p-n transistor?
(a) Collector and emitter terminals can be exchanged

(b) Collector is heavily doped, base width is small and emitter area is large
(c) Emitter, base and collector regions are equally doped
(d) Emitter is heavily doped, base width is small and collector area is large

[ESE Prelims : 2021]

4.9 'Positive and negative' ions by virtue of their net electrical charge, attract one another', these attractive bonding forces are

(a) Coulombic (b) Magnetic
(c) Electromagnetic (d) Non-magnetic

[ESE Prelims : 2022]

5. MAGNETIC PROPERTIES OF MATERIALS

5.1 Consider the following statements regarding water:

1. Water is a polar molecule
2. Water molecules align themselves in an electric field
3. Water molecules vary their alignment if the applied electric polarity changes with time
4. Water molecules align themselves in a magnetic field

Which of the above statements are correct?

(a) 1, 2 and 3 only (b) 1, 2 and 4 only
(c) 3 and 4 only (d) 1, 2, 3 and 4

[ESE Prelims : 2017]

5.2 On which of the following factors does hysteresis loss depend?

1. Magnetic field intensity
2. Frequency of the field
3. Volume of material
4. Neel temperature

(a) 1, 2 and 4 only (b) 1, 3 and 4 only
(c) 2, 3 and 4 only (d) 1, 2 and 3 only

[ESE Prelims : 2018]

5.3 The critical temperature above which ferromagnetic materials lose their magnetic property is called

(a) Kelvin point
(b) Curie point
(c) Recrystallization point
(d) Celsius point

[ESE Prelims : 2018]

5.4 Which one of the following types of materials is having high remanence, coercivity and saturation flux density as well as low permeability and high hysteresis energy losses?

(a) Soft magnetic materials
(b) Hard magnetic materials
(c) Hard electrical materials
(d) Soft electrical materials

[ESE Prelims : 2020]

6. CERAMICS

6.1 Consider the following assumptions made while developing the ionic packing theory:

1. Cations and anions are spherical but these spheres are not hard
2. Cations are always smaller than anions
3. Each cation would tend to be surrounded by the maximum number of anions permitted by geometry
4. Cations and anions do not touch each other

Which of the above assumptions are correct?

(a) 1 and 3 (b) 1 and 4
(c) 2 and 4 (d) 2 and 3

[ESE Prelims : 2017]

6.2 Consider the following statements:

1. Baking soda is used in fire extinguishers
2. Quick lime is used in the manufacture of glass

Which of the above statements is/are correct?

(a) 1 only (b) 2 only
(c) Both 1 and 2 (d) Neither 1 nor 2

[ESE Prelims : 2017]

6.3 **Statement (I) :** Metal carbides and carbon are used as refractories as they resist oxidation.

Statement (II) : Metal carbides and carbon are not particularly suitable for high temperature applications.

(a) Both Statement (I) and Statement (II) are individually true; and Statement (II) is the correct explanation of Statement (I)
(b) Both Statement (I) and Statement (II) are individually true; but Statement (II) is NOT the correct explanation of Statement (I)
(c) Statement (I) is true; but Statement (II) is false
(d) Statement (I) is false; but Statement (II) is true

[ESE Prelims : 2018]

6.4 **Statement (I) :** Abrasive ceramics are used to wear, grind, or cut away other material, which necessarily is softer.

Statement (II) : The prime requisite for abrasive ceramic group of materials is hardness or wear resistance and a high degree of toughness is essential to ensure that the abrasive particles do not easily fracture.

(a) Both Statement (I) and Statement (II) are individually true; and Statement (II) is the correct explanation of Statement (I)
(b) Both Statement (I) and Statement (II) are individually true; but Statement (II) is NOT the correct explanation of Statement (I)
(c) Statement (I) is true; but Statement (II) is false
(d) Statement (I) is false; but Statement (II) is true

[ESE Prelims : 2019]

6.5 In a stable ceramic crystal structure, a cation is surrounded by three anions in the form of a planar equilateral triangle. The ratio of the cation-anion radius for the crystal is nearly

(a) 0.16 (b) 0.24
(c) 0.32 (d) 0.41

[ESE Prelims : 2019]

6.6 **Statement (I):** The slip mode of deformation is the common mode in only one crystal at ambient and elevated temperature.

Statement (II): A slip plane and a slip direction that lies on it together constitute a slip system.

(a) Both Statement (I) and Statement (II) are individually true; and Statement (II) is the correct explanation of Statement (I)
(b) Both Statement (I) and Statement (II) are individually true; but Statement (II) is NOT the correct explanation of Statement (I)
(c) Statement (I) is true; but Statement (II) is false
(d) Statement (I) is false; but Statement (II) is true

[ESE Prelims : 2021]

7. POLYMERS

7.1 Phenol and formaldehyde are polymerized to a resultant product known as

(a) PVC (b) Bakelite
(c) Polyester (d) Teflon

[ESE Prelims : 2017]

7.2 The reason for low rate of plastic recycling are:

1. Scrap plastic has little value because virgin material is rather cheap.
2. Low density of plastic leads to high shipping and handling cost.
3. Recycling leads to a very severe public health hazard.

Which of the above reasons are correct?

(a) 1, 2 and 3 (b) 1 and 2 only
(c) 1 and 3 only (d) 2 and 3 only

[ESE Prelims : 2017]

7.3 How much sulphur is required per 100 kg of final rubber product to completely crosslink butadiene rubber?

(a) 17 kg (b) 27 kg
(c) 37 kg (d) 47 kg

[ESE Prelims : 2018]

7.4 **Statement (I) :** Long chain polymers are weaker than most ceramics and metals.

Statement (II) : The molecular chains in long chain polymers are bonded to each other with Van der Waals bonds.

(b) Both Statement (I) and Statement (II) are individually true, but Statement (II) is not the correct explanation of Statement (I)
(c) Statement (I) is true, but Statement (II) is false
(d) Statement (I) is false, but Statement (II) is true

[ESE Prelims : 2018]

7.5 Consider the following Repeat Unit Structure:

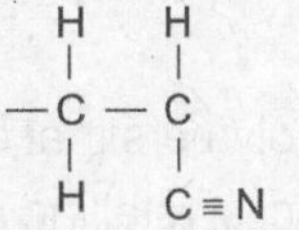

What is the above polymer?

(a) Poly(amide–imide)
(b) Polyacrylonitrile
(c) Polybutadiene
(d) Polyethylene

[ESE Prelims : 2019]

7.6 When two or more chemically different monomers are polymerized to form a cross link polymer along with some byproduct such as water, the process is known as

(a) Crystallographic polymerization
(b) Addition polymerization
(c) Copolymerization
(d) Condensation polymerization

[ESE Prelims : 2019]

7.7 **Statement (I)** : Cross linked polymers may be synthesized in which side-branch chains are connected to the main ones.

Statement (II) : Linear polymers are those in which the repeat units are joined together end to end in single chains.

(a) Both Statement (I) and Statement (II) are individually true; and Statement (II) is the correct explanation of Statement (I)
(b) Both Statement (I) and Statement (II) are individually true; but Statement (II) is NOT the correct explanation of Statement (I)
(c) Statement (I) is true; but Statement (II) is false
(d) Statement (I) is false; but Statement (II) is true

[ESE Prelims : 2019]

7.8 Polymers having strong primary bonds throughout, often formed by condensation polymerization, and their structure resembles one large molecule, are known as
(a) thermoplastic polymers
(b) thermosoftening polymers
(c) thermosetting polymers
(d) random polymers

[ESE Prelims : 2020]

7.9 Which one of the following is the trade name of polycarbonates?
(a) Alathon (b) Baylon
(c) Bexphane (d) Cycolac

[ESE Prelims : 2020]

7.10 Many bulk polymers that are crystallized from a melt, are semi crystalline and form which one of the following structures?
(a) Spherolite structure
(b) Spherelite structure
(c) Spherulite structure
(d) Spherilite structure

[ESE Prelims : 2022]

7.11 Some of the complex thermoplastic chains become so stiff that they act as rigid rods, even when heated above the melting point. These materials are
(a) Solid crystalline polymers
(b) Semi solid crystalline polymers
(c) Liquid crystalline polymers
(d) Copolymers **[ESE Prelims : 2022]**

8. COMPOSITES

8.1 Pneumatic structures are:
1. Membrane structures.
2. Stabilized by compressed air.
3. Characterized by internal pressure being less than external pressure.
4. Characterized by internal pressure being equal to or more than external pressure.

Select the correct answer using the codes given below:
(a) 1, 2 and 3 (b) 1, 2 and 4
(c) 2 and 4 only (d) 1 and 4 only

[ESE Prelims : 2017]

8.2 Consider the following statements:
1. Asphalt is a naturally occurring product having non-crystalline structure.
2. Rock Asphalt occurs in limestones or sandstones.
3. Tar is a residue left after distillation of crude oil.
4. Resins in Asphalt provide the stickiness to the product.

Which of the above statements are correct?
(a) 1, 2 and 3 only (b) 3 and 4 only
(c) 1, 2 and 4 only (d) 1, 2, 3 and 4

[ESE Prelims : 2017]

8.3 **Statement (I):** Physical properties of composite materials are generally isotropic.

Statement (II): The stiffness of composite panel will often depend upon the orientation of the applied forces and/or moments.

(a) Both Statement (I) and Statement (II) are individually true and Statement (II) is the correct explanation of Statement (I)
(b) Both Statement (I) and Statement (II) are individually true but Statement (II) is not the correct explanation of Statement (I)
(c) Statement (I) is true but Statement (II) is false
(d) Statement (I) is false but Statement (II) is true

[ESE Prelims : 2017]

9. MECHANICAL PROPERTIES OF MATERIALS

9.1 Fatigue resistance of materials can be improved by:
1. Shot peening
2. Polishing the surface

Which of the above is/are correct?
(a) 1 only (b) 2 only
(c) Both 1 and 2 (d) Neither 1 nor 2
[ESE Prelims : 2017]

9.2 **Statement (I):** Fatigue failure is a stress which is lower than the yield strength of a metal.
Statement (II): Repeated straining lowers the yield strength.
(a) Both Statement (I) and Statement (II) are individually true and Statement (II) is the correct explanation of Statement (I)
(b) Both Statement (I) and Statement (II) are individually true but Statement (II) is not the correct explanation of Statement (I)
(c) Statement (I) is true but Statement (II) is false
(d) Statement (I) is false but Statement (II) is true
[ESE Prelims : 2017]

9.3 During tensile testing of a material, if cross-sectional area of the specimen is doubled, the load required to produce the same elongation shall be
(a) Double (b) Half
(c) Same (d) Four times
[ESE Prelims : 2019]

9.4 The capacity of a material to absorb energy when it is deformed elastically and then, upon unloading, to have this energy recovered is called
(a) Ductility (b) Tensile strength
(c) Elasticity (d) Resilience
[ESE Prelims : 2019]

9.5 Which one of the following statements is correct regarding ductile fracture?
(a) Fractured surfaces are crystalline in appearance.
(b) There is virtually no reduction in cross-sectional area during fracture,
(c) Fracture takes place after necking with little sound.
(d) Percentage elongation is about 60% prior to fracture occurs.
[ESE Prelims : 2020]

9.6 Which of the following factors are affecting critical shear stresses?
1. Purity of metals reduces the critical shear stress.
2. Surface films greatly enhance the critical shear stress.
3. Rise in temperature.
4. Rate of deformation and the extent of initial deformation also help in raising the critical shear stress.

(a) 1, 2 and 3 only (b) 1, 2 and 4 only
(c) 1, 3 and 4 only (d) 1, 2, 3 and 4
[ESE Prelims : 2020]

9.7 The process by which plastic deformation is produced by dislocation motion is termed as
(a) Plane slit (b) Seepage
(c) Slip (d) Twinning
[ESE Prelims : 2022]

9.8 A relatively large plate of a glass is subjected to a tensile stress of 40 MPa. If the specific surface energy and modulus of elasticity for this glass are 0.3 J/m^2 and 69 GPa, respectively, what is approximate maximum length of a surface flaw that is possible without fracture?
(a) 6.2 μm (b) 8.2 μm
(c) 10.2 μm (d) 12.2 μm
[ESE Prelims : 2023]

9.9 A piece of copper originally 305 mm long is pulled in tension with a stress of 276 MPa. If the deformation is entirely elastic, what is the resultant elongation approximately?
(a) 3.3 mm (b) 0.33 mm
(c) 0.77 mm (d) 7.7 mm
[ESE Prelims : 2023]

9.10 What is the approximate value of ductility (%EL) of a cylindrical copper rod if it is cold worked such that the diameter is reduced from 15.2 mm to 12.2 mm? (Take the tensile strength from the curve for copper as 340 MPa).
(a) 7% (b) 3.56%
(c) 70% (d) 35.6%
[ESE Prelims : 2023]

10. FERROUS METALS

10.1 **Statement (I):** Manganese is always added to steels since it combines with the sulphur content to form manganese sulphide.

Statement (II): If manganese is not added, iron sulphide which is not harmful for steel, would form.

(a) Both Statement (I) and Statement (II) are individually true and Statement (II) is the correct explanation of Statement (I)
(b) Both Statement (I) and Statement (II) are individually true but Statement (II) is not the correct explanation of Statement (I)
(c) Statement (I) is true but Statement (II) is false
(d) Statement (I) is false but Statement (II) is true

[ESE Prelims : 2017]

10.2 Malleable cast iron is produced

1. By quick cooling of cast iron.
2. By adding magnesium to molten cast iron.
3. From white cast iron by annealing.

Which of the above statements is/are correct?

(a) 1 only (b) 2 only
(c) 3 only (d) 1, 2 and 3

[ESE Prelims : 2018]

10.3 **Statement (I)** : Mechanically, pearlite has properties intermediate between the soft ductile ferrite and the hard brittle cementite.

Statement (II) : Alpha iron can be made magnetic above 768°C.

(a) Both Statement (I) and Statement (II) are individually true and Statement (II) is the correct explanation of Statement (I)
(b) Both Statement (I) and Statement (II) are individually true, but Statement (II) is not the correct explanation of Statement (I)
(c) Statement (I) is true, but Statement (II) is false
(d) Statement (I) is false, but Statement (II) is true

[ESE Prelims : 2018]

10.4 The purpose of oil in a transformer is to

(a) protect the transformer from rusting
(b) avoid wear and tear of the transformer
(c) transfer heat from winding and core to the cooling surfaces of the transformer
(d) avoid noise in a transformer

[ESE Prelims : 2021]

10.5 Consider the following statements regarding effect of minor elements on steel properties:

1. Sulphur is present in steel either as iron sulphide or manganese sulphide and during the rolling or forging of steel, iron sulphide present in steel gets cracked/teared.
2. Silicon, in the form of ferro-silicon, is used widely as a deoxidant due to its low cost and high efficiency.
3. Silicon opposes the presence of iron oxide (FeO) which is very much detrimental to properties of steel.

Which of the above statements are correct?

(a) 1 and 2 only (b) 1 and 3 only
(c) 2 and 3 only (d) 1, 2 and 3

[ESE Prelims : 2024]

10.6 Consider the following statements regarding cast iron:

1. Gray cast irons can be classified depending on the shape of graphite that may be present in the form of either flakes or globules.
2. A class of cast iron known as Malleable obtained by treating molten metal by calcium silicide.
3. Meehanite cast irons have graphite nodules but are produced by heat treating white cast irons.

Which of the above statements are NOT correct?

(a) 1 and 2 only (b) 1 and 3 only
(c) 2 and 3 only (d) 1, 2 and 3

[ESE Prelims : 2024]

11. NON-FERROUS METALS AND ALLOYS

11.1 The material used in the production of bearings is

(a) Cast iron (b) Babbitt metal
(c) Pig iron (d) Steel

[ESE Prelims : 2018]

11.2 **Statement (I) :** High strength, super-duralumin alloys are adopted in the manufacture of aero engines.

Statement (II) : Precipitation heat treatment is adopted for duralumin products.

(a) Both Statement (I) and Statement (II) are individually true; and Statement (II) is the correct explanation of Statement (I)
(b) Both Statement (I) and Statement (II) are individually true; but Statement (II) is NOT the correct explanation of Statement (I)

(c) Statement (I) is true; but Statement (II) is false
(d) Statement (I) is false; but Statement (II) is true

[ESE Prelims : 2018]

11.3 Which of the following features of atoms determine the degree to which the solute atoms dissolve in the solvent atoms?
1. Atomic size factor
2. Crystal structure
3. Electronegativity

(a) 1 and 2 only (b) 1 and 3 only
(c) 2 and 3 only (d) 1, 2 and 3

[ESE Prelims : 2019]

11.4 In which one of the following phase transformations, there are no compositional alterations?
(a) Incongruent transformations
(b) Congruent transformations
(c) Non-equilibrium transformations
(d) Equilibrium transformations

[ESE Prelims : 2019]

Directions : Each of the next items consists of two statements, one labelled as 'Statement (I)' and the other as 'Statement (II)'. You are to examine these two statements carefully and select the answers to these items using the codes given below:

Codes:
(a) Both Statement (I) and Statement (II) are individually true; and Statement (II) is the correct explanation of Statement (I)
(b) Both Statement (I) and Statement (II) are individually true; but Statement (II) is NOT the correct explanation of Statement (I)
(c) Statement (I) is true; but Statement (II) is false
(d) Statement (I) is false; but Statement (II) is true

11.5 **Statement (I)** : The tie line is constructed across the two-phase region at the temperature of the alloy.
Statement (II) : The overall alloy composition is located on the tie line.

[ESE Prelims : 2019]

11.6 **Statement (I)** : Metals having same crystal structure will have greater solubility.
Statement (II) : Differences in crystal structure limits the solid solubility.

[ESE Prelims : 2019]

12. BASICS OF CORROSION, CREEP & HEAT TREATMENT

12.1 Which one of the following is termed as sacrificial protection of metal?
(a) Galvanization
(b) Tinning
(c) Organic coating
(d) Inorganic coating

[ESE Prelims : 2018]

12.2 The diffusion coefficient for copper in aluminium at 500°C and 600°C are $4.8 \times 10^{-14} m^2/s$ and $5.3 \times 10^{-3} m^2/s$ respectively. What is the approximate time at 500°C that will produce the same diffusion result (in terms of concentration of copper at some specific point in aluminium) as 10 h heat treatment at 600°C?
(a) 110.4 h (b) 152.4 h
(c) 210.4 h (d) 252.4 h

[ESE Prelims : 2023]

12.3 Consider the following statements regarding phase diagram rules for the crystal:
1. According to the Gibbs phase rule, the degree of freedom (F) = Number of components (C) + Number of phases (P) + 2.
2. The tie-line rule is applied to determine the compositions of two-coexisting phases in a binary phase diagram.
3. In the lever rule, the tie-line at the temperature of interest is treated as a lever arm, with the fulcrum at the overall composition.

Which of the above statement are correct?
(a) 1 and 2 only (b) 1 and 3 only
(c) 2 and 3 only (d) 1, 2 and 3

[ESE Prelims : 2024]

12.4 Consider the following statements regarding the gas carburizing.
1. Case depth can be obtained accurately.
2. More floor space is required than pack carburizing.
3. Process is rapid as less time is required than in pack carburizing.

Which of the above statements are correct?
(a) 1 and 2 only (b) 1 and 3 only
(c) 2 and 3 only (d) 1, 2 and 3

[ESE Prelims : 2024]

■■■■

Basics of Material Science and Engineering : Answer Key

1.1 (b)	2.1 (c)	2.2 (a)	2.3 (c)	2.4 (c)	2.5 (d)	2.6 (b)	2.7 (c)
2.8 (d)	2.9 (b)	3.1 (b)	3.2 (a)	3.3 (b)	3.4 (c)	3.5 (a)	3.6 (a)
3.7 (a)	3.8 (a)	3.9 (c)	4.1 (b)	4.2 (d)	4.3 (b)	4.4 (c)	4.5 (d)
4.6 (a)	4.7 (d)	4.8 (d)	4.9 (a)	5.1 (a)	5.2 (d)	5.3 (b)	5.4 (b)
6.1 (d)	6.2 (c)	6.3 (d)	6.4 (a)	6.5 (a)	6.6 (a)	7.1 (b)	7.2 (b)
7.3 (c)	7.4 (a)	7.5 (b)	7.6 (d)	7.7 (d)	7.8 (c)	7.9 (b)	7.10 (c)
7.11 (c)	8.1 (b)	8.2 (c)	8.3 (d)	9.1 (c)	9.2 (a)	9.3 (a)	9.4 (d)
9.5 (c)	9.6 (d)	9.7 (c)	9.8 (b)	9.9 (c)	9.10 (d)	10.1 (c)	10.2 (c)
10.3 (c)	10.4 (c)	10.5 (d)	10.6 (c)	11.1 (b)	11.2 (b)	11.3 (d)	11.4 (b)
11.5 (b)	11.6 (a)	12.1 (a)	12.2 (a)	12.3 (c)	12.4 (b)		

EXPLANATIONS

1.1 (b)

Material properties are dependent on their micro structure and response to force fields and surface interaction.

ADDITIONAL INFORMATION

Materials science deals with how a material changes with time (moves from non-equilibrium to equilibrium state) due to application of a certain field. It details the rate of various processes evolving in materials including shape, size composition and structure.

2.1 (c)

In the case of ionic bonding, the molecule is stable as long as the number of bonding electrons is equal to the number of antibonding electrons.

ADDITIONAL INFORMATION

Ionic bonding is the complete transfer of valence electron(s) between atoms and is a type of chemical bond that generates two oppositely charged ions.

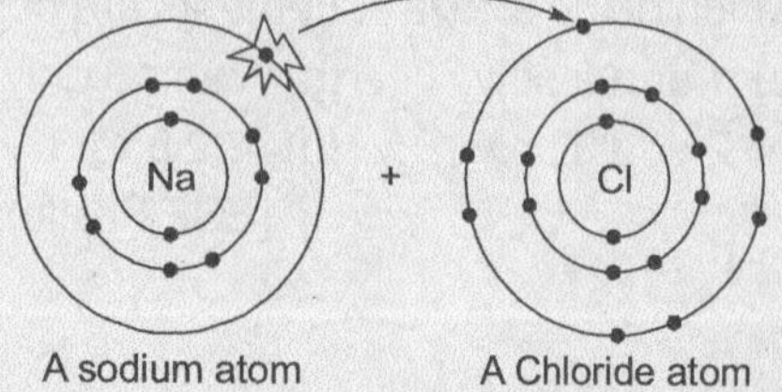

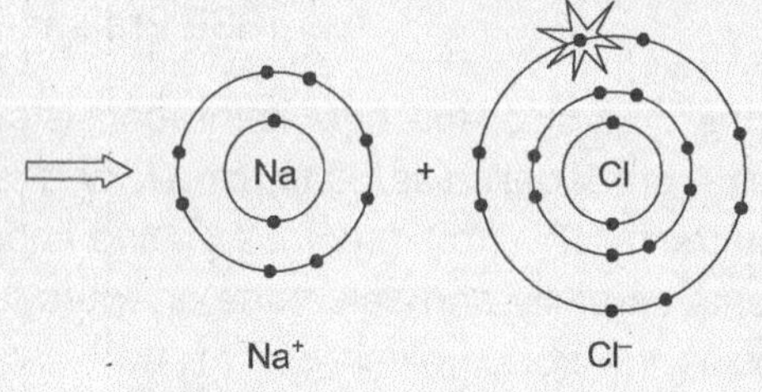

2.2 (a)

Characteristics of Alpha particles:

1. Alpha particles have large specific ionization values.
2. Alpha particles emitted by emitters are heavy elements.

ADDITIONAL INFORMATION

- Alpha particles consists of 2 protons and 2 neutrons.
- Mass of an alpha particle is 4 amu.
- Have low penetrating abilities.

2.3 (c)

Atoms can neither be created nor destroyed under similar conditions of temperature and pressure, equal volumes of gases contain an equal number of atoms.

ADDITIONAL INFORMATION

According to the Berzelius hypothesis, an equal volume of all gases under similar conditions of temperature and pressure contains an equal number of atoms.

2.4 (c)

Metallic bond is formed when atoms give up their electrons and become positive ions. So the bond is formed between free moving electrons and positive ions.

ADDITIONAL INFORMATION

- Covalent bonds also known as homopolar bonds.
- Electrostatic bonds are chemical bonds characterized by electrostatic attraction between ions and molecules.

2.5 (d)

Covalent compound: Whose atoms are bonds via a covalent bond.

Example:

- Carbon compounds are the most appropriate example of a covalent compound.
- Others are methane (CH_4), hydrogen (H_2), oxygen (O_2) etc.

ADDITIONAL INFORMATION

Covalent compounds are usually gaseous molecules at room temperature and pressure. They might also be liquids with low relatively low boiling points.

2.6 (b)

Thompson: The concept of converting mechanical work into heat.

James P.Joule: The energy equivalence between heat, work and electric power.

Max plank: The energy characteristics of light.

Albert Einstein : The theory of relativity.

2.7. (c)

Stereoisomerism: It denotes the situation in which atoms are linked together in the same order (head-to-tail) but differ in their spatial arrangement. For one stereoisomer, all the *R* groups are situated on the same side of the chain as follows:

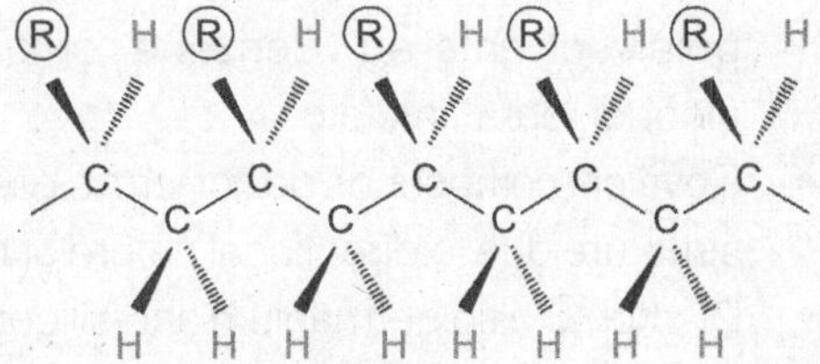

ADDITIONAL INFORMATION

Compounds which have their atoms connected in the same order but differ in 3-D orientation.

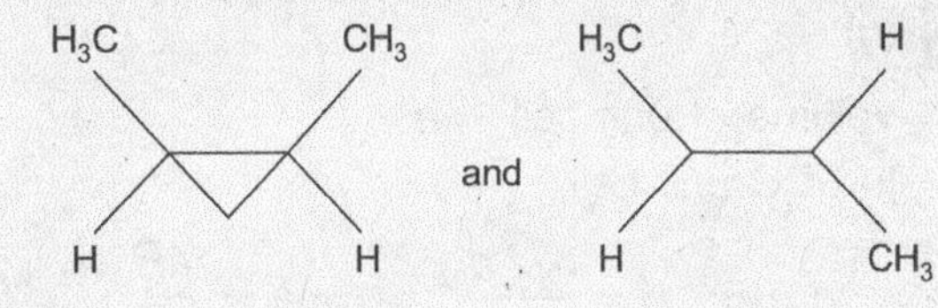

2.8 (d)

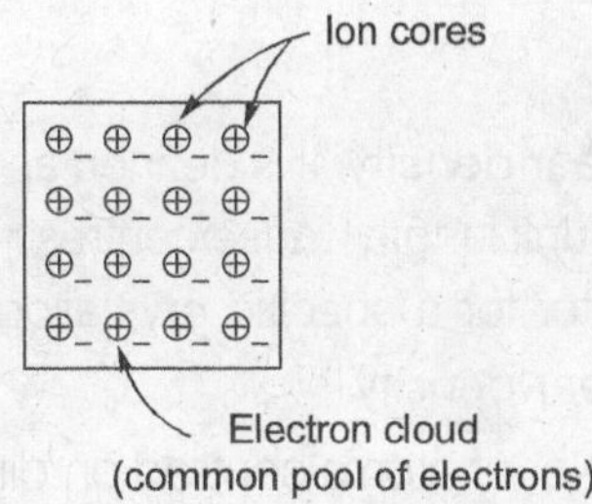

- In metallic bonding, the valence electrons are not bound to any particular atom in the solid and are more or less free to drift throughout the entire metal. They may be thought of belonging to the metal as a whole, or forming a "sea of electrons" or an "electron cloud".
- The free electrons shield the positively charged ion cores from mutually repulsive electrostatic forces. In addition, these electrons acts as a "glue" to hold the ion cores together.

In metallic bonding, the atoms can share the required number of electrons with their neighbours through the common pool to complete their octets. This sharing is not localized. The free nature of electron gas makes the bonding electrons resonate between different atoms and consequently, the mettalic sharing changes with time.

2.9 (b)

- Covalent compounds are insoluble in polar solvents like water but are soluble in non-polar solvents eg. Benzene, chloroform, alcohol, paraffins, etc.
- Covalent compounds do not form closed pack structure due to directional nature of bond.
- Diamond (also germanium and silicon) is the example of the simplest covalent structure, which is fairly open and empty and far from close-packed.

Hence, only statements 1 and 3 are correct.

3.1 (b)

Volume of unit cell = a^3

In FCC unit cell,

$$R = \frac{\sqrt{2}a}{4} \quad \Rightarrow a = \frac{4R}{\sqrt{2}}$$

$$\therefore \quad \text{Volume} = a^3 = \left(\frac{4R}{\sqrt{2}}\right)^3 = 16\sqrt{2}R^3$$

3.2 (a)

Linear density: It is defined as number of atoms per unit length whose centres lie on the direction vector for a specific crystallographic direction.

Linear density

$$= \frac{\text{No. of atoms centred on direction vector}}{\text{Length of direction vector}}$$

ADDITIONAL INFORMATION

- **Atomic density:** If atoms lose electrons, then their volume decreases and they have to give up energy to rebalance their atomic density.
- **Avagadro number:** The number of particles in a mole is called Avagadro number.

$$N_A = 6.02 \times 10^{23} \text{ particles/mole}$$

Mass of an individual atom can be calculated as

$$M_{atom} = \frac{\text{molar mass}}{N_A}$$

3.3 (b)

Stoichiometry may be defined as a state for any compounds where in there is the exact ratio of cations to anions as predicted by the chemical formula.

ADDITIONAL INFORMATION

- **Electroneutrality:** The principle of electroneutrality means that the overall number of positive and negative charges balances.
- **Frenkel defect:** When an atom or a small ion (usually a cation) leaves its place in the cell, it creates space and becomes a meeting place by staying close to one another.

3.4 (c)

Atomic packing factor

$$= \frac{\text{Sum of atomic volume in a unit cell}}{\text{Volume of a unit cell}}$$

$$\Rightarrow \quad \text{APF} = \frac{N \times \frac{4}{3}\pi r^3}{a^3}$$

For simple cubic,

$$N = \frac{1}{8} \times 8 = 1 \text{ atom/unit cell} \; ; \quad r = \frac{a}{2}$$

$$\therefore \quad \text{APF} = \frac{1 \times \frac{4}{3}\pi\left(\frac{a}{2}\right)^3}{a^3} = 0.52$$

ADDITIONAL INFORMATION

For simple cubic structure

Atom per unit cell : 1

Lattice constant, $a = 2r$

Co-ordination number, $CN = 6$

3.5 (a)

Schottky defect occurs when a pair of cation and anion is missing from crystal such that electrical neutrality is maintained.

ADDITIONAL INFORMATION

Defects or Imperfections in crystalline solid can be divided into four groups namely defects, point defects, volume defects and surface defects.

3.6 (a)

The factors characterize the formation of non-crystalline structure are:

- Non-formation of three-dimensional primary bond.
- Weak secondary bond.
- Open network of the atomic packing.

ADDITIONAL INFORMATION

- Non-crystalline solids do not have consistent arrangement of particles.
- Non-crystalline solids have well defined geometrical shape, have short range order.

3.7 (a)

There are two types of diffusion :
1. Microscopic diffusion
2. Macroscopic diffusion

In micro diffusion primarily atoms move a fraction of atomic distance.

ADDITIONAL INFORMATION

- **Vacancy Diffusion :** One vacancy position is shifting to another so there is no significant movement of atoms.
- **Surface diffusion :** Suppose two surfaces are in contact and still the atoms are moving from one surface to another, so there also diffusion is taking place from one or two atomic distances only.
- **Lattice diffusion:** It is taking place within the gain so here also the movement is not significant.
- **Inter-diffusion :** In case of bi-material alloys (such as Cu-Ni alloy) ; when we are mixing these materials atoms of one metal moves almost close to 100 to 200 atomic distances just to make the homogenization. Hence, in case of inter-diffusion, movement of atoms is quite significant so it will not fall under microscopic diffusion.

3.8 (a)

When a pair of one cation and one anion are absent from an ionic crystal, then the defect is called Schottky's defect.

ADDITIONAL INFORMATION

A stacking defect is a planar defect that can occur in crystalline materials. Crystalline materials form repeating patterns of layers of atoms. Errors can occur in the sequence of these layers and known as stacking faults.

3.9 (c)

Density of α-Fe = 7.87×10^3 kg/m^3

Atomic weight of Fe = 55.8

Avogadro's number,

$$N_A = 6.02 \times 10^{26} \text{ kg/mole}$$

Number of atoms per unit cell

$$= 2$$

$$\rho = \frac{nA}{V_C N_A}$$

where, ρ = density

n = No. of atoms per unit cell

V_C = Volume of unit cell

A = Atomic weight

N_A = Avogadro's number

$$\Rightarrow \quad V_C = \frac{nA}{\rho N_A}$$

$$= \frac{2 \times 55.8}{7.87 \times 10^3 \times 6.02 \times 10^{26}} \text{ m}^3$$

$$\Rightarrow \quad V_C = 2.355 \times 10^{-29} = a^3$$

(a = lattice constant or edge length of cubic unit cell)

$$\Rightarrow \quad a = 2.866 \times 10^{-10} \text{ m}$$

$$\Rightarrow \quad a = 2.866 \text{ Å}$$

4.1 (b)

$$R_T = R_0(1 + \alpha T)$$

$$\Rightarrow \quad 4 = R_0(1 + \alpha \times 40°) \quad \text{...(i)}$$

$$\Rightarrow \quad 6 = R_0(1 + \alpha \times 80°) \quad \text{...(ii)}$$

Solving equations (i) and (ii)

$$R_0 = 2 \ \Omega$$

4.2 (d)

Applications of Hall Effect:

1. it is used to determine whether a semiconductor is n-type or p-type.
2. to calculate carrier concentration.
3. to calculate mobility
4. to calculate drift velocity
5. to calculate the conductivity of specimen.
6. to calculate magnetic field *B*.

4.3 (b)

Photoelectric current does not depend upon photosensitivity of the non-metal.

ADDITIONAL INFORMATION

Photoelectric current increases with intensity, but reaches a saturation level for large change in voltages.

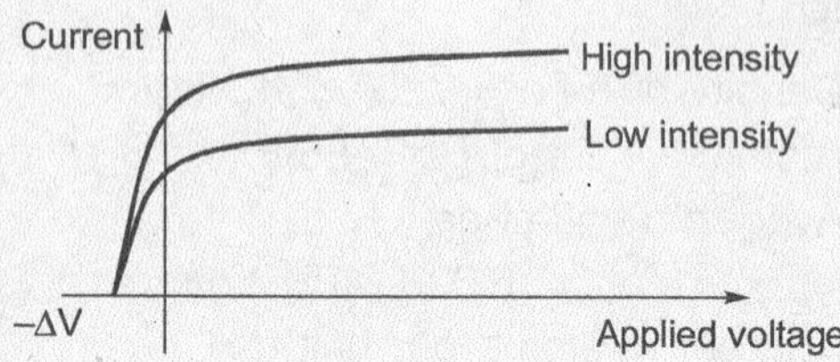

No current flows for voltages less than or equal to $-\Delta V$, the stopping potential. The stopping potential is independent of the radiation intensity.

4.4 (c)

MOSFET is a semiconductor device with output current controlled by an electric field and its current is carried predominantly by one type of carriers.

ADDITIONAL INFORMATION

An Insulated Gate Bipolar Transistor (IGBT) is a three-terminal power semiconductor device primarily forming an electronic switch. It was developed to combine with efficiency with fast switching. It consists of four alternating layers (P-N-P-N).

4.5 (d)

When Boron added with pure semiconductor it becomes p-type.

ADDITIONAL INFORMATION

When we add phosphorus, Arsenic and antimony is added to pure semiconductor it becomes n-type.

4.6 (a)

Impure semiconductor has more conductivity in contrast to pure semiconductor.

ADDITIONAL INFORMATION

An impure or extrinsic semiconductor material is made by doping trivalent or pentavalent elements with the pure semiconductor.

There are two types:

1. p-type; 2. n-type

4.7 (d)

Gettering problem is the major drawback of ion-implantation over diffusion process.

ADDITIONAL INFORMATION

Ion implantation is a low-temperature process by which ions of one element are accelerated into a solid target, there by changing the physical, chemical or electrical properties of the target.

Gettering is the common problem which frequently encounter in both silicon and in III – IV compounds.

4.8 (d)

For an n-p-n transistor, emitter is heavily doped, base width is small and collector area is large.

ADDITIONAL INFORMATION

A PNP transistor,

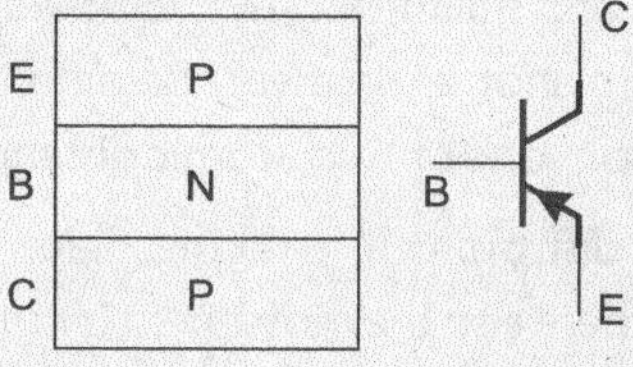

4.9 (a)

Positive ion (cation) and negative ion (anion) attract one another with electrostatic force (Coulombic force) of attraction to form ionic bond.

5.1 (a)

- Water is a polar molecule
- Water molecules align themselves in an electric field.
- Water molecules vary their alignment if the applied electric polarity changes with time.

ADDITIONAL INFORMATION

Water molecule:

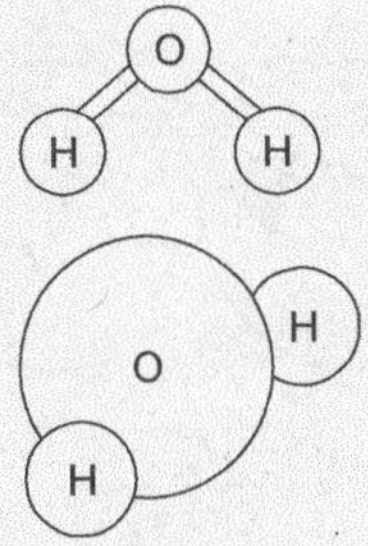

5.2 **(d)**

Neel temperature is the temperature at which antiferromagnetic material changes into paramagnetic material. So statement 4 is wrong, hence options (a), (b) and (c) are wrong; so the correct option is (d).

ADDITIONAL INFORMATION

Applications of hysteresis losses:

- Used in control systems for filtering of signals.
- This is used in aerodynamics when decreasing the angle of attack of the using after stall.
- Utilized in ferromagnetics in order to retain their memory.
- Used in respiratory physiology.

5.3 **(b)**

The critical temperature at which ferromagnetic material becomes paramagnetic is known as Curie temperature.

ADDITIONAL INFORMATION

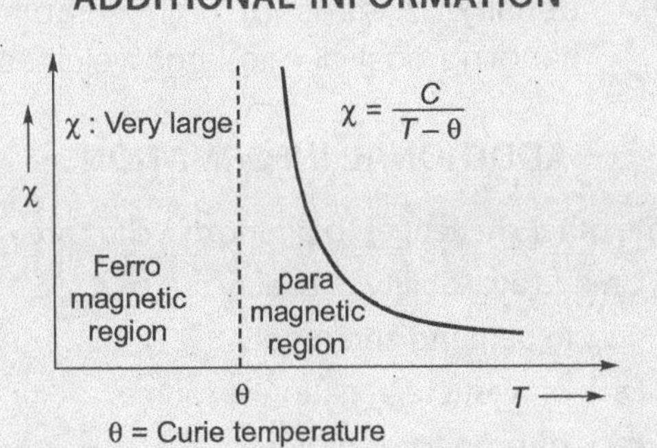

5.4 **(b)**

Properties of Hard magnetic material:

1. High retentivity
2. High coercivity
3. High hysteresis loss
4. Low permeability

ADDITIONAL INFORMATION

- For Soft magnetic material.

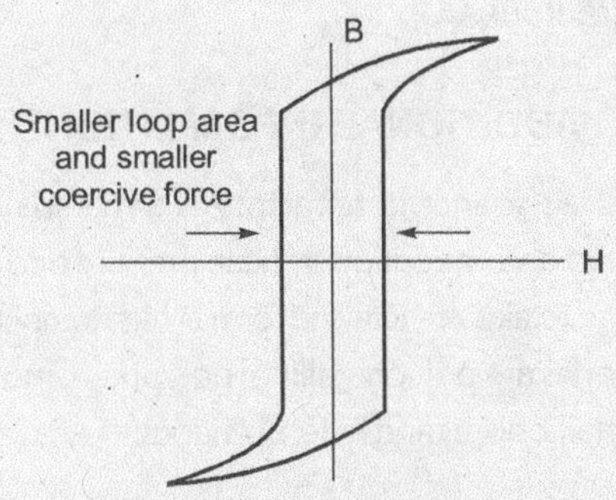

- For Hard magnetic material.

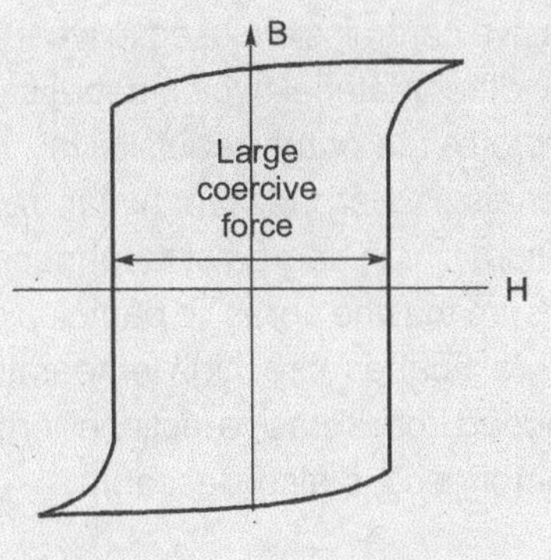

6.1 **(d)**

For stable structure in ionic material (like ceramics)

- Cations and anions are assumed hard sphere ions so statement 1 is wrong.
- Cations and anions must touch each other so statement 4 is wrong.

ADDITIONAL INFORMATION

Atomic packing factor

$$= \frac{\text{Total volume occupied by atoms in a unit cell}}{\text{Volume of the unit cell}}$$

6.2 (c)

Sodium bicarbonate commonly known as baking soda or bicarbonate of soda is a chemical compound used in fire extinguishers.
Quick lime is an alternate name for the chemical compound known as calcium oxide is used in manufacture of glass.

6.3 (d)

Metal carbides and carbon can be used as refractories.

ADDITIONAL INFORMATION

In materials science, a refractory is a material that is resistant to decomposition by heat, pressure or chemical attack and retains strength and form at high temperatures. Refractories are polycrystalline, polyphase, inorganic, non-metallic porus and heterogeneous.

6.4 (a)

Abrasive ceramics are used to wear, grind, or cut away other material, which necessarily is softer. Therefore, the prime requisite for this group of materials is hardness or wear resistance; in addition, a high degree of toughness is essential to ensure that the abrasive particles do not easily fracture. Furthermore, high temperatures may be produced from abrasive frictional forces, so some refractoriness is also desirable.

ADDITIONAL INFORMATION

An abrasive ceramic is a material, often a mineral, that is used to shape or finish a workpiece through rubbing which leads to part of the workpiece being worm away by friction. The ceramics which are used to cut, grind and polish other often softer materials are known as abrasives.

6.5 (a)

For stable crystal structure all the anions which are surrounding a cation should be in direct contact with that cation. This cation-anion contact is established for a specific value of r_C/r_A which is determined from geometrical considerations and assuming hard spherical ions as shown:

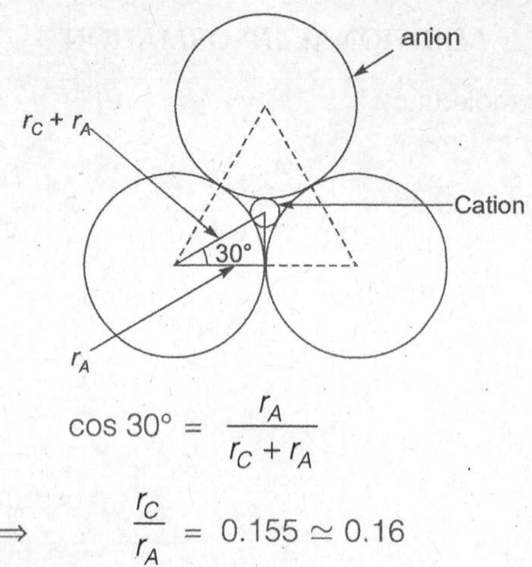

$$\cos 30° = \frac{r_A}{r_C + r_A}$$

$$\Rightarrow \quad \frac{r_C}{r_A} = 0.155 \simeq 0.16$$

7.1 (b)

Bakelite is made by condensation polymerisation of phenol and formaldehyde in an acid or alkaline medium. It is less brittle than many ceramics.

ADDITIONAL INFORMATION

- Polysters are manufactured by condensation polymerisation of dicarboxylic acid and dihydric alcohol e.g. ethylene glycol followed by curing with crosslinking agent (styrene).
- Addition polymerization is a process by which monomer units are attached one at a time in chain like fashion to form a linear macromolecule. There must be at least one double bond in the monomer. Generally thermoplastic are made by this process e.g. PVC, Teflon etc.

7.2 (b)

- Virgin plastics are typically cheaper than recycled plastics as they are cheaper to produce in mass.
- Since the plastic material has very low density, it leads to high shipping and handling cost due to large volumes.

ADDITIONAL INFORMATION

Some other challenges to plastic recycling are

- Limited availability of quality material.
- Gap in demand and supply.
- Lack of waste segregation at source.
- Lack of proper technology.
- Lack of proper industrial expertise.

7.3 (c)

As we know there is a one to one correspondence between the number of moles of sulphur and the number of moles of butadiene in the reaction and we have assumed only one sulphur atom is involved in each cross linking bond with butadiene. The average molecular weight of butadiene per mer is

= 6 Hydrogen atoms × 1 g/mole + 4 Carbon atoms × 12 g/mole = 54 g/mole

Fraction of sulphur used in butadiene rubber

$$= \frac{32}{32+54} = 0.372$$

(Atomic weight of sulphur = 32 g/mole)

For 100 kg of butadiene rubber, the required sulphur is 37.2 kg.

ADDITIONAL INFORMATION

The crosslinking process in elastomers is called vulcanization, which is achieved by non-reversible chemical reaction, ordinarily carried out at an elevated temperature. In most vulcanisation reactions, sulphur compounds are added to the heated elastomer, sulphur atoms bond with adjacent chains and cross links them.

7.4 (a)

Polymers are created by the process of polymerization, wherein their constituent elements called monomers, are reacted to form polymer chains. Polymers with Vander Waals forces linking chains are known to be weak and give the polymer a low melting point.

ADDITIONAL INFORMATION

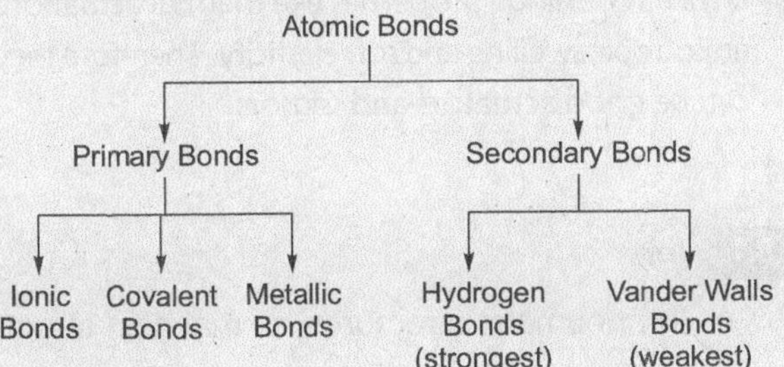

- Primary bonds are inter atomic in nature. These are stronger and more stable than secondary bonds. Hardness, strength and other properties are found in metals and ceramics due to primary bond.

7.5 (b)

$$\left[\begin{array}{cc} H & H \\ | & | \\ C - & C \\ | & | \\ C & C \equiv N \end{array} \right]_n \text{— Polyacrylonitrile (does not melt)}$$

ADDITIONAL INFORMATION

Polyacrylonitrile is used in form of fibre and because of strength and stability used as comonomer. Presence of high electronegative nitrile group on every other carbon atoms exerts mutual electrical repulsion and give stiff, rodlike structure. Because of this they form strong hydrogen bonding in fibre and give high strength and good resistance to moisture and solvents.

7.6 (d)

Condensation Polymerisation: It is the formation of polymer by stepwise inter molecular chemical reactions. There is usually a small molecular by product such as water that is eliminated. Generally, thermosetting are made by this process e.g. phenol formaldehyde (bakelite), amine formaldehyde, polyester, araldite etc.

ADDITIONAL INFORMATION

Thermoplastics are formed by addition polymerisation e.g. polyethylene, polyvinyl chloride (PVC), polypropylene etc.

7.7 (d)

In crosslinked polymers, adjacent chains are joined to one another at various positions by covalent bonds. The crosslinking process is either achieved during synthesis or by a non reversible chemical reactions that is carried out at elevated temperature.

ADDITIONAL INFORMATION

- In linear polymers, mer units are joined together end to end in single chains. These long chains are flexible. For linear polymer there are extensive Vander Waals and hydrogen bonding between the chains e.g. polyethylene, pvc, polyethylenes, PMMA, etc.

- In Branched polymers, side branches are connected to the main branch. These branches occur as a result of the side reaction that occur during synthesis of polymer.

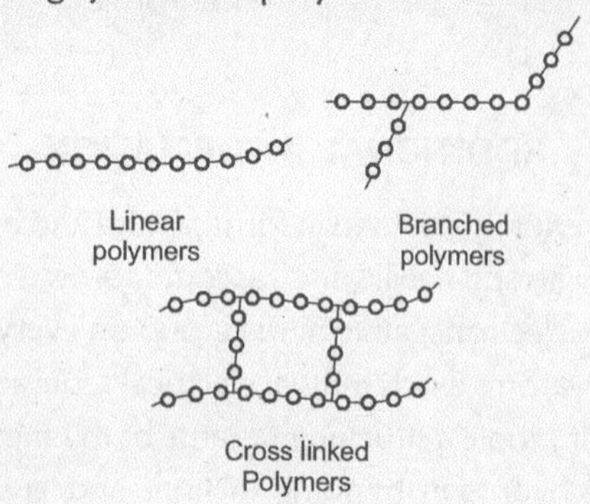

Linear polymers

Branched polymers

Cross linked Polymers

7.8 (c)

Thermosetting materials are those plastics which require heat and pressure to mould them into shape. They cannot be resoftened once they have set and hardened. These are formed by condensation polymerisation e.g.-Bakelite, Amine formaldehyde etc.

ADDITIONAL INFORMATION

Thermoplastics primarily include long chain straight or slightly branched molecules and the chains are held close to each other by secondary weak forces like Vander Waals forces. During heating, as the temperature increases the secondary forces are reduced and the sliding of these long chain molecular can easily occur one over the other at a reduced stress level.

7.9 (b)

Baylon is the trade name of polycarbonates.

ADDITIONAL INFORMATION

Two phenyl and two methyle group attached with same carbon atom in repeating structural unit produce stiff molecular structure, high impact energy due to carbonate linkage; high heat deflection temperature, good electrical insulating properties.

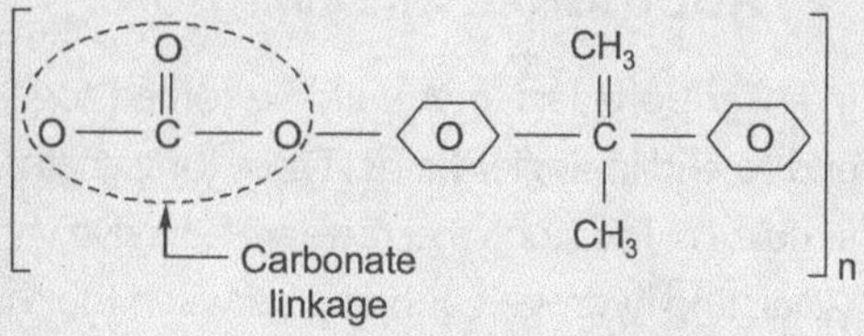

$$\left[-O-\overset{O}{\overset{\|}{C}}-O-\langle O \rangle-\underset{CH_3}{\overset{CH_3}{\overset{|}{\underset{|}{C}}}}-\langle O \rangle- \right]_n$$

Carbonate linkage

7.10 (c)

Many bulk polymers that are crystallized from a melt are semicrystalline and form a spherulite structure. As implied by the name, each spherulite may grow to be roughly spherical in shape. The spherulite consists of an aggregate of ribbon-like chain-folded crystallites (lamellae) approximately 10 nm thick that radiate outward from a single nucleation site in the center.

ADDITIONAL INFORMATION

Formation of spherulites affect many properties of the polymer material, in particular, crystallinity, density tensile strength, and Young's modulus of polymers increase during spherulization. This increase is due to the lamellae fraction within the spherulites, where the molecules are more densely packed than in the amorphous phase.

7.11 (c)

Liquid crystalline polymers are composed of extended, rod shaped and rigid molecules. In terms of molecular arrangement, these materials do not fall within any of the conventional liquid, amorphous, crystalline or semi-crystalline classifications, but may be considered as new state of matter, i.e., the liquid crystalline state being neither crystalline nor liquid.

ADDITIONAL INFORMATION

One of the most famous and classical applications for LCPs is Kevlar, a strong but light fiber with wide applications, notably bulletproof vests. In addition, LCPs have unique properties like thermal actuation, anisotropic swelling and soft elasticity. Therefore they can be good actuators and sensors.

8.1 (b)

In pneumatic structures, to maintain structural integrity, the structure must be pressurized such that the internal pressure equals or exceeds any external pressure being applied to the structure.

ADDITIONAL INFORMATION

Pneumatic structures are used in sports and recreation facilities, warehousing, temporary shelters and randomes. The structure can be either wholly, partially, or roof only air supported. A fully air supported structure can be intended to be a temporary or semi-temporary facility or permanent, whereas a structure with only air-supported roof can be built as a permanent building.

8.2 (c)

Asphalt is obtained either as a residue from the distillation of petroleum or from natural deposits. Asphalts consists of compounds of hydrogen and carbon with minor proportions of nitrogen, sulphur and oxygen. Asphalt is also referred to as bitumen whose main source is petroleum from where it is derived as the semi-solid compound and it is a thick, sticky, black, highly viscous liquid.

ADDITIONAL INFORMATION

When the crude oil undergoes fractional distillation where it is heated at a temperature of 525°C, the asphalt that is generated is referred to as "refined bitumen". The primary use of asphalt goes into the construction of road which accounts for 70% of the asphalt that is explored.

8.3 (d)

Composite's physical properties are not isotropic, but they are typically anisotropic i.e., different depending on the direction of the applied force or load. For instance, the composite's panel stiffness will usually depend upon the orientation of the applied forces and/or moments.

ADDITIONAL INFORMATION

Classification of composites are shown below :

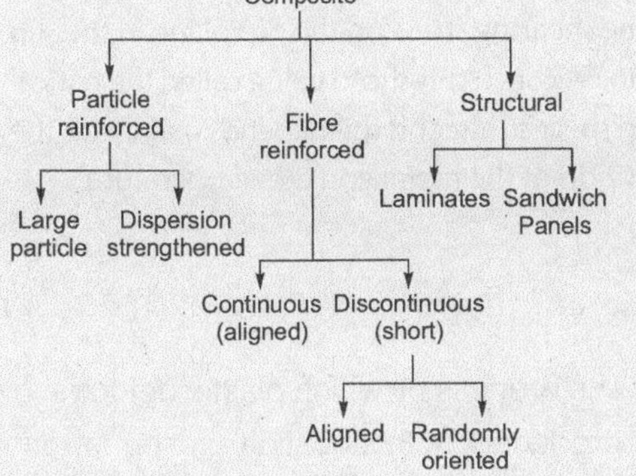

9.1 (c)

- Shot peening is a cold working process used to produce a compressive residual stress layer and modify the mechanical properties of metals and composites. Due to compressive residual stress on the surface, fatigue life of the component get improved.
- By polishing the surfaces, surface irregularities and cracks are improved. Hence, fatigue life get improved.

ADDITIONAL INFORMATION

There are several other way which improves the fatigue life. For e.g.

- Stress raisers such as sharp corners are avoided so that the stress concentration vanish or diminish considerably.
- Materials with finer grain sizes are utilized as they are stronger than coarse grain.
- The simplest and least expensive way of increasing the fatigue strength of carbon steels is by increasing the carbon content.

9.2 (a)

Fatigue fracture occurs at stress much lower that the tensile strength or even lower than yield stress for static loading. Cyclic stress initiate the crack, tensile stress cause the crack to grow through localized plastic strain. The fatigue failure, however depends upon a number of factors, such as the number of cycles, mean stress, stress amplitude, stress concentration, residual stresses.

ADDITIONAL INFORMATION

- Fatigue is the single largest cause or failure in metal estimated to comprise approximately 90% of all metallic failure.
- Fatigue strength of notched and unnotched grey cast iron specimens are same.

9.3 (a)

Elongation (δ) of a bar under tensile load is given by,

$$\delta = \frac{PL}{AE}$$

For same elongation (δ)

$$\delta_1 = \delta_2$$

$$\frac{P_1 L_1}{A_1 E_1} = \frac{P_2 L_2}{A_2 E_2}$$

$$\frac{P_2}{P_1} = \left(\frac{A_2}{A_1}\right)\left(\frac{E_2}{E_1}\right)\left(\frac{L_1}{L_2}\right)$$

$$\frac{P_2}{P_1} = \left(\frac{2A}{A}\right)\left(\frac{E}{E}\right)\left(\frac{L}{L}\right)$$

$$P_2 = 2P_1$$

ADDITIONAL INFORMATION

Following are the important parameters that can be determined from the stress-strain curve using original length, final length, original diameter and final diameter at fracture section.

1. Yield strength
2. Tensile strength
3. Percent elongation (for ductility)
4. Percent reduction in the area (for degree of cold working)

9.4 (d)

The ability of a material to absorb energy when deformed elastically and to return the same when unloaded is termed as resilience. This energy up to the elastic limit (σ_e) per unit volume is known as Modulus of resilience.

Strain energy per unit volume

$U = \frac{1}{2}\sigma \in$, at any stage upto elastic limit

Modulus of resilience,

$$u = \frac{\sigma_e^2}{2E}$$

ADDITIONAL INFORMATION

- Tensile strength, is the ability of the material to resist external load causing tensile stress, without fracture.
- Elasticity is defined as the ability of the material to regain its original shape and size instantaneously after the deformation, when the external forces are removed.
- Ductility is the ability of material to deform to a greater extent before the sign of crack, appears when it is subjected to tensile force.

9.5 (c)

Ductile fracture occurs after prolonged plastic deformation and is accompanied by necking. It requires higher strain energy. Due to necking warning sign can be easily sensed. Ductile fracture generally takes place along a plane on which shear stress is maximum.

ADDITIONAL INFORMATION

Brittle fracture occurs with little or no plastic deformation. Thus strain energy required for it is lower. Thus failure sign cannot be marked. Brittle failure takes place along a plane on which normal stress is maximum.

9.6 (d)

Critical resolved shear stress is

$$\tau_R = \frac{F}{A}\cos\phi \cos\alpha$$

where F is the tensile force, A is the area of cross-section of the specimen, ϕ is the angle between normal to slip plane and tensile area, and α is the angle of slip direction with tensile axis.

Factors influencing critical shear stress:

- Purity in metals reduces the critical stress.
- Surface effects like surface films greatly increase the critical shear stress.
- Temperature increases thermal mobility and hence decreases the critical shear stress.

ADDITIONAL INFORMATION

The amount of slip taking place in a single crystal depends on the magnitude of the shearing stress produced by an external force, geometry of the crystal structure and the orientation of the active slip planes with respect to the shearing stress. Slip begins when the shearing stress on the slip plane in the slip direction reaches a threshold value called the critical resolved shear stress, and this depends mainly on the composition of the metal and the temperature.

9.7 (c)

- The process by which plastic deformation is produced by dislocation motion is termed as slip.

- In twinning, a portion of the crystal takes up an orientation that is related to the orientation of the rest of the untwinned lattice, in a definite symmetrical way. The twinned portion of the crystal is a mirror image of the parent crystal. The plane of symmetry between the two portions is called twinning plane.

ADDITIONAL INFORMATION

Dislocations do not move with the same degree of ease on all crystallographic planes of atoms and in all crystallographic directions. Slip is more difficult in low-density planes because atoms are far apart and slip distances are large.

9.8 (b)

Stress (σ) required for crack propagation in a brittle material is described by the expression.

$$\sigma = \left(\frac{2E\gamma_s}{\pi a}\right)^{1/2}$$

$E \rightarrow$ Modulus of elasticity
$\gamma_s \rightarrow$ Specific surface energy
$a \rightarrow$ Length of a surface flaw

Rearrangement of this expression such that 'a' is dependent variable, and realizing that σ = 40 MPa, γ_s = 0.3 J/m^2, E = 69 GPa leads to

$$a = \frac{2E\gamma_s}{\pi\sigma^2} = \frac{2\times(69\times10^9)(0.3)}{\pi\times(40\times10^6)^2}$$

$= 8.2 \times 10^{-6}$ m = 8.2 μm

ADDITIONAL INFORMATION

There are three types of fractures:

(a) Shear fracture due to extensive slip on active slip planes.

(b) Fibrous fractured surface caused by shear cleavage, and

(c) Ductile fracture exhibiting considerable plastic deformation.

9.9 (c)

Since deformation is entirely elastic.

So, $\sigma = \in E$

where $\sigma \rightarrow$ stress; $\in \rightarrow$ strain; $E \rightarrow$ Modulus of elasticity

$$\Rightarrow \quad \sigma = \frac{\Delta l}{l_o}E$$

$$\Rightarrow \quad \Delta l = \frac{\sigma l_o}{E} = \frac{(276 \text{ MPa})\times(305 \text{ mm})}{110\times10^3 \text{ MPa}}$$

(For Cu, E = 110 GPa)

Δl = 0.756 mm

$\Delta l \simeq$ 0.77 mm

where $\Delta l \rightarrow$ elongation; $l_o \rightarrow$ original length

ADDITIONAL INFORMATION

The elongation of specimen is uniform along the gauge length upto the maximum load, corresponding to ultimate tensile strength. Beyond this point, necking begins and deformation is no longer uniform along the length of the specimen.

9.10 (d)

Percentage cold work (%CW) is defined as

$$\%CW = \left(\frac{A_o - A_d}{A_o}\right)\times100 = \frac{\pi\left(\frac{15.2}{2}\right)^2 - \pi\left(\frac{12.2}{2}\right)^2}{\pi\left(\frac{15.2}{2}\right)^2}\times100$$

= 35.6%

For copper, ductility (%EL) versus % CW curve is given as

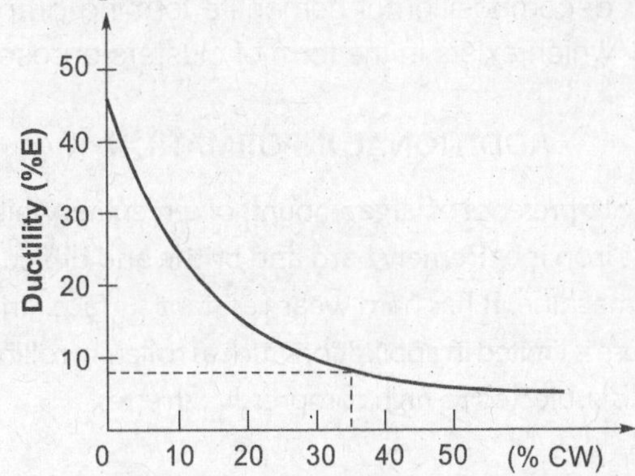

ADDITIONAL INFORMATION

Effects of cold working on parts:

- Better surface finish.
- Increase in strength due to work hardening.
- Ductility and Malleability reduces after cold working.

10.1 (c)

Every solid and liquid matter in the nature will have some impurity of sulphur. Sulphur after reacting with iron produces iron sulphide (FeS). FeS gets accumulated on the grain boundaries and is having lower melting point. Upon hot working the metal FeS melts out producing number of cracks on the grain boundaries. As a result of that material will fail as brittle fracture. This phenomenon is called "Hot shortness". To avoid this phenomenon, 'Mn' is added into the material 'Mn' captures sulphur before it can react with iron. Also a large number of 'MnS' pockets will be present in the materials which improves the machinability because MnS is having how shear strength.

ADDITIONAL INFORMATION

If 'Mn' is around 12%, the strength of material appears to be very high called "Hadfield steel" which is used to make bull dozer rolls, jaw crusher plates, nuts and bolts etc.

10.2 (c)

By a heat treatment process, white cast iron is converted into a malleable iron. Heating white cast iron at temperature between 800°C and 900°C for prolonged time period, and in neutral atmosphere (to prevent oxidation) causes decomposition of cementite forming graphite, which exists in the form of clusters or rosettes.

ADDITIONAL INFORMATION

Due to presence of large amounts of cementite, white cast iron is extremely hard and brittle and difficult to machine. It has hard wear-resistant surface, and its use is limited in applications such as rollers in rolling mills subjected to high compressive stresses.

10.3 (c)

- In pearlite, there are alternate strips (layers) of ferrite and cementite. Ferrite is soft but cementite is harder and more brittle than ferrite. Rather, cementite reinforces the ferrite.
- At 768°C (this temperature is called curie temperature) only magnetic property of iron is disappeared and structure remains same. Above curie point magnetic properties does not exist in iron.

ADDITIONAL INFORMATION

The property of iron changes when temperature changes i.e. it exists in different forms at different temperature.

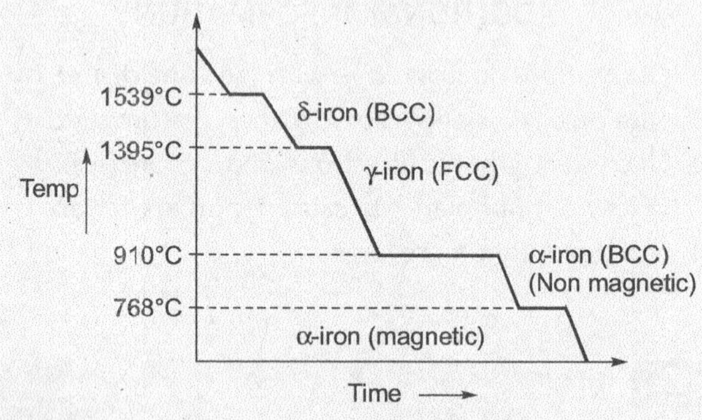

10.4 (c)

Functions of oil in a transformer:

1. **Insulation:** Transformer oil has a much higher dielectric strength than air. The insulating material is immersed in oil, which not only improves insulation strength but also protects it from moisture.
2. **Heat dissipation:** Transformer oil has a large specific heat and is often used as a coolant.
3. **Arc suppression:** Because the transformer oil has good thermal conductivity, and can separate a large amount of gas under the action of the high temperature of the arc, and generate a large pressure, the arc extinguishing performance of the medium is improved and the arc is quickly extinguished.

ADDITIONAL INFORMATION

- The density of transformer oil should be as small as possible to facilitate the precipitation of moisture and impurities in the oil.
- The viscosity should be moderate, too large will affect the convection heat dissipation, too small will reduce the flash point.

10.6 (c)

- The class of cast iron obtained by treating molten metal by calcium silicide is popularly known as Meehanite.
- Malleable cast irons, like nodular cast irons, have graphite nodules but are produced by heat treating white cast irons.

Hence, statements 2 and 3 are incorrect.

11.1 (b)

Babbitt metal is used as the lining for bearing shells of cast iron, steel and bronze. The Babbitt lining prevents friction and wear that is common when a lubricant is not able to prevent the bearing's moving part from welding together. The babbitt therefore extends the life of the bearing.

ADDITIONAL INFORMATION

A best bearing material should have low coefficient of friction, high compressive strength, high fatigue strength, and low thermal expansion.

11.2 (b)

- Duralumin is the first commercially heat treatable alloy extensively used in aircraft industry for making frames, ribs, propeller blades and in automobiles. Composition of duralumin is Cu4, Mg0.5, Si0.7 and rest Al.
- Precipitation heat treatment involves heating alloys upto an appropriate temperature and holding that specific temperature for an adequate time to introduce one or several constituents into the solid solution, and then undergo rapid cooling to have these enclosed with the solution. Precipitation hardening is used for non-ferrous alloys e.g. duralumin, aluminium alloysets.

ADDITIONAL INFORMATION

Aluminium lithium alloys are used in aerospace industries due to their low density, high specific strength, excellent fatigue strength and low temperature toughness.

11.3 (d)

Following conditions are required to form complete substitutional solid solution.

(i) Two elements must have similar crystal structure.
(ii) The difference in their atomic radii should be less than 15%.
(iii) Two elements must have same valency.
(iv) The solute and solvent should have similar electronegativity.

ADDITIONAL INFORMATION

- When two or more element mixed and they do not form a chemical bond then it is termed as alloy.
- When two or more elements are mixed and they are bound together by chemical bond then it is termed as compound.

11.4 (b)

There are three types of phase transformations:

- Diffusion dependent phase transformation in which the number or the composition of the phases is invariant. This is also termed as congruent phase transformation.
- Diffusion dependent phase transformation in which there is a change in composition or even in the number of phases present. This is termed as incongruent phase transformation.
- Transformation in which there is no diffusion. This type of transformation result in a metastable phase e.g. martensitic phase transformation.

ADDITIONAL INFORMATION

Phase diagram is a plot between temperature and composition of a system which indicates the changes that takes place during heating and cooling of an alloy series.

11.5 (b)

As per lever rule, to find the fractions of liquid phase and solid phase in the binary alloy:

1. The tie line is constructed across the two phase region at the temperature of the alloy.
2. The overall alloy composition is located on the tie line.
3. The fraction of a phase is computed by taking the length of tie line from the overall alloy. Composition to the phase boundary for the other phase and dividing by the total tie line length.

ADDITIONAL INFORMATION

Eutectic point is the point in the phase diagram at which the liquidus and the solidus lines meet, or it is defined as the point at which three phases, a liquid phase, and two solid phase coexist. The number of degrees of freedom at the eutectic point is zero.

11.6 (a)

Following conditions are required to form complete substitutional solid solution.

(i) Two elements must have similar crystal structure.

(ii) The difference in their atomic radii should be less than 15%.

(iii) Two elements must have same valency.

(iv) The solute and solvent should have similar electronegativity.

ADDITIONAL INFORMATION

- When two or more element mixed and they do not form a chemical bond then it is termed as alloy.
- When two or more elements are mixed and they are bound together by chemical bond then it is termed as compound.

12.1 (a)

Galvanization or galvanizing is the process of applying a protective zinc coating to steel or iron to prevent rusting.

ADDITIONAL INFORMATION

The other methods of corrosion prevention are:

1. Alloying the metal.
2. Use of inhibitors.
3. Use of metallic and non-metallic coatings.

12.2 (a)

The diffusion equation for given case is

$$\frac{x^2}{Dt} = \text{Constant}$$

where D = Diffusion coefficient; t = Time; x = Diffusion depth

Given that the composition in both diffusion situations will be equal at the same position (i.e., x is also a constant)

So, Dt = Constant (at both temperatures)

$$D_{500}t_{500} = D_{600}t_{600}$$

$$\Rightarrow \quad t_{500} = \frac{D_{600}t_{600}}{D_{500}}$$

$$= \frac{5.3\times10^{-3}\times\frac{m^2}{s}\times10h}{4.8\times10^{-14}\frac{m^2}{s}}$$

$$= 1.104 \times 10^{12} \text{ h}$$

Note : There is error in question, D at 600°C should be equal to 5.3×10^{-13} m²/s, then

$$t_{500} = \frac{5.3\times10^{-13}\times10}{4.8\times10^{-14}}$$

$$t_{500} = 110.4 \text{ h}$$

So, option (a) is correct.

ADDITIONAL INFORMATION

A natural tendency of the molecule or atoms to flow from higher concentration to lower concentration is called diffusion, it is a time and temperature dependent process.

12.3 (c)

Gibb's phase rule,

P + F = C + N

F → Degree of freedom

P → Number of phases

C → Number of components forming the system.

N → System variable

N = 2, if variables are temp and pressure

N = 1, if variable is temp only

Let's take, N = 2 (for alloys)

$P + F = C + 2$

$\Rightarrow F = C - P + 2$

Phase diagrams are graphical representations of the physical states of a material (such as liquid, solid, gas) at various conditions of temperature, pressure, or composition. Tie-lines play a significant role in the Lever Rule application. Also known as the 'isotherm', it's a line drawn at the temperature of interest on the phase diagram. This line connects the compositions of two phases in equilibrium at that temperature. The tie-line helps to identify the composition of each phase in equilibrium and allows the determination of the amounts of each phase (using the Lever Rule).This tie-line's lengths on either side of the overall composition act as 'arms' of a lever, with the overall composition acting as the balancing point.

Hence statements-1 is wrong, so options (a), (b) and (d) cannot be true. Hence, correct option is (c).

12.4 (b)

The advantages of gas carburizing over pack carburizing are as below:

- Possibility of better regulation to obtain accurate case depth and hardness through controlling the gas composition.
- Less time is required, since there is no need to pack boxes filled with low-heat conductive substances.

It needs only half as much floor space as required for pack carburizing.

Hence, statements 1 and 3 are correct.

■■■■

Information and Communication Technologies

CONTENTS

Information and Communication Technologies

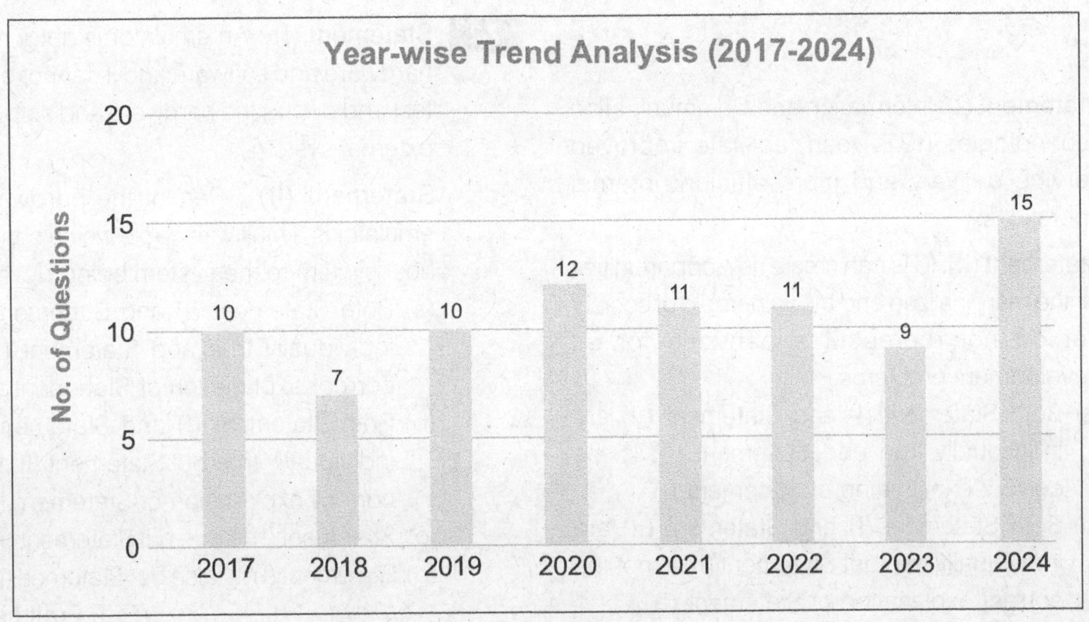

TOPIC-WISE TREND ANALYSIS (2017-2024)

Sl. No.	Topics	Total Number of Questions
1.	Introduction to ICT	2
2.	ICT Based Tools	10
3.	Communication	2
4.	Networking	40
5.	e-Governance	16
6.	Technology Based Education	9
7.	Recent Development Related to ICT	6

Information and Communication Technologies

1. INTRODUCTION TO ICT

1.1 **Statement (I):** Information and Communication Technologies (ICTs) can facilitate improved service delivery and more efficient internal operations.

Statement (II): ICTs can create new opportunities for the marginalized and the vulnerable of society but do not represent a panacea for all development problems.

(a) Both Statement (I) and Statement (II) are individually true and Statement (II) is the correct explanation of Statement (I)
(b) Both Statement (I) and Statement (II) are individually true but Statement (II) is not the correct explanation of Statement (I)
(c) Statement (I) is true but Statement (II) is false
(d) Statement (I) is false but Statement (II) is true **[ESE Prelims : 2020]**

1.2 Which one of the following involves for the examination of Internet records to track down the identity of someone who posted in a discussion forum on one Website might search for clues to the poster's identity on Facebook, Twitter, and other online sources?

(a) Pornography (b) Internet filter
(c) Doxing (d) Internet censorship

[ESE Prelims : 2024]

2. ICT BASED TOOLS

2.1 Which type of output device creates coloured images which look and feel like photographs?

(a) Electrostatic plotter
(b) Laser printer
(c) Dye sublimation printer
(d) Inkjet plotter

[ESE Prelims : 2018]

2.2 **Statement (I) :** An emulator is not a mixture of hardware and software and it cannot be used to test and debug the hardware and software of an external system.

Statement (II) : Part of the hardware of an emulator is a multiwire cable which connects the host system to the system being developed.

(a) Both Statement (I) and Statement (II) are individually true and Statement (II) is the correct explanation of Statement (I)
(b) Both Statement (I) and Statement (II) are individually true but Statement (II) is not the correct explanation of Statement (I)
(c) Statement (I) is true but Statement (II) is false
(d) Statement (I) is false but Statement (II) is true

[ESE Prelims : 2019]

2.3 The core elements of high-level programming languages are:

(a) Keywords, Expressions and Punctuation
(b) Functions, Keywords and Operators
(c) Keywords, Operators and Punctuation
(d) Functions, Expressions and Operators

[ESE Prelims : 2019]

2.4 Which one of the following systems is used when there are rigid time requirements on the operation of a processor or the flow of data, and thus is often used as a control device in a dedicated application?

(a) A real-time system
(b) A distributed system
(c) A parallel system
(d) A serial system **[ESE Prelims : 2022]**

2.5 A device which exhibits irregular or unpredictable response times is called

(a) Asynchronous (b) Synchronous
(c) Sharable (d) Non-sharable

[ESE Prelims : 2023]

2.6 Which one of the following tables is used by operating system to keep the track of many I/O requests at the same time?
(a) File allocation table
(b) Device – status table
(c) Memory – status table
(d) Interrupt driven table
[ESE Prelims : 2023]

2.7 A stream of a video image that is one-quarter the size of a standard TV image; that is, it has a resolution of 352 by 240 pixels. If each pixel is represented by 24 bits of information, as would be the case for 24-bit color, then what is the approximate size of each frame?
(a) 247.5 KB (b) 352.5 KB
(c) 417.5 KB (d) 532.5 KB
[ESE Prelims : 2023]

2.8 Which one of the following features is/are used when a website is complex, consisting of many pages produced by multiple authors working for the same company, often desirable to have a way to prevent a different page from having a different appearance?
(a) Checkbox (b) Style sheets
(c) Table (d) Forms
[ESE Prelims : 2024]

2.9 Which one of the languages is used to develop the web pages in the structured and for automated processing?
(a) eXtensible Markup Language
(b) Hypertext Markup Language
(c) eXtended Hyper Text Markup Language
(d) Markup Language
[ESE Prelims : 2024]

2.10 Which one of the following is a harmful program that resides in the active memory of the computer and duplicates itself without human intervention, often sending copies of themselves to other computers by email?
(a) Worms (b) Viruses
(c) Bugs (d) Spam
[ESE Prelims : 2024]

3. COMMUNICATION

3.1 What does CDMA stand for?
(a) Code Division Mobile Access
(b) Code Division Multiple Access
(c) Code Division Multiple Applications
(d) Code Division Mobile Applications
[ESE Prelims : 2018]

3.2 Which one of the following transmission media is/are used for the remote control communication for televisions, VCRs, and stereos etc.?
(a) Fiber optics
(b) Fiber cables
(c) The electromagnetic spectrum
(d) Unguided infrared and millimeter waves
[ESE Prelims : 2024]

4. NETWORKING

4.1 Consider the following statements on 'firewall' used in computing systems:
1. It controls and monitors the data traffic flow between inside and outside network
2. It protects and secures the inside network from any outside network
3. It can be implemented in software or hardware or a combination of both

Which of the above statements are correct?
(a) 1 and 2 only (b) 1 and 3 only
(c) 2 and 3 only (d) 1, 2 and 3
[ESE Prelims : 2017]

4.2 Consider the following statements regarding Holostore:
1. It is a device that reads and writes data in an optical form
2. It is a computer storage device
3. It refers to Institutions where holography is taught

Which of the above statements are correct?
(a) 1, 2 and 3 (b) 1 and 3 only
(c) 1 and 2 only (d) 2 and 3 only
[ESE Prelims : 2017]

4.3 Consider the following statements regarding internet of things (IoT):
1. IoT extends the communication via internet to all the things that surround us
2. IoT is M2M communication

3. IoT uses only wireless technology
4. The major objectives for IoT are the creation of smart environments/spaces and self-aware things.

Which of the above statements are correct?

(a) 1 and 3 (b) 2 and 3
(c) 1 and 4 (d) 2 and 4

[ESE Prelims : 2017]

4.4 Consider the following statements concerning e-governance and technology:

1. Rich Site Summary (RSS) is not a very useful tool for working on Web-service technology
2. Myspace is one of the most visited networks in the world
3. Facebook.com is not the most useful social networking site of choice for most students
4. Wiki- is a group collaboration software tool working on Web-service technology.

Which of the above statements are correct?

(a) 1 and 4 (b) 2 and 4
(c) 1 and 3 (d) 2 and 3

[ESE Prelims : 2017]

4.5 The quality of service provided in a computer network is

(a) Presentation layer issue
(b) Session layer issue
(c) Network layer issue
(d) Data link layer issue **[ESE Prelims : 2019]**

4.6 In a computer network, a point-to-point transmission, with one sender and one receiver is called

(a) Unicasting (b) Multicasting
(c) Broadcasting (d) Internetworking

[ESE Prelims : 2019]

4.7 The Protocol (http), the DNS name of the host, and the file name is identified through

(a) Uniform Resource Locator
(b) Web Browser
(c) Web Server
(d) IP address **[ESE Prelims : 2019]**

4.8 The traditional way to handle forms and other interactive Web pages is a system called

(a) Graphical User Interface
(b) Common Gateway Interface
(c) Text Based User Interface
(d) Command Line Interface

[ESE Prelims : 2019]

4.9 Pretty Good Privacy (PGP) which encrypts the data by using a block cipher is used in

(a) FTP security (b) e-mail security
(c) Browser security (d) Bluetooth security

[ESE Prelims : 2019]

4.10 Which of the following are the advantages of coding audiovisual objects?

1. It allows interaction with the content
2. It improves reusability and coding the content
3. It allows content-based scalability

(a) 1 and 2 only (b) 1 and 3 only
(c) 2 and 3 only (d) 1, 2 and 3

[ESE Prelims : 2019]

4.11 The transmission of real-time streams across networks uses Bandwidth Allocation Mechanism (BAM), which is based on

(a) Stream peak rate
(b) Bucket rate
(c) Token bucket depth
(d) Packet size **[ESE Prelims : 2019]**

4.12 The Pre-echo PE distortions in audio signal represents the

(a) Theoretical limit on compressibility of particular signals
(b) Imaginary components of a signal
(c) Critical band analysis of a signal
(d) Histogram of the signals

[ESE Prelims : 2019]

4.13 The Ethernet designed by IEEE to compete with LAN protocols which can transmit data ten times faster at a rate of 100 Mbps is

(a) fast Ethernet
(b) bridged Ethernet
(c) switched Ethernet
(d) full-duplex Ethernet **[ESE Prelims : 2020]**

4.14 IEEE standard protocol which defines a wireless Personal Area Network (PAN) operable in a room is

(a) Wi-Fi (b) Bluetooth
(c) Infrared (d) Wireless LAN

[ESE Prelims : 2020]

4.15 Which one of the following points is a private switching station that connects the national internet service provider's network and operates at a high data rate up to 600 Mbps?

(a) Locking point (b) Peering point
(c) Hub point (d) Modem point

[ESE Prelims : 2020]

4.16 Which one of the following protocols is used to address the true routing decisions problems?
(a) Exterior Gateway Protocol
(b) Border Gateway Protocol
(c) Open Shortest Path First Protocol
(d) Interior Gateway Routing Protocol
[ESE Prelims : 2021]

4.17 Which one of the following standards is used in vehicular communication system?
(a) IEEE 802.11a (b) IEEE 802.11p
(c) IEEE 802.11g (d) IEEE 802.11h
[ESE Prelims : 2021]

4.18 Which of the following network metrics are used to evaluate the performance of a network?
(a) Throughput and Delay
(b) Reliability and Security
(c) Topology and Type of connection
(d) Portability and Security
[ESE Prelims : 2021]

4.19 Which of the following things are defined by uniform resource locator for specifying the information on the internet?
(a) protocol, host computer, throughput and delay
(b) host computer, destination computer and delay
(c) throughput, delay, port and path
(d) protocol, host computer, port and path
[ESE Prelims : 2021]

4.20 Which one of the following documents are created and handled by the Common Gateway Interface (CGI) technology?
(a) Dynamic documents
(b) Static documents
(c) Tampered documents
(d) Linked documents
[ESE Prelims : 2021]

4.21 Which one of the following services does NOT come under category of Cloud computing?
(a) IaaS (Infrastructure as a Service)
(b) Saas (Software as a Service)
(c) PaaS (Platform as a Service)
(d) BDaaS (Big data as a Service)
[ESE Prelims : 2021]

4.22 What is the key size of Data Encryption Standard algorithm in cryptography?
(a) 56 bit (b) 62 bit
(c) 168 bit (d) 128 bit
[ESE Prelims : 2021]

4.23 Which one of the following protocols is designed to provide privacy between two communicating applications viz a client and a server?
(a) Data link layer protocol
(b) Physical layer protocol
(c) Secure socket layer protocol
(d) Session layer protocol
[ESE Prelims : 2022]

4.24 Which one of the following schemes is used for radio stations within the same region, where each radio station has its own frequency?
(a) Space division multiplexing
(b) Frequency division multiplexing
(c) Time division multiplexing
(d) Code division multiplexing
[ESE Prelims : 2022]

4.25 The AES key expansion algorithm takes as input a 4-word (16-byte) key and produces a linear array of
(a) 50 words (200 bytes)
(b) 44 words (176 bytes)
(c) 40 words (160 bytes)
(d) 35 words (140 bytes)
[ESE Prelims : 2022]

4.26 Which one of the following are the well-known routing attacks on IoT?
(a) Clone Id and Sybil attacks
(b) Selective-reversing attacks
(c) Packet reversing attacks
(d) Frame selective wired attacks
[ESE Prelims : 2022]

4.27 Which one of the following layers in the OSI reference model is concerned with transmission of unstructured bit stream over physical medium; deals with the mechanical, electrical, functional, and procedural characteristics to access the physical medium?
(a) Transport layer (b) Network layer
(c) Data link layer (d) Physical layer
[ESE Prelims : 2022]

4.28 Which one of the following servers is a tool that allows an information provider to prepare indexes of unstructured documents, and allows users to search these indexes with natural language questions?
(a) Name server
(b) Terminal server
(c) Wide area information server
(d) File server

[ESE Prelims : 2022]

4.29 Which one of the following statements is related to frequency hopping spread spectrum?
(a) It is a spread spectrum technique which allows for the coexistence of multiple networks in the same area by separating different networks using different hopping sequences.
(b) It is a spread spectrum technique which allows for the coexistence of multiple networks in the different area by separating different networks using different hopping sequences.
(c) It is a spread spectrum technique which does not allow for the coexistence of multiple networks in the same area by separating same networks using different hopping sequences.
(d) It is a spread spectrum technique which allows for the coexistence of single network in the different area by separating different networks using same hopping sequence.

[ESE Prelims : 2023]

4.30 What is the approximate effective throughput, if user wants to fetch a 1-MB file across a 1-Gbps network with a round-trip time of 100 ms?
(a) 50.1 Mbps (b) 74.1 Mbps
(c) 84.1 Mbps (d) 90.1 Mbps

[ESE Prelims : 2023]

4.31 In a network, a transcontinental channel with a one-way latency of 50 ms and a bandwidth of 45 Mbps is able to hold how many bits that fit in the pipe approximately?
(a) 2.25×10^6 bits
(b) 1.25×10^6 bits
(c) 50.00×10^6 bits
(d) 45.00×10^6 bits

[ESE Prelims : 2023]

4.32 Consider the following statements for significance of prominence in the Internet architecture:
1. Programmers are free to define new channel abstractions or applications that run on top of any of the existing protocols.
2. It defines a common method for exchanging packets among a wide collection of networks.
3. It allows someone to propose a new protocol to be included in the architecture.

Which of the above statements is/are correct?
(a) 1 only (b) 2 and 3 only
(c) 1 and 3 only (d) 1, 2 and 3

[ESE Prelims : 2023]

4.33 Consider the following statements regarding the failure in the network:
1. Bit errors typically occur because outside forces, such as lightning strikes, power surges, and microwave ovens, interfere with the transmission of data.
2. One of the main difficulties in dealing with lost packets is distinguishing between a packet that is indeed lost and one that is merely late in arriving at the destination.
3. The failure can be caused by software that crashes, a power failure, or a reckless backhoe operator.

Which of the above statement(s) is/are correct?
(a) 1 only (b) 1 and 2 only
(c) 2 and 3 only (d) 1, 2 and 3

[ESE Prelims : 2023]

4.34 Which one of the following protocols, is used to wrap IP packets with the additional feature of multiplexing and de-multiplexing multiple processes using a single IP address?
(a) User Datagram Protocol
(b) Transport Control Protocol
(c) Internet Protocol
(d) Point-to-Point Protocol

[ESE Prelims : 2024]

4.35 Which one of the following code modules is/are used where the browser fetches from a special directory on the disk and installs as an extension to itself?
(a) Uniform Resource Locators
(b) Browser

(c) Plug-in
(d) Client server

[ESE Prelims : 2024]

4.36 Which one of the following interfaces is used to allow web servers to talk to back-end programs and scripts that can accept input and generate HTML pages in response?
(a) Application Programming Inter- face
(b) User Interface
(c) Application Interface Marker
(d) Common Gateway Interface

[ESE Prelims : 2024]

4.37 Which one of the following status code responses gives the internal server error?
(a) 200 (b) 500
(c) 100 (d) 300

[ESE Prelims : 2024]

4.38 Which one of the following attacks is one in which a malicious hacker takes over computers via the Internet and causes them to flood a target site with demands for data and other small tasks?
(a) Rootkits
(b) Distributed Denial-of-Service
(c) Phishing
(d) Smishing

[ESE Prelims : 2024]

4.39 Which one of the following is a set of programs that enables its user to gain administrator-level access to a computer without the end user's consent or knowledge?
(a) Distributed Denial-of-Service
(b) Phishing
(c) Smishing
(d) Rootkit

[ESE Prelims : 2024]

4.40 Which one of the following is software and/or hardware that monitor system and network resources and activities, and notify network security personnel when it detects network traffic that attempts to circumvent the security measures of a networked computer environment?
(a) An intrusion detection system
(b) A protection of evidence and activity logs system
(c) A critical internet security threats system
(d) An illusion detection system

[ESE Prelims : 2024]

5. e-GOVERNANCE

5.1 **Statement (I):** Digital India is a program to transform India into a digitally empowered society.
Statement (II): Digital India is a program to institute smart cities.
(a) Both Statement (I) and Statement (II) are individually true and Statement (II) is the correct explanation of Statement (I)
(b) Both Statement (I) and Statement (II) are individually true but Statement (II) is not the correct explanation of Statement (I)
(c) Statement (I) is true but Statement (II) is false
(d) Statement (I) is false but Statement (II) is true.

[ESE Prelims : 2017]

5.2 Consider the following statements:
Vision of Digital India launched by the Government of India is centered on
(i) Digital infrastructure as a utility only to senior citizens.
(ii) Governance and Services on demand
(iii) Digital Empowerment of every citizen
Which of the above statements are Correct?
(a) (i), (ii) and (iii) (b) (i) and (ii) only
(c) (i) and (iii) only (d) (ii) and (iii) only

[ESE Prelims : 2017]

5.3 PRAGATI is the acronym
(a) Pro-active governance and transparency in India
(b) Pro-active governance and timely implementation
(c) Primary governance for transparency in India
(d) Promoting and accomplishing governance with assured and timely implementation

[ESE Prelims : 2017]

5.4 The main goal of G2B (Government to Business) is:
1. To increase productivity by giving more access to information
2. To lower the cost of doing business
3. To create more transparency
Select the correct answer using the codes given below:

(a) 1 and 2 only (b) 1 and 3 only
(c) 2 and 3 only (d) 1, 2 and 3

[ESE Prelims : 2017]

5.5 Which of the following are the benefits of e-governance system?

1. Simplicity, efficiency and accountability.
2. Quality service to citizens.
3. Better access to information.
4. Expanded reach to governance.

(a) 1, 2 and 3 only (b) 1, 2 and 4 only
(c) 3 and 4 only (d) 1, 2, 3 and 4

[ESE Prelims : 2018]

5.6 Consider the following statements:

1. National Agricultural Portal, eNAM, is designed to create a unified national market for agricultural commodities.
2. Farmers can showcase their produce online from the nearest market and the buyer can quote his price from anywhere.

Which of the above statements is/are correct?

(a) 1 only (b) 2 only
(c) Both 1 and 2 (d) Neither 1 nor 2

[ESE Prelims : 2018]

5.7 **Statement (I) :** Information and communication technologies can play a key role in the development and economic growth of rural India.

Statement (II) : Successful ICT application in e-governance giving respective one-stop solutions for rural communities is an absolute need of the hour.

(a) Both Statement (I) and Statement (II) are individually true and Statement (II) is the correct explanation of Statement (I)
(b) Both Statement (I) and Statement (II) are individually true, but Statement (II) is not the correct explanation of Statement (I)
(c) Statement (I) is true, but Statement (II) is false
(d) Statement (I) is false, but Statement (II) is true

[ESE Prelims : 2018]

5.8 Which one of the following is the nodal department to implement public internet access program and rural internet connectivity by converting its offices as multi-service centres?

(a) Department of Electronics and Information Technology
(b) Department of Information and Broadcasting
(c) Department of Telecommunication
(d) Department of Posts

[ESE Prelims : 2020]

5.9 Which one of the following is **not** the vision area of Digital India as a program to transform India into a digitally empowered society and knowledge economy?

(a) Infrastructure as utility to every citizen
(b) Governance and services on demand
(c) Free Wi-Fi access
(d) Digital empowerment of citizens

[ESE Prelims : 2020]

5.10 Infrastructure aspects provided by the Government of India in formation of National e-Governance Plan for storage of data and hosting applications, network connectivity and capacity building respectively are

(a) SDC, SWAN and NISG
(b) SWAN, SDC and NISG
(c) SDC, NISG and SWAN
(d) SWAN, NISG and SDC

[ESE Prelims : 2020]

5.11 Which one of the following is not the characteristic of Good Governance and e-Governance that are closely linked and depend on each other?

(a) Accountable (b) Transparent
(c) Consciousness (d) Consensus-oriented

[ESE Prelims : 2020]

5.12 **Statement (I):** Long-term sustainability of e-Governance projects does not depend on financial viability, especially if they are to be implemented in the Public-Private Partnership (PPP) mode.

Statement (II): Front-end e-services are possible without back-end computerization.

(a) Both Statement (I) and Statement (II) are individually true and Statement (II) is the correct explanation of Statement (I)
(b) Both Statement (I) and Statement (II) are individually true but Statement (II) is not the correct explanation of Statement (I)
(c) Statement (I) is true but Statement (II) is false
(d) Statement (I) is false but Statement (II) is true.

[ESE Prelims : 2020]

5.13 Which one of the following frameworks is developed to assess the value of the increasing investments made on e-governance projects in terms of service orientation, technology architecture, replicability and sustainability in various states across the country?

(a) eTechnology Group@IMRB
(b) e-Governance Assessment Framework
(c) Sustainable Access in Rural India
(d) e-Governance Action Plan

[ESE Prelims : 2021]

5.14 Which one of the following is the latest in series being organized as the largest virtual gathering to create dialogues, and accelerate innovation in agriculture?

(a) Agri-India Hackathon 2020
(b) National Agriculture Higher Education Project
(c) ENSURE
(d) National Mission for Sustainable Development

[ESE Prelims : 2021]

5.15 **Statement (I):** Major e-Governance projects bear fruit only when application of IT is preceded by process re-engineering.

Statement (II): Initiatives which save the citizens' time, money and effort are able to succeed even when back-end computerization is not done.

(a) Both Statement (I) and Statement (II) are individually true and Statement (II) is the correct explanation of Statement (I)
(b) Both Statement (I) and Statement (II) are individually true but Statement (II) is not the correct explanation of Statement (I)
(c) Statement (I) is true but Statement (II) is false
(d) Statement (I) is false but Statement (II) is true.

[ESE Prelims : 2021]

5.16 **Statement (I)** : Content is the heart of any IT project.

Statement (II) : Implementation and maintenance of e-government projects through IT professional hired from the market is likely to result in failure of the project as the organization is bound to disown such outsiders.

(a) Both Statement (I) and Statement (II) are individually true and Statement (II) is the correct explanation of Statement (I)
(b) Both Statement (I) and Statement (II) are individually true, but Statement (II) is not the correct explanation of Statement (I)
(c) Statement (I) is true, but Statement (II) is false
(d) Statement (I) is false, but Statement (II) is true

[ESE Prelims : 2022]

6. TECHNOLOGY BASED EDUCATION

6.1 Consider the following as advantages of ICT tools in educational systems:

1. Increased capacity and cost effectiveness of the educational system
2. Achievement by target groups that had limited access to traditional education
3. Support for improvement of the quality and relevance of existing structures of education
4. Provision of links between various educational institutions for knowledge sharing

Which of the above statements are correct?

(a) 1, 2 and 3 only (b) 1, 2 and 4 only
(c) 3 and 4 only (d) 1, 2, 3 and 4

[ESE Prelims : 2017]

6.2 Which of the following sets of free software tools are suitable for ICT-based education as well as an open source?

(a) Scilab, Osdag, PHP and Latex
(b) Java, LibreOffice, Audacity and Matlab
(c) Scilab, Arduino, LibreOffice and Latex
(d) Scilab, Octave, Netduino and Latex

[ESE Prelims : 2018]

6.3 Which one of the following is **not** a Creative Commons license which users can choose to apply when publishing their work?

(a) Attribution
(b) Share-Alike
(c) Copyright Infringement
(d) No Derivative Works [ESE Prelims : 2020]

6.4 Which one of the following is **not** the skill needed in the workplace of the future for inventive thinking using information and communication technology in education?

(a) Adaptability
(b) Responsibility
(c) Curiosity and creativity
(d) Risk-taking [ESE Prelims : 2020]

6.5 The pedagogy which involves productive learning and finding new solutions to problems, where manipulation of existing information and creation of real-world products are possible with ICT, is called
(a) collaborative pedagogy
(b) creative pedagogy
(c) integrative pedagogy
(d) evaluative pedagogy [ESE Prelims : 2020]

6.6 Which one of the following learnings uses web technology to conduct conventional classes with distant learners?
(a) Learner-led e-learning
(b) Instructor-led e-learning
(c) Telementoring and e-coaching
(d) Facilitated e-learning
[ESE Prelims : 2021]

6.7 Which one of the following is an advantage of branched or Intrinsic programming type or style in ICT based teaching and learning process?
(a) Large frames reduce the time of learning
(b) There is a possibility of guess-work
(c) It is very expensive to provide so many audio-visual aids
(d) Revise/redesign at frequent intervals is difficult and expansive
[ESE Prelims : 2022]

6.8 Which one of the following learnings is a teaching approach that engages students in sustained, collaborative real-world investigations?
(a) Project-based learning
(b) Cooperative learning
(c) Collaborative learning
(d) Outcome based learning
[ESE Prelims : 2022]

6.9 Which of the following issues were addressed while establishing an IT Policy?
1. Respect of the intellectual rights of others, including trade secrets, copyrights, patents, and trade- marks
2. Inappropriate use of IT resources, such as Web surfing, blogging, personal emailing, and other use of computers for purposes other than business.
3. The need to protect the security of IT resources through adherence to good security practices, such as not sharing user IDs and passwords, using hard-to-guess passwords, and frequently changing passwords.
4. The use of the computer to intimidate, harass, or insult others through abusive language in emails and by other means.

Select the correct answer using the code given below:
(a) 1 and 3 only (b) 2 and 3 only
(c) 2, 3 and 4 only (d) 1, 2, 3 and 4
[ESE Prelims : 2024]

7. RECENT DEVELOPMENT RELATED TO ICT

7.1 Consider the following statements regarding Holostore:
1. It is a device that reads and writes data in at optical form
2. It is a computer storage device
3. It refers to Institutions where holography is taught.

Which of the above statements are correct?
(a) 1, 2 and 3 (b) 1 and 3 only
(c) 1 and 2 only (d) 2 and 3 only
[ESE Prelims : 2017]

7.2 In a radar system, the term 'Rat-Race' is used in connection with
(a) Modulator
(b) Pulse characteristics
(c) Receiver Bandwidth
(d) Duplexer [ESE Prelims : 2018]

7.3 Consider the following statements regarding Cybersecurity Multi-Donor Trust Fund:
1. The World bank has launched a new 'Cybersecurity Multi-Donor Trust Fund' to better roll out cybersecurity development agenda in a systematic manner.
2. The new fund has been developed as an associated trust fund under the broader Digital Development Partnership (DDP) umbrella program.
3. World Bank has partnered with four countries, namely Estonia, Japan, Germany, and the Netherlands, to launch the fund.

Which of the above statements are correct?
(a) 1 and 2 only (b) 2 and 3 only
(c) 1 and 3 only (d) 1, 2 and 3

[ESE Prelims : 2022]

7.4 Which one of the statements is NOT relevant to quantum computing?
(a) Quantum computing is that much more powerful functions may be computed using qubits and quantum gates.
(b) Quantum operations are well adapted to describe discrete state changes, that is, transformations between an initial state and final state, without explicit reference to the passage of time.
(c) Quantum computation does not support entanglement and measurements of a quantum computer's registers can yield only a small, discrete set of values.
(d) Quantum computing is the use of quantum phenomena such as superposition and entanglement to perform the computation.

[ESE Prelims : 2023]

7.5 Which one of the following Acts mandates schools and libraries in India to use some form of technological protection to block computer access to obscene material, pornography, and anything else considered harmful to minors?
(a) Telecommunications Act
(b) Child Online Protection Act
(c) Children's Internet Protection Act
(d) Communications Decency Act

[ESE Prelims : 2024]

7.6 Which of the following Acts is required for the commercial emailers in sending out messages that advertise a commercial product or service?
(a) Controlling the Assault of Non- Solicited Pornography Marketing Act
(b) Communications Assistance for Law Enforcement Act
(c) Communications Act of 1934
(d) Communications Decency Act

[ESE Prelims : 2024]

■■■■

Information and Communication Technologies (ICT) : Answer Key

1.1 (b)	**1.2** (c)	**2.1** (c)	**2.2** (d)	**2.3** (c)	**2.4** (a)	**2.5** (a)	**2.6** (b)
2.7 (a)	**2.8** (b)	**2.9** (a)	**2.10** (a)	**3.1** (b)	**3.2** (d)	**4.1** (d)	**4.2** (c)
4.3 (d)	**4.4** (a)	**4.5** (c)	**4.6** (a)	**4.7** (a)	**4.8** (b)	**4.9** (b)	**4.10** (d)
4.11 (a)	**4.12** (a)	**4.13** (a)	**4.14** (b)	**4.15** (b)	**4.16** (c)	**4.17** (b)	**4.18** (a)
4.19 (d)	**4.20** (a)	**4.21** (d)	**4.22** (a)	**4.23** (c)	**4.24** (b)	**4.25** (b)	**4.26** (a)
4.27 (d)	**4.28** (c)	**4.29** (a)	**4.30** (b)	**4.31** (a)	**4.32** (d)	**4.33** (d)	**4.34** (b*)
4.35 (c)	**4.36** (d)	**4.37** (b)	**4.38** (b)	**4.39** (d)	**4.40** (a)	**5.1** (c)	**5.2** (d)
5.3 (b)	**5.4** (d)	**5.5** (d)	**5.6** (c)	**5.7** (a)	**5.8** (d)	**5.9** (c)	**5.10** (a)
5.11 (c)	**5.12** (d)	**5.13** (b)	**5.14** (a)	**5.15** (b)	**5.16** (d)	**6.1** (d)	**6.2** (a, c)
6.3 (c)	**6.4** (b)	**6.5** (b)	**6.6** (b)	**6.7** (b)	**6.8** (a)	**6.9** (d)	**7.1** (c)
7.2 (d)	**7.3** (d)	**7.4** (c)	**7.5** (c)	**7.6** (a)			

EXPLANATIONS

1.1 (b)

Statement (I) brings out a broad general utility or positive outcome of Information and Communication Technology (ICT), while Statement (II) puts-forth a real limitation of ICT despite some important utility. So, both statements are only individually true.

1.2 (c)

Doxing is a way of tracing someone or getting information about an individual using sources on the internet and social engineering techniques.

2.3 (c)

Individual statements that are used in a program in a High-Level Language (HLL) are called statements, and a programming statement nearly in all high-level programming languages (say Application Programming Interfaces or APIs) includes keywords, operators and punctuation. While Machine Level Languages (MLLs) are binary based say Operating System (OS), High Level Languages (HLLs) say multiple Application Programs are based on Alpha-Numerals.

2.4 (a)

A real time system is used when there are rigid time requirements on the operation of a processor or the flow of data, and thus is often used as a control device in a dedicated application.

2.6 (b)

Device status table contain entry for each I/O device indicating its type, address and state.

2.8 (b)

To maintain a consistent appearance across multiple pages on a complex website produced by multiple authors, the feature commonly used is CSS (Cascading Style Sheet). CSS allows to define and apply consistent styles (such as colors, font, and layouts) across various pages, ensuring a unified look and feel throughout the website.

2.9 (a)

eXtensible Markup Language (XML) allows for the representation of structured data in a format that is easy to understand and process by both humans and automated systems. XML It is commonly used for developing web pages, especially when the data needs to be structured for automated processing or interchange between different systems or applications.

2.10 (a)

A computer worm is a standalone malware computer program that replicates itself in order to spread to other computers. It often uses a computer network to spread itself, relying on security failures on the target computer to access it. It will use this machine as a host to scan and infect other computers.

3.2 (d)

Infrared waves are used for short range communication like the remote control for televisions, VCRs, wireless mouse, keyboard, printer, etc. It uses unbounded wireless media that does not require any physical medium to transmit electromagnetic signals.

4.5 (c)

Network Layer is layer of internet just above Data Link Layer and below Transport Layer in 7-Layered Open System Interconnection (OSI) Model. In a computer network issues like delay, transit time, jitter, routing, etc affect the quality of network services and these are in normal course part of Network Issue. Network Layer includes routing services, IP addressing, internet traffic control, etc. Though Application Layer – the top-most layer serves as the main layer for application services or Application Programming Interface (API).

4.6 (a)

A transmission between a single transmitter and a single receiver is a unicast transmission, thus in it packets are sent to a single network destination in a particular address (one-to-one). Debra Cameron, '*'Internet2' – The Future of the Internet and Next-generation Initiatives*' "Unicast transmission is sent point to point from one sender to one receiver". While, multicast transmission denotes transmission of information between one or several transmitters and several receivers (one-to-many). Whereas in broadcast it is one-to-all that is to all possible destinations.

4.7 **(a)**

Domain name of a website is hosted on some IP Address, while Uniform Resource Locator (URL) is the unique web address of a web page. Thus, multiple web pages in a website have unique web addresses in that domain name. Hyper Text Transfer Protocol (HTTP) or its more secure version Hyper Text Transfer Protocol Secure (HTTPS) is used for data transmission in the web traffic. Uniform Resource Locator – main form of Uniform Resource Identifier (URI) – obvious, is made of retrieval method/protocol (HTTP here), the host server and the domain name and that web page address/location.

4.8 **(b)**

Common Gateway Interface (CGI) has traditionally been used (i.e., in web 1.0) for interactive web pages or communication in then World Wide Web (WWW). Active Servers Pages (ASP) are used for more interactivity of web pages. Common Gateway Interface (CGI) specifications govern the interaction of web servers and external programs.

4.9 **(b)**

Pretty Good Privacy (PGP) is a program that uses encryption to protect the privacy of e-Mail or even other data files. Now it is more popular email encryption software used for digitally signing emails or other data thus enhancing security features.

4.10 **(d)**

Coding audio-video information for example music, videos, movies, etc (for example there is Moving Pictures Expert Group - MPEG) has led to multiple benefits. As every object has binary mask thus more interaction, modification and re-use is possible. There is also use of compression standards which obviously leads to better scalability of the content. It has also led to the growth of interactive web-based creative multi-media content and Digital TV. Coded multimedia audio-video digital content can be better synchronized and integrated.

4.11 **(a)**

Bandwidth Allocation Mechanism (BAM) is based on the Stream Peak Rate, and it is used inside Premium Service Architecture in order to support real-time stored stream transmission. Thus, it helps in checking the need for additional bandwidth allocation in multi-media transmission in any communication channel.

4.12 **(a)**

Pre-echo, also referred to as forward echo, can occur when transient sound can cause a type of distortion, and it denotes theoretical limit on compressibility of a particular signal. Pre-echo occurs when signals with sharp attack begin near the end of transform block immediately following a region of low energy.

4.13 **(a)**

Fast Ethernet created by IEEE - under the name 802.3u - can transmit data 10 times faster at the rate of 100 Mbps. 100 BASE-T or IEEE 802.3u aims also include competing with Fiber Channel, compatibility with Standard Ethernet, etc. However, the length of the cable - range - in copper based Fast Ethernet is lesser as compared to Fiber Mode.

4.14 **(b)**

Bluetooth or IEEE 802.15.1 Standards came up as cost-effective (Wireless Personal Area Network (W-PAN) Technology-Type with one Primary Node and maximum Seven Secondary Nodes, named after Herald Blatand. While IEEE 802.11 standards or Wi-Fi is primarily a solution for Wireless Local Area Network (W-LAN). Frequency range for Bluetooth is 2.400 GHz to 2.485 GHz while for Wi-Fi it is 2.40 GHz to 5.0 GHz.

4.15 **(b)**

National Internet Service Providers (ISPs) are connected together through Private Switching Stations which are referred to as Peering Points or even Internet Exchange Points (IXPs) normally operating at high data rate say upto 600 Mbps. And the benefits in the direct interconnection through them include high data rate, better routing efficiency and cost-effectiveness.

4.16 (c)

In packet switch network data is transmitted in the form of multiple smaller packets through multiple paths while routing is a process of selecting a path for traffic in a network but, in packet switching networks routing is the higher level decision making that directs network packets from source to destination. The protocol used to address the true routing decision problem is "Open Shortest Path First Protocol". Routing decisions are made by routing information exchange between routers.

4.17 (b)

802.11 is a WLAN (Wireless Local Area Network) standard developed by IEEE (Institute of Electrical and Electronics Engineers). 802.11 x refers to a family of specifications developed by the IEEE for wireless LAN technology where IEEE 802.11 P is used in vehicular communication system.

4.18 (a)

Performance of a network is the measurement of service quality of a network. The characteristics that measure the performance of a network are :

Bandwidth : Amount of bandwidth allocated to network.

Throughput : Data sent successfully per unit time.

Latency (Delay) : Time taken by data to reach from sender to receiver.

Jitter : Disturbance in the network.

4.19 (d)

URL (Uniform Resource Locator) contains the name of the protocol needed to access a resource. Second part identifies IP address or domain name and subdomain. After this, URL also specifies path to a specific page or file within a domain. A network port to use to make the connection.

e.g. :https://upsc.gov.in/notification **protocol domain path.**

4.20 (a)

Common Gateway Interface (CGI) that acts like a middleman that passes user's request to external databases and receives the processed data. Hence, in HTTP and www CGI handles 'Dynamic Documents'.

4.21 (d)

Cloud computing is primarily broadly categorised as:

1. Infrastructure-as-a-Service (IaaS),
2. Platform-as-a-Service (PaaS), and
3. Software-as-a-Service (SaaS). Even though Big Data as a Service (BDaaS) makes use of cloud computing infrastructure and services but as such it is not categorised as the three main categories of cloud computing models.

4.22 (a)

Data Encryption Standard (DES) is a type of block cipher that encrypts data in blocks of size 64 bits each. But key length is 56 bits because in DES process every 8th bit of key is discarded. In 64 bits a total of 8 bits are discarded.

Total key length in DES : 64 – 8 = 56 bits.

4.23 (c)

In SSC protocol, communication between applications i.e. client and server over a network remains secure. SSL works by using certificate for authentication of client and server and uses encryption/decryption with unique key.

4.24 (b)

A broadcasting method in radio where frequency modulation is used. In this method, each radio station has its own frequency.

4.25 (b)

In AES key expansion algorithm a word key is taken to produce a linear array of 44 words. Each round uses 4 of these words and each word contains 32 bytes which means each subkey is 128 bit long.

4.26 (a)

Network attacks are those attacks that damage an entire system by orchestrating the IoT network. Some well known routing attacks on IoT are: Selective forwarding attacks, sink hole attack Hellow flood attack, Wormhole attack, clone Id, etc.

4.27 (d)

Physical layer is responsible for the communication of unstructured raw data streams over a physical medium. It defines a range of aspect that includes electrical, mechanical, and physical system and networking device that includes specifications such as cable size, signal frequency, voltage, etc.

4.28 (c)

WAIS (Wide Area Information Server) is an internal system in which specialized subject databases are created at multiple server locations, kept track of by a directory of servers at one location and made accessible for searching by users with WAIS client programme.

4.29 (a)

Frequency hopping : Keep changing from one frequency to another to avoid eaves dropping. Spread spectrum means co-existence of multiple network in the same area by separating different network using different hopping sequence.

4.30 (b)

Throughput = data size/(t.t + 2p.t)

2 × p.t = RTT = 100 millisec

Transmission time = data size/BW

= 8 M bits/1000 Mbps = 8 milli sec

Throughput = 8 × 1 Mbits/(8 + 100) milli sec

= 8000/108 = 74.07 Mbps

4.31 (a)

1 sec = 45 × 106 bits

50 ms = 45 × 106 × 50 × 10–3 bits

= 2.25 × 106 bits

4.34 (b*)

UDP (User Datagram Protocol) and TCP (Transmission Control Protocol) perform the demultiplexing and multiplexing jobs by including two special fields in the segment headers: the source port number field and the destination port number field. The port number field helps to differentiate between data streams from multiple applications through a single IP address.

4.35 (c)

In the context of web browsers, plug-ins are often used to extend browser capabilities by allowing the browser to interpret and display certain types of content that it wouldn't natively support. These plug-ins can be fetched from special directories on the disk and installed as extensions to the browser.

4.36 (d)

The interface used to allow web servers to communicate with backend programs and scripts, which can accept input and generate HTML pages in response, is typically CGI. Common Gateway Interface (CGI) is a standard protocol that enables web servers to execute external program or scripts and include their output in HTTP responses. This allows web servers to add dynamic content to web pages, such as forms, generated images, and other interactive features.

4.37 (b)

The HTTP status code that indicates an internal server error is 500. This status code is a generic message indicating that the server has encountered an unexpected condition and cannot fulfill the request.

4.38 (b)

A distributed denial-of-service (DDoS) attack is a malicious attempt to disrupt the normal traffic of a targeted server, service or network by overwhelming the target or its surrounding infrastructure with a flood of Internet traffic which may consists of data requests, pings, or other small tasks, causing the target site's servers to become overloaded and unable to respond to legitimate requests.

4.39 (d)

A rootkit is a set of programs that enables its users to gain administrator-level access to a computer without the end user's consent or knowledge. It is a malicious software that hides its presence from detection by antivirus programs or other security tools.

4.40 (a)

The software/hardware designed to monitor system and network resources, detect anomalous activities and notify network security personnel when it identifies attempts to circumvent security measures of a networked computer environment is: An intrusion detection system (IDS).

IDS tools monitor network or system activities for signs of unauthorized access, attack or policy violation.

5.1 (c)

The Indian Government launched the Digital India campaign to make government services available to citizens electronically by online infrastructure improvement and also by enhancing internet connectivity. It also aims to empower the country digitally in the domain of technology. Prime Minister Narendra Modi launched the campaign on 1st July 2015.

5.3 (b)

PRAGATI is an information and communications technology-based platform for, as the name reveals, pro-active governance and timely implementation.

1. It was launched in 2015.
2. It is an interactive and integrated platform chiefly for addressing the grievances of the common man, while at the same time, monitoring government projects and programmes/schemes.
3. The platform was designed by the PMO in coordination with the National Informatics Centre.
4. The chief aims of the PRAGATI platform: • Project implementation

- Project monitoring
- Grievance redressal

The platform combines three technologies namely,

- Digital data management
- Video conferencing
- Geo-spatial technology

The platform also brings together the different secretaries of the GOI and the States' Chief Secretaries, along with the PM.

The PM discusses the issues related to the projects and is also presented with the ground-level situation.

5.6 (c)

- e-NAM or the e-trading platform (online trading portal) for the National Agriculture Market (NAM) was launched by the Prime Minister of India, on April 2016.
- e – National Agriculture Market (e-NAM) is a pan-India electronic trading portal that nets the prevailing Agricultural Produce Market Committees (APMC) Mandis for making a united national market for agricultural commodities.
- The e-NAM project would operate via the online portal that is linked to the states' Mandis (Wholesale markets). All the participating states will be providing the software (Website and Mobile Application) for e-NAM at no cost.

5.8 (d)

"A total of 150,000 Post Offices are proposed to be converted into multi service centres. Department of Posts would be nodal department to implement Public Internet Access Programme".

5.9 (c)

Three key vision area of Digital India are:

1. Digital Infrastructure as a Core Utility to Every Citizen.
2. Governance and Services on Demand, and
3. Digital Empowerment of Citizens.

5.10 (a)

State Data Centres (SDCs): Secure Data Storage, hosting applications and online delivery of services, remote management and service integration. etc.

State Wide Area Networks (SWAN): As the converged backbone network for data, voice and video communication throughout any State/UT, and

National Institute for Smart Government (NISG): Capacity building, consulting services, project management and talent acquisition.

5.11 (c)

The World Bank earlier identified four basic elements of Good Governance:

1. Accountability
2. Participation
3. Predictability, and
4. Transparency

Then, there came expansion of these or addition to these either by World Bank itself or by even other institutions like UNDP which include consensus oriented, responsive and inclusive, Rule of Law, effective and efficient, etc. Online-participation also much helps in consensus-building.

5.12 (d)

Long-term sustainability of e-Governance Projects without financial viability is neither possible nor feasible. While, Front-end e-services at-times are possible without back-end computerisation, though not that feasible, or without integrative approach e-service cannot work well. For example, even if some policy approvals at back end are not computerised, even then some front-end service applications are possible to be brought.

5.13 (b)

In India there is 'National eGovernance Service Delivery Assessment Framework', globally too there is UN e-Government Assessment Framework for an assessment of e-Government readiness of the member countries. In India, in order to undertake assessment of projects and enable comparison of project performance projects are assessed on Assessment Framework.

5.14 (a)

- Agri-India Hackathon is the largest virtual gathering to create dialogues and accelerate innovations.
- Virtual Agri-hackathon 2020 is organised by the Department of Agriculture and Cooperation and Farmers' Welfare in association with IARI, Pusa, New Delhi.

5.15 (b)

Second Administrative Reforms Commission (SARC) - set up in 2005 - brought out Fifteen Reports, and its Eleventh Report (brought in December 2008) is on e-Governance (Promoting e-Governance - The SMART Way Forward).

Both the statements given in this question are given in this Report.

The Main aim of Process Reengineering is removing unnecessary process and modernising and simplifying essential ones making use of modern technology for better efficiency and service delivery.

The Report also observes that for complete transformation of governance there has to be an end-to-end ICT enablement coupled with process re-engineering.

5.16 (d)

- Once the automation starts, the primary function can not be changed. Hence statement (I) is wrong.
- To provide information about the product to the designer and production people is the primary function.

6.4 (b)

Skills needed for innovative thinking in Digital Age:

1. Adaptability/Ability to manage complexity
2. Curiosity-desire to know as fuel for life-long learning.
3. Creativity-using the imagination to develop new and original things, and
4. Risk taking-the willingness to place something valued in a position or situation where it could be exposed to damage or loss (say for quantum leaps).

6.5 (b)

Creative Learning: Promotes manipulation of existing information and creation of real-world products.

Integrative Learning: Overcomes artificial separation between disciplines and links theory and practice, or even thematic learning.

Evaluative Learning: Student directed and diagnostic rather than static approach to learning.

Active Learning: Small group of learners actively involve in activities and pace of learning, and

Collaborative learning: More collaboration and interaction between students, teachers, experts, etc.

6.6 (b)

e-Learning is a method where learner can learn from anywhere and anytime. Generally, e-learning does not require a presence of instruction. But in above question it is asking about conducting conventional, classes using web technology. Hence, presence of virtual instructor is there. Therefore, answer is (b).

6.7 (b)

In branching or intrinsic programming student is provided a piece of information, and a situation with multiple choice. A student who responds incorrectly will either be returned to the original frame and a student who responds correctly will advance to the next frame. Hence, there is a scope of guess work in it.

6.8 (a)

Projects are organized around a driving question. A teaching method or approach that engages student in sustained, collaborative, real world investigation is project based learning.

6.9 (d)

Issues addressed in IT Policy:

- Security concern
- Data Privacy
- Uses Policy
- Access control
- Licensing
- Network infrastructure
- Cloud computing
- Ethical use

7.3 (d)

- The World Bank has launched a new Cybersecurity Multi-Donor Trust Fund aimed at better defining and systematically rolling out the cybersecurity development agenda to help ensure a more substantial reflection of cybersecurity considerations across its programs and financing.
- Developed as an associated trust fund under the World Bank's broader Digital Development Partnership Umbrella (DDP), the Cybersecurity Multi-Donor Trust Fund's launch was made possible with donor contributions from Estonia, Germany, Japan, and the Netherlands.
- The new Cybersecurity Multi-Donor Trust Fund will help ensure that countries and their citizens can safely and securely take full advantage of the ongoing digital transformation and development

7.4 (c)

Facts:

- Quantum computing uses phenomena in quantum physics to create new ways of computing.
- Quantum computing involves qubits.
- Unlike a normal computer bit, which can be either 0 or 1, a qubit can exist in a multidimensional state.
- The power of quantum computers grows exponentially with more qubits.
- Classical computers that add more bits can increase power only linearly.

7.5 (c)

The Children's Internet Protection Act (CIPA) was enacted in 2000 to address concerns about children's access to obscene or harmful content over the Internet. This Act requires the schools and libraries to install filters on computers used by minors and adults.

7.6 (a)

The CAN-SPAM Act [Controlling the Assault of Non- Solicited Pornography Marketing Act] of USA is a law that sets rules for commercial email, establishes requirements for commercial messages, gives recipients the right to have you stop emailing them, and spells out tough penalties for violations. The Federal Trade Commission (FTC) enforces the CAN-SPAM Act.

■■■■

Ethics and Values in Engineering Profession

CONTENTS

Ethics and Values in Engineering Profession

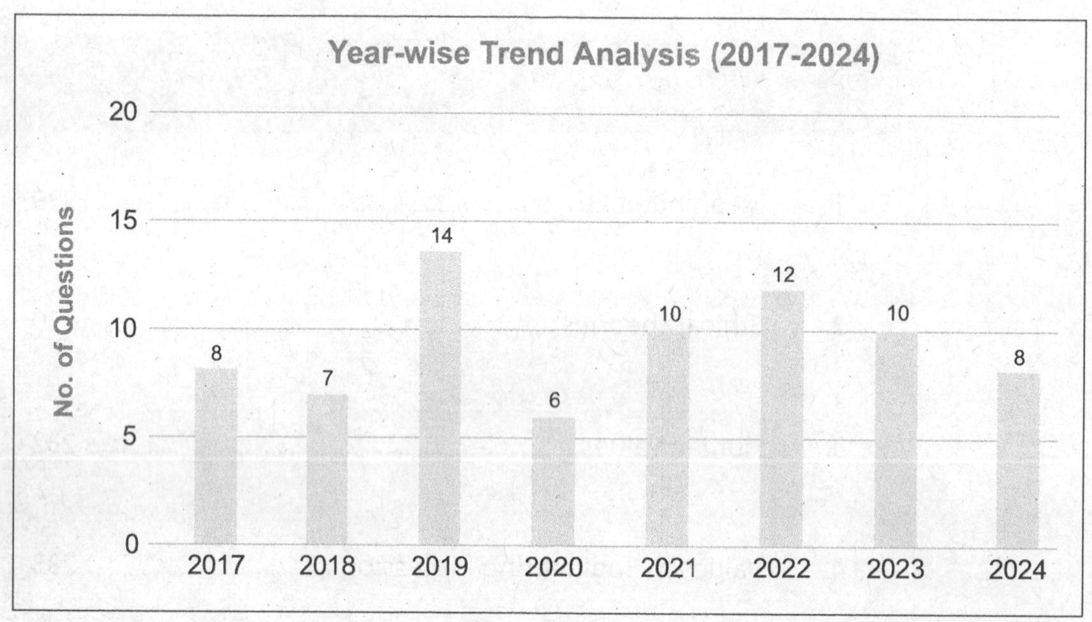

TOPIC-WISE TREND ANALYSIS (2017-2024)		
Sl. No.	**Topics**	**Total Number of Questions**
1.	Meaning of Ethics	9
2.	Ethical Theories	18
3.	Human Values	9
4.	Values in Engineering Profession	14
5.	Ethical issues, Dilemmas, Value Conflicts & Unethical Behaviour in Engineering	11
6.	Responsibilities of Engineers	8
7.	Codes of Conduct	6

Ethics and Values in Engineering Profession

1. MEANING OF ETHICS

1.1 The Ethical skills are:
(a) Care of possessions, modesty, posture, self-reliance, tidy appearance
(b) Good behaviour, good manners, good relationships, good environment
(c) Attention, calmness, concentration, self-confidence, self-esteem
(d) Code of conduct, responsibility, efficiency, perseverance, punctuality

[ESE Prelims : 2017]

1.2 Ethical issues that can affect an Engineer's professional and personal life are termed as
(a) Macro-ethics (b) Micro-ethics
(c) Morals (d) Rights

[ESE Prelims : 2017]

1.3 Which one of the following statement is correct with respect to the '*societal development*'?
(a) Behaviour grows into habits, habits into tradition and tradition becomes custom
(b) Customs grow into mores and mores grow into custom
(c) Behaviours grow into customs and customs grow into traditions
(d) Folkways grow into tradition and traditions grow into customs

[ESE Prelims : 2019]

1.4 One branch of ethical philosophy-claims that it is possible to know right from wrong or good from bad in a very clear and objective manner, is called
(a) Non-Cognitivism (b) Ethical Pluralism
(c) Cognitivism (d) Utilitarianism

[ESE Prelims : 2019]

1.5 Consider the following statements regarding '*Engineering Ethics*':
1. It is the activity of understanding moral values
2. It resolves the moral issues and justifies moral judgments
3. It would refer to the set of specifically moral problems and issues related to Engineering

Which of the above statements are correct?
(a) 1, 2 and 3 (b) 1 and 2 only
(c) 1 and 3 only (d) 2 and 3 only

[ESE Prelims : 2019]

1.6 The philosophical study of beliefs and knowledge is better known as
(a) Ontology (b) Epistemology
(c) Entomology (d) Etymology

[ESE Prelims : 2019]

1.7 The basic difference between a professional and an amateur is
(a) a professional is someone who is connected with a job that needs special training or skill, while an amateur is someone who works in multi-dimensions without any specialization
(b) a professional is clear in thinking and focused on the job, while an amateur is confused and distracted from the job
(c) a professional does high quality work/job in a specific area, while an amateur is associated with specific area with lowest pay
(d) a professional remains positive and achieves despite facing grievances, while an amateur does work efficiently due to many imagined grievances

[ESE Prelims : 2020]

1.8 'Euthanasia' refers to the
(a) loyalty of the people that take pride in being part of their organization and care for the organization above their own well-being
(b) ills in the society that are caused by ignorance and lack of respect for the laws of the land
(c) emotional intelligence to understand how people perform various functions
(d) killing of a terminally ill person suffering acutely with no hope of survival

[ESE Prelims : 2020]

1.9 Which one of the following does NOT come under business ethics?
(a) Avoid breaking the law
(b) Avoid actions that are bad for one's image
(c) Avoid action
(d) Avoid conflict [ESE Prelims : 2022]

2. ETHICAL THEORIES

2.1 Consider the following factors in making ethical judgement:
1. The motive from which the action springs
2. The nature of the act itself, including the means adopted
3. The resulting consequences

Which of the above factors are correct?
(a) 1 and 2 only (b) 1 and 3 only
(c) 1, 2 and 3 (d) 2 and 3 only
[ESE Prelims : 2017]

2.2 **Statement (I):** What is legal may not always be ethical.
Statement (II): Ethical standards and the law, share the same theme, i.e., what is permissible and impermissible.
(a) Both Statement (I) and Statement (II) are individually true and Statement (II) is the correct explanation of Statement (I)
(b) Both Statement (I) and Statement (II) are individually true, but Statement (II) is not the correct explanation of Statement (I)
(c) Statement (I) is true, but Statement (II) is false
(d) Statement (I) is false, but Statement (II) is true
[ESE Prelims : 2018]

2.3 Which one of the following statement is correct with respect to '*the convergence theory*' on social change?
(a) The societal adaptive culture is changing more slowly
(b) As societies become modernized, they begin to resemble one another more closely
(c) The developed countries show more growth in social changes than the less developed countries
(d) Strong opposition by old people retards the social change
[ESE Prelims : 2019]

2.4 'Utilitarianism' in the professional ethics is
(a) an acquired habit that helps to lead a rational life
(b) a skill to solve a current ethical problem by comparing it with similar problems from the past and their outcome
(c) a right of activists to decide their own duties
(d) a judgment of an action by the consequences of that action
[ESE Prelims : 2020]

2.5 The basic ethical principle of 'Beneficence' states that
(a) all our thoughts and actions must be directed to ensure that others benefit from these thoughts and actions
(b) our actions must result in the least harm to the others
(c) we should not impose our views on others
(d) our actions must be fair to everyone
[ESE Prelims : 2020]

2.6 Match the following:
List-I
A. Act Utilitarian Theory
B. Rule Utilitarian Theory
C. Duty Ethics Theory
D. The Rights Theory

List-II
1. John Locke
2. Immanuel Kant
3. Richard Brandt
4. J.S. Mill

Select the correct matching using code given below:

	A	B	C	D
(a)	4	3	1	2
(b)	4	3	2	1
(c)	1	2	3	4
(d)	1	2	4	3

[ESE Prelims : 2021]

2.7 'Groupthink', a noteworthy feature of the organizational settings within which engineers work and deliberate in groups, has been suggested by
(a) Abraham Maslow (b) Irving Janis
(c) B.F. Skinner (d) Christopher Meyers
[ESE Prelims : 2021]

2.8 Carol Gilligan is associated with
(a) the natural justice
(b) the responsibility
(c) the principle of loyalty
(d) the ethics of care [ESE Prelims : 2021]

2.9 **Statement (I):** Moral pluralists maintain that there are moral truths, but they do not form a body of coherent and consistent truths in the way that one finds in science or mathematics.
Statement (II): Moral truth are real, but partial.
(a) Both Statement (I) and Statement (II) are individually true and Statement (II) is the correct explanation of Statement (I)
(b) Both Statement (I) and Statement (II) are individually true, but Statement (II) is not the correct explanation of Statement (I)
(c) Statement (I) is true, but Statement (II) is false
(d) Statement (I) is false, but Statement (II) is true
[ESE Prelims : 2021]

2.10 Which one of the following is NOT a type of virtue for responsible professionalism?
(a) Public-spirited virtue
(b) Team work virtue
(c) Self-realization virtue
(d) Self-direction virtue [ESE Prelims : 2022]

2.11 Arrange the following in hierarchical order as suggested by Carroll in the four-part model of corporate social responsibility?
1. Economic 2. Philanthropic
3. Legal 4. Ethical
Select the correct answer using the code given below:
(a) 2, 4, 3, 1 (b) 4, 3, 1, 2
(c) 2, 1, 3, 4 (d) 1, 3, 4, 2
[ESE Prelims : 2022]

2.12 Unlike newspapers in the old days, today's newspapers and televised news programs are full of stories about murders and assaults in our city. One can only conclude from the change that violent crime is now out of control, and, to be safe from personal attack, one should not leave one's home except for absolute necessities.
Which one of the following, if true, would cast the most serious doubt on the conclusion?
(a) Newspapers and televised news programs have more comprehensive coverage of violent crime than newspapers did in the old days.
(b) National data show that violent crime is out of control everywhere, not just in the author's city.
(c) Police records show that people experience more violent crimes in their own neighborhoods than they do outside their neighborhoods.
(d) Murder comprised a larger proportion of violent crimes in the old days than it does today.
[ESE Prelims : 2022]

2.13 Consider the following objectives of the study on professional ethics :
1. Forming consistent viewpoints based on facts.
2. Searching beyond obvious the alternative responses to issues and being receptive to creative solutions
3. Comprehending, assessing different views.
Which of the above objective(s) is/are correct?
(a) 2 and 3 only (b) 1, 2 and 3
(c) 2 only (d) 1 and 3 only
[ESE Prelims : 2023]

2.14 The normative sense of engineering ethics does NOT include:
(a) Knowing moral values, finding accurate solutions to moral problems and justifying moral Judgements in engineering practices.
(b) Generating alternate courses of action to resolve the dilemma.
(c) Study of decisions, policies, and values that are morally desirable in the engineering practice and research.
(d) Using codes of ethics and standard and applying them in their transactions by engineers.
[ESE Prelims : 2023]

2.15 Which one of the following characteristic features distinguishes Carol Gilligan's theory from Kohlberg's theory with regard to the moral development.
(a) Transactional approach
(b) Logic and rule centric
(c) More of caring
(d) Justice
[ESE Prelims : 2023]

2.16 Which one of the following theorists and philosophers in NOT associated with the 'Duty Ethics' ?
(a) Immanuel Kant (b) John Locke
(c) John Rawl (d) C.W.D. Ross
[ESE Prelims : 2023]

2.17 **Statement (I):** Ethics involves the discipline of systematic enquiry into moral norms of standards of behavior and their underlying values and justification.
Statement (II): Applied ethics looks into the way in which moral value can be applied to particular areas of concern such as business.
(a) Both Statement (I) and Statement (II) are individually true and Statement (II) is the correct explanation of Statement (I).
(b) Both Statement (I) and Statement (II) are individually true but Statement (II) is **not** the correct explanation of Statement (I).
(c) Statement (I) is true but Statement (II) is false.
(d) Statement (I) is false but Statement (II) is true.
[ESE Prelims : 2023]

2.18 Consider the following statements: The philosopher W. D. Ross, who constructed a list of basic duties or obligations, which he called prima facie duties. His lists of prima facie duties are given below:
1. Duties resting on previous acts.
2. Duties of gratitude, Duties of justice.
3. Duties of beneficence, Duties of self-improvement.
4. Duties to injure others, unexceptional to be widely practiced.

Which of the above statements are correct?
(a) 1, 2 and 4 (b) 1, 3 and 4
(c) 1, 2 and 3 (d) 2, 3 and 4
[ESE Prelims : 2024]

3. HUMAN VALUES

3.1 The famous statements "The weak can never forgive. Forgiveness is the attribute of the strong" is given by
(a) Swami Vivekananda
(b) Mahatma Gandhi
(c) Martin Luther
(d) Sri Aurobindo [ESE Prelims : 2021]

3.2 Which philosopher suggested Wisdom, Courage, Temperance and Justice as four 'cardinal virtues'?
(a) Aristotle (b) Aquinas
(c) Socrates (d) Plato
[ESE Prelims : 2021]

3.3 Select inappropriate statement about integrity
(a) It involves the discovery and communication of the truth
(b) It leads to a concern for the whole situation in decision-making, including an awareness of the professional's own attitudes, standards and value systems
(c) It is simply truthfulness or avoidance of lying
(d) It ensures that the professional does not accept 'moral distance'
[ESE Prelims : 2021]

3.4 Which one of the following is NOT included in Nussbaum's basic human functional capabilities?
(a) Being able to live a human life of normal length
(b) Being able to use senses, imagine, think, and reason
(c) Being able to laugh, play, and enjoy recreational activities
(d) Being able to earn livelihood and live peacefully
[ESE Prelims : 2022]

3.5 **Statement (I) :** Social involvement discourages additional government regulation and intervention.
Statement (II) : Social involvement can create a weakened international balance of payments situation.
(a) Both Statement (I) and Statement (II) are individually true and Statement (II) is the correct explanation of Statement (I)
(b) Both Statement (I) and Statement (II) are individually true, but Statement (II) is not the correct explanation of Statement (I)
(c) Statement (I) is true, but Statement (II) is false
(d) Statement (I) is false, but Statement (II) is true
[ESE Prelims : 2022]

3.6 Some argue that laws are instituted at least in part to help establish a particular moral fabric in society. But the primary function of law is surely to help order society so that its institutions, organizations, and citizenry can work together harmoniously, regardless of any further moral aims of the law. Indeed, the highest courts have on occasion treated moral beliefs based on conscience or religious faith as grounds for making exceptions in the application of laws.

The statements above, if true, most strongly support which one of the following?

(a) The manner in which laws are applied sometimes takes into account the belief of the people governed by those laws.
(b) The law has an one of its functions the ordering of society but is devoid of moral aims.
(c) Actions based on religious belief or on moral conviction tend to receive the protection of the highest courts.
(d) The way a society is ordered by law should not reflect any moral convictions about the way society ought to be ordered.

[ESE Prelims : 2022]

3.7 Which one of the following statements is NOT correct regarding human values?

(a) Values mean an in-built mechanism which distinguishes the right from the wrong.
(b) Values provide us with a unique, personal and moral template that we use subconsciously to assess and judge the intensions and actions of others and ourselves.
(c) Values serve the process of 'becoming' in the sense of transformation of the level of consciousness to purer, higher levels.
(d) Values are essentially objective while skills are subjective.

[ESE Prelims : 2023]

3.8 Which one of the following statements is NOT correct?

(a) Notions or beliefs about manners, tastes, customs and towards laws are few examples of morality.
(b) Morality is more general and prescriptive based on customs and traditions; whereas ethics is specific and descriptive.
(c) Morality thrusts on judgement and punishment, in the name of God or by laws; whereas ethics, thrust is on influence, education, training through codes, guidelines and correction.
(d) Morality is more concerned with the results of wrong action, when done; whereas ethics is with the results of a right action, when not done.

[ESE Prelims : 2023]

3.9 What are the types of Moral Judgments?

(a) Permissible, Intent, Obligatory, Standpoint
(b) Professional, Impermissible, Obligatory, Supererogatory
(c) Permissible, Impermissible, Obligatory, Supererogatory
(d) Professional, Impermissible, Obligatory, Standpoint

[ESE Prelims : 2024]

4. VALUES IN ENGINEERING PROFESSION

4.1 Consider the following statements:

(i) The failure of many structures
(ii) Accidents on major highways
(iii) Discharge of effluents which pollute rivers
(iv) Dangerous gas emissions

These are attributable to:

1. Improper design
2. Not adhering to proper maintenance practices
3. Lack of proper inspection and quality control during construction
4. Corruption amongst several concerned individuals
5. Lack of public concern

Select the most appropriate answer using the codes given below:

(a) 1, 2 and 3 only (b) 1, 2 and 4 only
(c) 3, 4 and 5 only (d) 1, 2, 3, 4 and 5

[ESE Prelims : 2017]

4.2 For which of the following sectors do comprehensive safety and health statutes, for regulating Occupational Safety and Health (OSH) at work places, exist at present in India?

(a) Mining, Factories, Ports and Agriculture
(b) Factories, Mining Agriculture and Construction

(c) Mining, Factories, Ports and Construction
(d) Factories, Ports, Agriculture and Construction
[ESE Prelims : 2017]

4.3 Consider the following statements regarding safety during demolition of a multistory building:
1. Demolition need not proceed storey by storey.
2. Floor openings, not used for material chutes, should be enclosed with adequate guard rails.
3. Before demolition starts all lath and loose plaster shall be stripped off throughout the building.
4. Adequate and well-stiffened lateral bracing shall be provided for walls.

Which of the above statements are correct?
(a) 1, 2 and 3 only (b) 1 and 4 only
(c) 2, 3 and 4 only (d) 1, 2, 3 and 4
[ESE Prelims : 2017]

4.4 Consider the goals of the safety policy of a company, at once designating also the responsibilities and authority for their achievement:
1. Safety of employees and the public at large
2. Efforts to be made to involve all managers, supervisors and employees in the development and implementation of safety procedures.
3. Clearing of all caveats that may tell on keeping the morale of the employees high

Which of the above statements are correct?
(a) 1, 2 and 3 (b) 2 and 3 only
(c) 1 and 2 only (d) 1 and 3 only
[ESE Prelims : 2017]

4.5 Professionals who breach the '*duty of care*' are liable for injuries their negligence causes. This liability is commonly referred to as
(a) Professional offense
(b) Professional negligence
(c) Professional misdeed
(d) Professional malpractice
[ESE Prelims : 2019]

4.6 In the professional ethics, the degree of safety proposed to be attained varies with
(a) design, duration and product
(b) cost of risk, design and utility
(c) cost of risk, perception and utility
(d) product, perception and cost of risk involved
[ESE Prelims : 2020]

4.7 Which one of the following principles refers to whom may be affected by the actions of the company that affect health, safety, or the environment and refrain from taking reprisals against employees who report dangerous incidents to management or appropriate authorities?
(a) Safe product and service
(b) Informing the public
(c) Environmental restoration
(d) Risk reduction
[ESE Prelims : 2022]

4.8 Consider the following statements regarding engineers as responsible experimenters:
1. It includes a conscientious commitment to live by moral values.
2. It can be accountable for the results of the project.
3. It restricts free-personal involvement in all steps of the project or product development.
4. It includes constant awareness of the progress of the experiment and readiness to monitor the side effects, if any.

Which of the above statements are correct?
(a) 1 and 2 only (b) 3 and 4 only
(c) 1, 2 and 4 only (d) 1, 2, 3 and 4
[ESE Prelims : 2022]

4.9 Which one of the following is NOT Davis' eight moral tests?
(a) Harm test (b) Acceptability test
(c) Virtue test (d) Professional test
[ESE Prelims : 2022]

4.10 The 'work ethics' is aimed at NOT ensuring which of the following?
(a) The economy and productivity.
(b) Safety and privacy.
(c) Consumption and distribution.
(d) Health and hygiene. **[ESE Prelims : 2023]**

4.11 A very compassionate man, Engineer Bernard Amadei in 2001 was pro-foundly affected by the poor living conditions in underdeveloped countries, such as the absence of clean. water. He founded EWB-USA in 2001 for improving the living condition of the poor. Engineering students in EWB are responsible for many projects throughout the world that have enhanced human well-being. What is the full form of the term EWB?

(a) Economically Water Boys
(b) Engineers Well to do Boys
(c) Engineers Without Borders
(d) Engineers Water Boys

[ESE Prelims : 2024]

4.12 Which of the following tests are suggested by Philosopher Michael Davis that rely on our commonsense morality, but also reflect some of the concepts in moral theories or approaches?
(a) Harm Test, Publicity Test, Defensibility Test, Reversibility Test, Virtue Test, Professional Test, Colleague Test and Organization Test
(b) Defensibility Test, Reversibility Test, Virtue Test, Professional Test, Colleague Test, Heat Test, Organism Test and Purity Test
(c) Purity Test, Defensibility Test, Reversibility Test, Professional Test, Colleague Test, Heat Test Organism and Virtue Test
(d) Defensibility Test, Heat Test, Principal Test, Reversibility Test, Virtue Test, Professional Test, Organism Test and Colleague Test

[ESE Prelims : 2024]

4.13 If a utilitarian approach requires that we maximize well-being, how should we go about determining the criteria we should use in seeking this maximization? One approach that has appeal from the engineering perspective is CBA, which holds that the course of action that produces the greatest benefit or utility relative to cost should be chosen. What is the full form of the term CBA?
(a) Cost Benefit Approach
(b) Competitive Benefit Approach
(c) Competitive Benefit Analysis
(d) Cost Benefit Analysis

[ESE Prelims : 2024]

4.14 Joshua B. Kardon presents "an engineer is not liable, or responsible, for damages for every error. Society has decided, through case law, that when you hire an engineer, you buy the engineers normal errors. However, if the error is shown to have been worse than a certain level of error, the engineer is liable". That level, the line between non-negligent and negligent error is called
(a) Engineering Judgement
(b) Normal Distribution
(c) Standard of Care
(d) Performance Relative Standard

[ESE Prelims : 2024]

5. ETHICAL ISSUES, DILEMMAS, VALUE CONFLICTS & UNETHICAL BEHAVIOUR IN ENGINEERING

5.1 What is meant by 'Conflict of Interest'?
(a) Being interested in many subjects.
(b) Hobbies interfering in education.
(c) Least interest in the job taken up or assigned.
(d) A conflict between the private interests and the official responsibilities of a person in a position of trust.

[ESE Prelims : 2018]

5.2 What is 'Nepotism'?
(a) Undermining the morale of workers.
(b) Harassment of women workers.
(c) Being autocratic in decision-making.
(d) Hiring friends or relatives and showing favouritism in work.

[ESE Prelims : 2018]

5.3 A Whistleblower is someone who
(a) Whistles classical music.
(b) Informs on any illegal, unethical or corrupt activity going on in the organisation.
(c) Is adept in whistling.
(d) Boasts about himself/herself.

[ESE Prelims : 2018]

5.4 IPR protects the use the information and ideas that are of
(a) Ethical value (b) Moral value
(c) Social value (d) Commercial value

[ESE Prelims : 2018]

5.5 Information used in a business, generally unknown to the public, that the company has taken strong measures to keep confidential is called
(a) A patent (b) A copyright
(c) A trade secret (d) A trade mark

[ESE Prelims : 2019]

5.6 A situation where very high prices are charged from customers for a limited period of time is known as
(a) Gouging (b) Zipping
(c) Bamboozing (d) Hoodwinking
[ESE Prelims : 2019]

5.7 Consider the following steps for an individual regarding preparation for disclosure of unethical behaviour:
1. Study and document the facts and formulate a plan for an appeal.
2. Take up the matter with higher management.
3. Discuss the matter with immediate supervisor.
4. If the internal appeal does not resolve the conflict, then he should notify the company that he intends to continue with an external review of the problem.

What is the correct sequence of order of the above steps?
(a) 2, 3, 1 and 4 (b) 1, 3, 2 and 4
(c) 3, 2, 4 and 1 (d) 1, 2, 3 and 4
[ESE Prelims : 2019]

5.8 What are the core qualities of a professional practitioner?
1. Integrity both with themselves and with others
2. Independence-to be free of secondary interests with other parties
3. Competence
4. Discretion-care with communications

Select the correct answer using the codes given below:
(a) 1, 2, 3 and 4 (b) 1, 2 and 3 only
(c) 1 and 3 only (d) 3 and 4 only
[ESE Prelims : 2019]

5.9 When should whistle blowing be attempted?
1. There must be a clear and great harm that can be avoided
2. The whistleblower must be in a clear position to report on the problem
3. The whistleblower must have a reasonable chance of success in stopping the harmful activity
4. The whistleblower feels that all other lines of action within the context of the organization have been explored and shut off

Select the correct answer using the codes given below:
(a) 1, 2, 3 and 4 (b) 1, 2 and 4 only
(c) 1, 3 and 4 only (d) 2 and 3 only
[ESE Prelims : 2019]

5.10 Whistleblowing in an organizational set up affects:
1. Peer professional relationships
2. Relationships with management
3. Family relations

Which of the above statement(s) is/are correct?
(a) 1 and 2 only (b) 2 only
(c) 1, 2 and 3 (d) 1 only
[ESE Prelims : 2021]

5.11 Hooch & Bootlegging refers to
(a) the prohibition law on unethical practices related to liquor
(b) laws against giving money to beggars as a generous act
(c) food adulteration rules
(d) laws on checking the illegal business of duplicate goods
[ESE Prelims : 2021]

6. RESPONSIBILITIES OF ENGINEERS

6.1 **Statement (I)** : Increasingly, employers have generally tended to expect engineers to possess both hard skills and soft skills.
Statement (II) : Soft skills mean the knowledge of software.
(a) Both Statement (I) and Statement (II) are individually true and Statement (II) is the correct explanation of Statement (I)
(b) Both Statement (I) and Statement (II) are individually true, but Statement (II) is not the correct explanation of Statement (I)
(c) Statement (I) is true, but Statement (II) is false
(d) Statement (I) is false, but Statement (II) is true
[ESE Prelims : 2018]

6.2 Which of the following are the attributes of a profession?
1. The work requires sophisticated skills, use of judgment and exercise of discretion
2. Membership in the profession does not require extensive formal education as well as practical training
3. There are set standards for admission to the profession and conduct for members

4. Significant public good results from practice of the profession

Select the correct answer using the codes given below:

(a) 1, 2 and 3 only (b) 1, 2 and 4 only
(c) 1, 3 and 4 only (d) 2, 3 and 4 only

[ESE Prelims : 2019]

6.3 Which of the following are the salient features of the Patent Act 1970 ?

1. It codifies inventions which are not patentable
2. It provides for endorsement of patent with the words 'license of right'
3. It provides for revocation of patents in public interest
4. It has provision for validity period also for the patents

Select the correct answer using the codes given below:

(a) 1, 2, 3 and 4 (b) 1, 2 and 4 only
(c) 1, 3 and 4 only (d) 2 and 3 only

[ESE Prelims : 2019]

6.4 **Statement (I) :** Agency-loyalty is acting to fulfill one's contractual duties to an employer.

Statement (II) : Agency-loyalty is entirely a matter of actions, whatever its motives.

(a) Both Statement (I) and Statement (II) are individually true; and Statement (II) is the correct explanation of Statement (I)
(b) Both Statement (I) and Statement (II) are individually true; but Statement (II) is NOT the correct explanation of Statement (I)
(c) Statement (I) is true; but Statement (II) is false
(d) Statement (I) is false; but Statement (II) is true

[ESE Prelims : 2019]

6.5 Which one of the following is **not** the skill needed in the workplace of the future for inventive thinking using information and communication technology in education?

(a) Adaptability
(b) Responsibility
(c) Curiosity and creativity
(d) Risk-taking

[ESE Prelims : 2020]

6.6 The Engineers for a Sustainable World (ESW) meant for using the professional talents to create a more sustainable world was founded in

(a) 1999 (b) 2000
(c) 2001 (d) 2002

[ESE Prelims : 2022]

6.7 Which one of the following is NOT included under the categories of civic virtues as indispensable for a self-governing administration?

(a) Self-reflection
(b) Self-restraint
(c) Self-reliance
(d) Self-assertion

[ESE Prelims : 2023]

6.8 What are the three stages in the Development of Professional Identity?

(a) Possessing Knowledge, Professional Services, Self-Defining or Integrated Professional
(b) Independent Operator, Professional Services, Self-Defining or Integrated Professional
(c) Possessing Knowledge, Team Oriented Idealist, Self-Defining or Integrated Professional
(d) Independent Operator, Team Oriented Idealist, Self-Defining or Integrated Professional

[ESE Prelims : 2024]

7. CODES OF CONDUCT

7.1 **Statement (I):** Engineers shall hold paramount the safety, health and welfare of the public while performing their professional duties.

Statement (II): Engineers shall continue their professional development throughout their careers.

(a) Both Statement (I) and Statement (II) are individually true and Statement (II) is the correct explanation of Statement (I)
(b) Both Statement (I) and Statement (II) are individually true but Statement (II) is not the correct explanation of Statement (I)
(c) Statement (I) is true but Statement (II) is false
(d) Statement (I) is false but Statement (II) is true

[ESE Prelims : 2017]

7.2 Consider the following statements regarding the code of ethics for Engineers:

1. The safety, health and welfare of the public are of paramount importance.
2. Perform services only in the area of their competence.
3. Issue public statements strictly in an objective and truthful manner.
4. Avoid deceptive acts.

Which of the above statements are correct?

(a) 1, 2 and 3 only (b) 1, 2 and 4 only
(c) 3 and 4 only (d) 1, 2, 3 and 4

[ESE Prelims : 2018]

7.3 Which one of the following statements is NOT correct about the codes of conduct?

(a) These cover general guiding principles
(b) Their purpose is to regulate the conduct of members on various transactions
(c) These are the broader sets of principles that are designed to inform specific laws or government actions
(d) These translate the values into specific behavioral standards, keeping in mind the possible reflection on the stakeholders' interest

[ESE Prelims : 2021]

7.4 Which one of the following is NOT covered under the International Labour Organization declaration on fundamental principles and rights at work (1998)?

(a) Freedom of association and the right to collective bargaining
(b) The elimination of forced and compulsory labour
(c) The abolition of child labour
(d) The global compact on migration

[ESE Prelims : 2022]

7.5 Spirituality is promoted in the work-place by adhering to the following activities:

1. Verbally respect the individuals as humans and recognize their values in all decisions and actions.
2. Support causes outside the business.
3. Do unto others as you would have them do unto you.
4. Realization of the self-potential through meditative acts.

Which of the above activities are correct?

(a) 2, 3 and 4 only
(b) 1, 2, 3 and 4
(c) 2 and 3 only
(d) 1, 2 and 3 only [ESE Prelims : 2023]

7.6 The first of the Fundamental Canons of the code of the National Society of Professional Engineers says that engineers shall hold

(a) paramount the safety, health, and welfare of the public
(b) devotion to clients as the first responsibility
(c) devotion to his employer
(d) devotion to the public

■■■■

Ethics and Values in Engineering Profession : Answer Key

1.1 (d)	**1.2** (b)	**1.3** (a)	**1.4** (c)	**1.5** (a)	**1.6** (b)	**1.7** (a)	**1.8** (d)
1.9 (c)	**2.1** (c)	**2.2** (b)	**2.3** (b)	**2.4** (d)	**2.5** (a)	**2.6** (b)	**2.7** (b)
2.8 (d)	**2.9** (a)	**2.10** (c)	**2.11** (d)	**2.12** (a)	**2.13** (a)	**2.14** (b)	**2.15** (c)
2.16 (b)	**2.17.** (b)	**2.18.** (c)	**3.1** (b)	**3.2** (d)	**3.3** (a)	**3.4** (d)	**3.5** (b)
3.6 (a)	**3.7** (d)	**3.8** (a)	**3.9** (c)	**4.1** (d)	**4.2** (c)	**4.3** (c)	**4.4** (a)
4.5 (b)	**4.6** (b)	**4.7** (b)	**4.8** (c)	**4.9** (b)	**4.10.** (c)	**4.11** (c)	**4.12** (a)
4.13 (d)	**4.14** (c)	**5.1** (d)	**5.2** (d)	**5.3** (b)	**5.4** (d)	**5.5** (c)	**5.6** (a)
5.7 (b)	**5.8** (a)	**5.9** (b)	**5.10** (a)	**5.11** (a)	**6.1** (c)	**6.2** (c)	**6.3** (a)
6.4 (b)	**6.5** (b)	**6.6** (d)	**6.7** (d)	**6.8** (c)	**7.1** (a)	**7.2** (d)	**7.3** (a)
7.4 (d)	**7.5** (d)	**7.6** (a)					

EXPLANATIONS

1.2 (b)

Micro-ethics is about unawareness of the consequences of individual behavior. Micro-ethics understands that there is a wide variety of actions that are all within the law, but that have a wide variety of consequences, some of which are socially desirable and others are not.

1.4 (c)

- The above statement defines Cognitivism.
- Cognitivism is perhaps best defined as the denial of non-cognitivism. Cognitivists think that moral sentences are apt for truth or falsity, and that the state of mind of accepting a moral judgment is typically one of belief.

1.5 (a)

- All the chosen statements are in coherence with scope of Engineering Ethics.
- Engineering ethics are principles and guidelines engineers follow to ensure their decision-making is aligned with their obligations to the public, their clients, and the industry.

1.6 (b)

- The above statement forms the definition of Epistemology.
- It is the theory of knowledge, especially with regard to its methods, validity, and scope, and the distinction between justified belief and opinion.

1.8 (d)

Euthanasia is the practice of intentionally ending life to eliminate pain and suffering. Different countries have different euthanasia laws.

1.9 (c)

- Business Ethics doesn't include Avoid Action, as it is essential pre-requisite for Ethics.
- Business ethics is the implementation of policies and procedures regarding topics such as fraud, bribery, discrimination, and corporate governance.

2.3 (b)

- Convergence theory presumes that as nations move from the early stages of industrialization toward becoming fully industrialized, they begin to resemble other industrialized societies in terms of societal norms and technology.
- The characteristics of these nations effectively converge. Ultimately, this could lead to a unified global culture if nothing impeded the process.
- Convergence theory has its roots in the functionalist perspective of economics which assumes that societies have certain requirements that must be met if they are to survive and operate effectively.

ADDITIONAL INFORMATION

Convergence theory became popular in the 1960s when it was formulated by the University of California, Berkeley Professor of Economics Clark Kerr.

2.4 (d)

Utilitarianism is an ethical theory that determines right from wrong by focusing on outcomes. It is a form of consequentialism. Utilitarianism holds that the most ethical choice is the one that will produce the greatest good for the greatest number.

2.7 (b)

- This term was coined is 1952 by William.H. Whyle Jr. and most of the initial on Groupthink was conducted by Irving Janis.
- Groupthink is phenomenon that occurs when a group of individuals reaches a consensus w/o critical reasoning or evaluation of the consequences or alternatives.

2.10 (c)

Professionalism responsibility/Types of virtues:

1. Public spirited virtues
2. Team work virtue
3. Self-direction virtue
4. Proficiency virtue

Hence, Self-realization virtues is answer.

2.11 (d)

CSR, in the form that we see today, became popular after it was defined by Archie Carroll's "Pyramid of Corporate Social Responsibility" in 1991. Its simplicity, yet ability to describe the idea of CSR with four areas, has made the pyramid one of the most accepted corporate theories of CSR since.

Carroll's pyramid suggests that corporate has to fulfil responsibility at four levels – Economic, Legal, Ethical and Philanthropic.

1. **Economic Responsibility :** The lowest level of the pyramid represents a business's first responsibility, which is to be profitable. Without profit, the company would not be able to pay their workers, employees will lose their jobs even before the company starts CSR activities.
2. **Legal Responsibility :** The second level of the pyramid is the business's legal obligation to obey the law. This is the most important responsibility out of the four levels as this will show how companies conduct their business in the marketplace.
3. **Ethical Responsibility :** A company should not only be obeying the law, but it should also do their business ethically. Unlike the first two levels, this is something that a company is not obligated to do.
4. **Philanthropic Responsibility :** At the top of the pyramid, occupying the smallest space is philanthropy.

Hence, option (d) i.e. 1,3,4,2 is correct order.

According to UPSC answer is option (a).

2.15 (c)

- Carol Gilligan states that the post-conventional level of moral thinking can be dealt based on the two types of thinking. Gilligan's theory is based on the two main ideas, the care-based morality (usually found in women) and the justice-based morality (usually found in men).
- Kohlberg's theory of moral development is a theory that focuses on how children develop morality and moral reasoning. Kohlberg's theory suggests that moral development occurs in a series of six stages and that moral logic is primarily focused on seeking and maintaining justice.

3.4 (d)

Martha Nussbaum identified **an Aristotelian set of ten universal, normative capabilities** that act as freedoms that are **generally protected by law**. These may be viewed as needs, although they are also related to values.

1. **Life :** Being able to live a complete and satisfying life into old age. Not having life cut short or being made such that it hardly seems worth living.
2. **Bodily Health :** Living with good health, and not in a state where ill health seriously affects the quality of life. Having access to medical help as needed. To have good food and be able to exercise in ways that sustain health.
3. **Bodily Integrity :** Being able to go where you want to go. Being free from attack and abuse of any kind. Being able to satisfy healthy bodily needs.
4. **Senses, Imagination and Thought :** Being able to use all of one's senses. Being free to imagine, think and reason.
5. **Emotion :** Being able to become attached to other things and people outside of ourselves, loving and caring for them. Experiencing grief, longing, gratitude and justified anger.
6. **Practical Reason :** Being able to consider and develop understanding of good and evil, and to think critically about the world and one's own place in it.
7. **Affiliation :** Being able to associate with others, living with them and acting for them. Showing concern for people in general and interacting with others.
8. **Other Species :** Being able to live with the full range of creatures and plants that inhabit the world around us. To be able to enjoy nature and appreciate its beauty.
9. **Play :** Being able to laugh, play games and generally have fun. Not having one's enjoyment and recreation criticized or prevented.

10. **Control Over One's Environment :** Being able to participate in political activities, making free choice and joining with others to promote political views.

Hence, Option (d) i.e. Being able to earn livelihood and live peacefully is not included.

4.2 (c)

The Constitution of India contains specific provisions on the occupational safety and health (OSH) of workers which currently covers only four industry sectors: factories, mines, ports and construction.

4.3 (c)

Demolition of exterior walls and floors must begin at the top of structure and proceed downward.

4.8 (c)

Responsible Experimentalists:

1. **Conscientiousness :** Protect safety knowledge, respect right of consent of public
2. **Relevant Information or Comprehensive perspective :** Awareness of experimental nature of projects, forecasting, monitoring
3. **Moral autonomy :** Personally engaged, thoughtful, involvement in project
4. **Accountability :** Accept responsibility for results of a project (avoid fragmentation, diffusion, time pressures)

Hence, statement 3 i.e. It restricts free-personal involvement in all steps of the project or product development is wrong.

Hence, Correct Option is 1, 2 and 4 only.

4.9 (b)

Michael Davis guide to ethical decision-making include following eight moral tests. Some of these include:

(a) **Harm test:** Does this option do less harm than the alternatives?

(b) **Publicity test:** Would I want my choice of this option published in the newspaper?

(c) **Defensibility test:** Could I defend my choice of this option before a congressional committee or committee of peers?

(d) **Reversibility test:** Would I still think this option was a good choice if I were adversely affected by it?

(e) **Colleague test:** What do my colleagues say when I describe my problem and suggest this option as my solution?

(f) **Professional test:** What might my profession's governing body for ethics say about this option?

(g) **Organization test:** What does my company's ethics officer or legal counsel say about this?

Hence, **Acceptability test** is not mentioned by Michael Davis.

5.3 (b)

Workplace whistleblowing occurs when an individual reports wrongdoing in an organisation, such as financial misconduct or discrimination. This person is often an employee but can also be a third-party such as a supplier or customer. Internal whistleblowing is when someone makes a report within an organisation

5.5 (c)

- The above given statement is the definition of a 'trade secret'.
- Trade secrets are intellectual property (IP) rights on confidential information which may be sold or licensed.

5.6 (a)

- The above statement forms the definition of Gouging.
- Price gouging occurs when a seller increases the prices of goods, services, or commodities to a level much higher than is considered reasonable or fair.

5.7 (b)

The order of the statement chosen form the logical sequence with respect to an individual regarding preparation for disclosure of unethical behaviour.

6.2 (c)

The chosen statements are in line with the definition of profession.

6.6 (d)

Engineers for a Sustainable World (ESW) is an international non-profit network of students & professionals united by their shared passion for technical sustainability.

What is ESW's vision & mission?

It strives to empower students and professionals to tackle sustainability challenges. The ESW works towards a sustainable world supported by a network of passionate engineers. Its goals are as follows:

1. Design and implement hands-on sustainability projects across the globe.
2. Educate students and practicing engineers on sustainability best practices.
3. Unite those passionate about sustainability in a productive international network.

7.4 (d)

ILO Declaration on Fundamental Principles and Rights at Work

Adopted in 1998, the Declaration commits Member States to respect and promote principles and rights in four categories, whether or not they have ratified the relevant Conventions. These categories are:

1. Freedom of association and the effective recognition of the right to collective bargaining,
2. The elimination of forced or compulsory labour,
3. The abolition of child labour and
4. The elimination of discrimination in respect of employment and occupation.

■■■■

Current Issues of National & International Importance

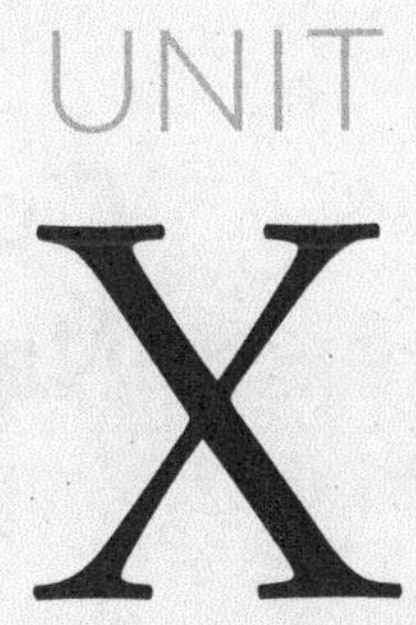

CONTENTS

Current Issues of National & International Importance

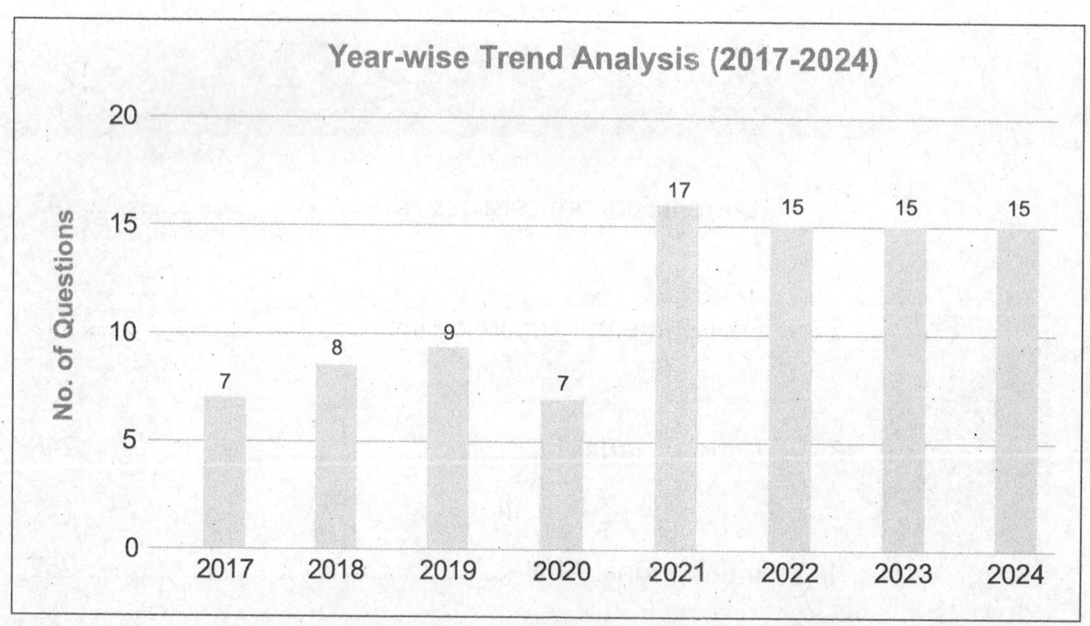

TOPIC-WISE TREND ANALYSIS (2017-2024)		
Sl. No.	**Topics**	**Total Number of Questions**
1.	Current Economics Issues	26
2.	Environmental and Wild Life	3
3.	National Affairs	28
4.	International Affairs	11
5.	Defence in News	6
6.	Science and Technology	8
7.	Miscellaneous	13

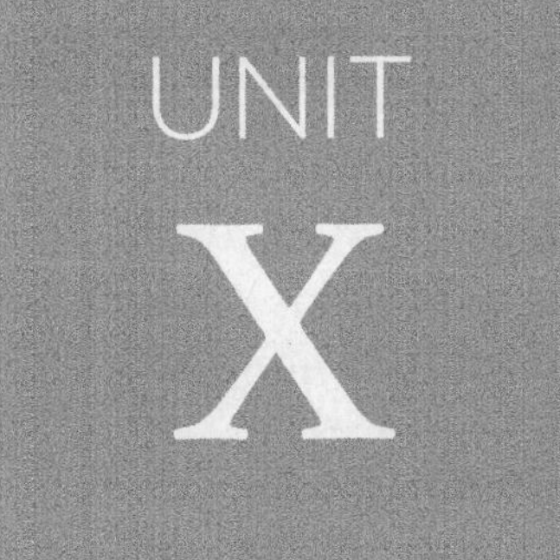

Current Issues of National & International Importance

1. CURRENT ECONOMICS ISSUES

1.1 Which of the following methods are considered as most favourable to production organizations liable for taxation?

1. Straight line method of depreciation.
2. Declining balance method of depreciation.
3. Sum of the years digits methods of depreciation.
4. Sinking fund method of depreciation.

Select the correct answer using the codes given below:

(a) 1 and 2 (b) 2 and 3
(c) 3 and 4 (d) 4 and 1

[ESE Prelims : 2017]

1.2 Consider the following statements regarding repo rate:

1. It is the rate at which RBI lends money to Commercial Banks generally against Government securities.
2. It is the rate at which RBI borrows money from commercial banks generally against government securities.
3. It is the rate at which commercial banks repo deposits with RBI.

Which of the above statements is/are correct?

(a) 1 only (b) 2 only
(c) 3 only (d) 1, 2 and 3

[ESE Prelims : 2017]

1.3 Consider the following statements regarding GST:

1. The GST Bill 2014 has the purpose to improve the value added tax on goods and services.
2. It can be imposed differently in different states.
3. It is a comprehensive tax imposed nation wide irrespective of any state concerned.
4. It is a significant step in the reform of indirect taxation in India.

Which of the above statements are correct?

(a) 1, 2 and 3 (b) 1, 2 and 4
(c) 2, 3 and 4 (d) 1, 3 and 4

[ESE Prelims : 2017]

1.4 Consider the following statements:

Non-performing assets (NPAs) decline in value when

1. Demand revives in the economy.
2. Capacity utilization increases.
3. Capacity utilization, through substantive, is yet sub-optimal.
4. Capacity utilization decreases consequent upon merger of units.

Which of the above statements are correct?

(a) 1, 3 and 4 only (b) 1, 2 and 4 only
(c) 1, 2 and 3 only (d) 1, 2, 3 and 4

[ESE Prelims : 2018]

1.5 What is Crowdfunding?

(a) Money collected for public welfare projects by levying an entry fee to exhibitions, shows, etc.
(b) Money collected by charitable organizations by placing a donation box at prominent locations.
(c) Money raised by innovators and inventors by launching their products and services through the Internet.
(d) Money raised by individuals by passing the hat around to onlookers at a street performance.

[ESE Prelims : 2018]

1.6 Which one of the following is **not** a component of 'Capital Receipts'?

(a) Market borrowings including special bonds

(b) External loans raised by the Central Government from abroad

(c) Receipts from taxes on property and capital transactions

(d) Provident Funds (State Provident Funds and Public Provident Fund)

[ESE Prelims : 2019]

1.7 Which one of the following statement is correct with respect to the '*societal development*'?

(a) Behaviour grows into habits, habits into tradition and tradition becomes custom

(b) Customs grow into mores and mores grow into custom

(c) Behaviours grow into customs and customs grow into traditions

(d) Folkways grow into tradition and traditions grow into customs

[ESE Prelims : 2019]

1.8 Which one of the following statement is correct with respect to '*the convergence theory*' on social change?

(a) The societal adaptive culture is changing more slowly

(b) As societies become modernized, they begin to resemble one another more closely

(c) The developed countries show more growth in social changes than the less developed countries

(d) Strong opposition by old people retards the social change

[ESE Prelims : 2019]

1.9 With respect to the conduct and performance of a company, '*perfect competition*' refers to

(a) Large number of small firms producing differentiated products

(b) Complete freedom in economic life and absence of rivalry among firms

(c) Many companies selling similar products with free entry

(d) Sole producer selling a distinct product

[ESE Prelims : 2019]

1.10 The cheapest method of marketing of securities with the only cost incurred being on sending '*letters of rights*' to existing holders is

(a) Public issue through prospectus method

(b) Offer for sale method

(c) Rights issue

(d) Subscription by inside coterie method

[ESE Prelims : 2019]

1.11 'Fiscal policy' means

(a) Balancing the revenue collection and expenditure

(b) Establishing equilibrium between demand and supply of goods and services

(c) Use of taxation, public borrowing and public expenditure by Government for purposes of 'stabilization' or 'development'

(d) Deficiency as an instrument of growth

[ESE Prelims : 2019]

1.12 Which of the following come under the offerings of '*MUDRA*' Bank?

1. Portfolio Credit Guarantee
2. Credit for large industries
3. MUDRA Card
4. Credit Enhancement

Select the correct answer using the codes given below:

(a) 1, 2 and 3 only (b) 1, 3 and 4 only
(c) 1, 2 and 4 only (d) 2, 3 and 4 only

[ESE Prelims : 2019]

1.13 Which of the following is/are the key reasons for encouraging start up Entrepreneurship?

1. Innovations
2. Focusing on service industry
3. Bringing the values of proactivity into the society

Select the correct answer using the codes given below:

(a) 1 only (b) 2 only
(c) 1 and 3 only (d) 1, 2 and 3

[ESE Prelims : 2019]

1.14 Which of the following are the main objectives of Gold Monetization Scheme launched in the country?

1. To monetize gold holdings in the country
2. To increase export of gold from the country
3. To reduce India's import bill
4. To meet the targets of reduction in fiscal deficit

Select the correct answer using the codes given below:
(a) 1 and 4 only (b) 2 and 4 only
(c) 2 and 3 only (d) 1 and 3 only

[ESE Prelims : 2019]

1.15 What is TADF?
(a) Technology Acquired Desired Firm
(b) Technologically Advanced Direct Fund
(c) Technologically Accomplished Direct Fund
(d) Technology Acquisition and Development Fund

[ESE Prelims : 2020]

1.16 Which one of the following is a measure of sustainable income level that can be secured without decreasing the stock of natural assets?
(a) Natural Capital Stock
(b) Environmental Value
(c) Green Accounting
(d) Social Discount Rate

[ESE Prelims : 2020]

1.17 Which of the following is a resource allocation as per Chenery's development process?
(a) Investment
(b) Structure of domestic demand
(c) Labour allocation
(d) Government revenue

[ESE Prelims : 2020]

1.18 Which one of the following ratios is referred to as everything that has been invested in the past and to the whole income?
(a) Capital-output ratio
(b) Average capital-output ratio
(c) Incremental capital-output ratio
(d) Marginal ratio

[ESE Prelims : 2020]

1.19 Which one of the following methods of planning is an attempt to work out the implications of the development effort in terms of factor allocations and product yields so as to maximize income and employment?
(a) Perspective planning
(b) Physical planning
(c) Financial planning
(d) Indicative planning

[ESE Prelims : 2020]

1.20 Which one of the following reflects an intrinsic or true value for factors or products?
(a) Price inflation
(b) Economy pricing
(c) Penetration pricing
(d) Shadow price [ESE Prelims : 2020]

1.21 Which one of the following control policies leaves no freedom to private enterprise to buy plant, machinery, raw materials from the country of its choice?
(a) Import control
(b) Export control
(c) Exchange control
(d) Physical control [ESE Prelims : 2020]

1.22 **Statement (I):** Issue of shares is the most common method especially to raise long-term funds.

Statement (II): The equity shareholders are residual owners who have restricted claim on income as dividend.
(a) Both Statement (I) and Statement (II) are individually true and Statement (II) is the correct explanation of Statement (I)
(b) Both Statement (I) and Statement (II) are individually true but Statement (II) is not the correct explanation of Statement (I)
(c) Statement (I) is true but Statement (II) is false
(d) Statement (I) is false but Statement (II) is true [ESE Prelims : 2021]

1.23 Which one of the following statements is NOT correct regarding the Pension Fund Regulatory and Development Authority (PFRDA)?
(a) It has increased the entry age for the National Pension System (NPS) from 60 years to 65 years
(b) Earlier the eligible age to invest in NPS was 18-65 years which has now been revised to 18-70 years
(c) As per the revised norms, any Indian Citizen, resident or non-resident and Overseas Citizen of India (OCI) between the age of 65-70 years can join NPS
(d) Subscribers can continue or defer their NPS Account up to the age of 75 years

[ESE Prelims : 2022]

1.24 Consider the following statements :
1. India's Goods and Services Tax collection for the month of June 2023 reached R. 1.61 trillion, according to the Ministry of Finance.
2. India received its highest-ever FDI inflow of US $83.57 billion in the fiscal year 2021-2022.
3. The net direct tax collection in the current fiscal year has witnessed a significant growth of 16%, reaching Rs. 4.75 lakh crore, indicating a surge in economic activity.

Which of the above statements are correct?
(a) 1 and 2 only (b) 1 and 3 only
(c) 2 and 3 only (d) 1, 2 and 3

[ESE Prelims : 2024]

1.25 Consider the following economic activities:
1. Public administration
2. Financial services
3. Mining and quarrying

Which of the above economic activities fall under the tertiary sector?
(a) 1 and 3 only (b) 1, 2 and 3
(c) 2 and 3 only (d) 1 and 2 only

[ESE Prelims : 2023]

1.26 IMF raises its projection for economic growth in 2021-22 to
(a) 11.3% (b) 12.5%
(c) 10.2% (d) 8.4%

[ESE Prelims : 2023]

2. ENVIRONMENTAL AND WILD LIFE

2.1 The Thirteenth Meeting of the Conference of the Parties to the Convention on the Conservation of Migratory Species of Wild Animals (CMS COP 13) in 2020 was held in
(a) Brazil (b) India
(c) Indonesia (d) Canada

[ESE Prelims : 2021]

2.2 Energy services for sustainable development are directly linked to:
1. Poverty 2. Lifestyles
3. Women 4. Deforestation

Select the correct answer using the codes given below:
(a) 1, 2 and 3 only (b) 1, 2 and 4 only
(c) 2, 3 and 4 only (d) 1 and 3 only

[ESE Prelims : 2021]

2.3 Consider the following statements regarding Forum for Decarbonizing Transport:
1. NITI Aayog and World Resources Institute (WRI), India, jointly launched the 'Forum for Decarbonizing Transport' in India.
2. NITI Aayog is the implementing partner for India.
3. The aim of the project is to bring down the peak level of GHG emissions (transport sector) in Asia (in line with a well below 2-degree pathway), resulting in problems like congestion and air pollution.

Which of the above statements are correct?
(a) 1 and 2 only (b) 2 and 3 only
(c) 1 and 3 only (d) 1, 2 and 3

[ESE Prelims : 2022]

3. NATIONAL AFFAIRS

3.1 Consider the following statements:
Vision of Digital India launched by the Government of India is centered on
1. Digital infrastructure as a utility only to senior citizens.
2. Governance and Services on demand
3. Digital Empowerment of every citizen

Which of the above statements are correct?
(a) 1, 2 and 3 (b) 1 and 2 only
(c) 1 and 3 only (d) 2 and 3 only

[ESE Prelims : 2017]

3.2 Which one of the following Committees was set up by the Government of India in September 2014 to restructure the Railways and to suggest ways for resource mobilization?
(a) Bibek Debroy Committee
(b) C. Rangarajan Committee
(c) Parthasarathi Shome Committee
(d) Sundar Committee [ESE Prelims : 2017]

3.3 PRAGATI is the acronym
(a) Pro-active governance and transparency in India
(b) Pro-active governance and timely implementation
(c) Primary governance for transparency in India
(d) Promoting and accomplishing governance with assured and timely implementation

[ESE Prelims : 2017]

3.4 Government of India had introduced the Consumer Protection Bill, 2015, in the Lok Sabha. The Bill gives the right to consumers to

1. Seek redressal against unfair or restrictive trade practices.
2. File a complaint for overcharging or deceptive charging.

Which of the above is/are included in the Bill?

(a) 1 only (b) 2 only
(c) Both 1 and 2 (d) Neither 1 nor 2

[ESE Prelims : 2018]

3.5 Technology Promotion, Development and Utilization Programme implemented by Department of Scientific and Industrial Research has which of the following components?

1. Industrial R&D Promotion Programme
2. Flagship Programme
3. Information Technology and e-Governance

(a) 1 and 2 only (b) 1 and 3 only
(c) 2 and 3 only (d) 1, 2 and 3

[ESE Prelims : 2018]

3.6 Consider the following statements:

1. IPDS strengthens the distribution network in urban areas while DDUGJY does the same in rural areas.
2. DELP focuses to substitute LED bulbs for incandescent bulbs.

Which of the above statements is/are correct?

(a) 1 only (b) 2 only
(c) Both 1 and 2 (d) Neither 1 nor 2

[ESE Prelims : 2018]

3.7 Consider the following statements:

1. National Agricultural Portal, eNAM, is designed to create a unified national market for agricultural commodities.
2. Farmers can showcase their produce online from the nearest market and the buyer can quote his price from anywhere.

Which of the above statements is/are correct?

(a) 1 only (b) 2 only
(c) Both 1 and 2 (d) Neither 1 nor 2

[ESE Prelims : 2018]

3.8 Match the following:

List-I

A. Utkarsh Bangla Scheme
B. Placement Linked Skill Training Programme
C. SURYA Scheme
D. Employment Linked Skill Training Programme

List-II

1. West Bengal
2. Rajasthan
3. Haryana
4. Assam

Select the correct matching using the code given below:

	A	B	C	D
(a)	2	3	4	1
(b)	2	3	1	4
(c)	1	4	3	2
(d)	1	4	2	3

[ESE Prelims : 2021]

3.9 With a view to encourage and promote Indian artisans and their handicraft, Hunar Haat offers an effective platform. Where was the 22nd Hunar Haat held?

(a) Jaipur (b) Ferozpur
(c) Rampur (d) Bharatpur

[ESE Prelims : 2021]

3.10 Which one of the following statements is NOT correct regarding the National Education Policy 2020 in India?

(a) It proposes sweeping changes in the education system from pre-primary to PhD and skill development
(b) It states that universities from among top 100 in the world will be able to set up campuses in India
(c) It expects that India will achieve 60% GER by 2030.
(d) It suggests NAAC to be merged with UGC and AICTE

[ESE Prelims : 2021]

3.11 According to the National Institutional Ranking Framework 2020, which institute was on the top in overall ranking?
(a) Indian Institute of Technology, Madras
(b) Indian Institute of Science, Bengaluru
(c) Indian Institute of Technology, Delhi
(d) Indian Institute of Technology, Bombay

[ESE Prelims : 2021]

3.12 Consider the following statements with respect to the schemes initiated by the Government of India in 2020.
1. NISHTHA is a teachers training program.
2. SVANidhi is a scheme to facilitate artisans to access affordable working capital loan.
3. SATYABHAMA is a scheme to promote research and development in science and technology.
4. Manodarpan is a scheme to promote tourism in rural parts of India.

Which of the above statements is/are correct?
(a) 1 and 3 only (b) 2 and 4 only
(c) 3 and 4 only (d) 2 only

[ESE Prelims : 2021]

3.13 Which one of the following statements is NOT correct about the Atal Tunnel?
(a) It is the highest altitude tunnel in the world.
(b) It was inaugurated on 03 October 2020 in Rohtang.
(c) It connects Manali to Lahauli-Spiti valley.
(d) It is capable of handling 5000 cars and 2500 trucks per day with maximum speed of 80 kmph.

[ESE Prelims : 2021]

3.14 Which one of the following is the latest in series being organized as the largest virtual gathering to create dialogues, and accelerate innovation in agriculture?
(a) Agri-India Hackathon 2020
(b) National Agriculture Higher Education Project
(c) ENSURE
(d) National Mission for Sustainable Development

[ESE Prelims : 2021]

3.15 Which one of the following national parks has become the first national park in India to be equipped with satellite phones?
(a) Kaziranga National Park in Assam
(b) Sundarbans National Park in West Bengal
(c) Desert National Park In Rajasthan
(d) Indravati National Park in Chhattisgarh

[ESE Prelims : 2022]

3.16 Which one of the following is NOT important initiatives under EASE 4.0?
(a) Smart lending for aspiring India
(b) New age 24 × 7 banking with resilient technology
(c) Collaborative banking for synergistic outcomes
(d) Parameters of FI-Index

[ESE Prelims : 2022]

3.17 World's largest star sapphire cluster has been found in
(a) Rajkot, India
(b) Ratnapura, Sri Lanka
(c) Pretoria, South Africa
(d) Brisbane, Australia

[ESE Prelims : 2022]

3.18 Which one of the following ministries has repealed the Unmanned Aircraft Systems (UAS) Rules, 2021 and replaced the same with the liberalized Drone Rules, 2021?
(a) Ministry of Home Affairs
(b) Ministry of Defence
(c) Ministry of Science and Technology
(d) Ministry of Civil Aviation

[ESE Prelims : 2022]

3.19 What one of the following is NOT an objective of Mahila Kisan Sashaktikaran Pariyojana (MKSP)?
(a) To create sustainable agricultural livelihood opportunities for women in agriculture.
(b) To ensure food and nutrition security at the household and the community level.
(c) To enable women to have better access to inputs and services of the government and other agencies.
(d) To help women educate the rural folk and improve their living condition.

[ESE Prelims : 2023]

3.20 Consider the following statements regarding the aim of Jal Jeevan Mission to provide every rural household of the country with adequate tap water of prescribed quality on regular basis:

1. It seeks to ensure 'ease of living' which leads to healthier as well as hygienic living conditions in rural areas.
2. It aims to establish water tanks in good numbers with the slogan 'Har Ghar Jal'.
3. By ensuring community participation at village-level, it will help in developing local leadership based no Gandhiji's philosophy of 'Gram Swarajya'.
4. The Mission seeks to achieve its goal by 2024.

Which of the above statements are correct?

(a) 1,3 and 4 only (b) 1, 2, 3 and 4
(c) 1 and 3 only (d) 2 and 4 only

[ESE Prelims : 2023]

3.21 Select the State/s and/or UTs of India which have been ranked first as 'Zero Hunger' as per SDG : India Agenda for Development:

(a) Tamil Nadu and Delhi
(b) Kerala and Chandigarh
(c) Gujarat and Delhi
(d) Goa and Lakshadweep

[ESE Prelims : 2023]

3.22 Consider the following statements:

1. Ford India will stop manufacturing vehicles in India but will retain the engine-making an technology services business as part of restructuring its India operations.
2. Zee Entertainment Enterprises Ltd announces a merger with Sony pictures Networks India.
3. Yashoda Hegde is the new CEO of Coffee Day Enterprises Ltd.

Which of the above statements is/are correct?

(a) 1 and 3 only (b) 1 only
(c) 1 and 2 only (d) 2 only

[ESE Prelims : 2023]

3.23 Consider the following statements regarding Hemis Festival :

1. The Hemis Festival in Ladakh is a renowned religious celebration.
2. The Hemis Festival is dedicated to the birth anniversary of Lord Padmasambhava.
3. Hemis Festival offers a mesmerizing experience of Tibetan Tantric Buddhism.

Which of the above statements are correct?

(a) 1 and 2 only (b) 1 and 3 only
(c) 2 and 3 only (d) 1, 2 and 3

[ESE Prelims : 2024]

3.24 Which of the following cities achieved the Guinness World Records by constructing Single Lane Bituminous Concrete Road and longest Double Decker Viaduct with Highway Flyover and Metro Rail?

(a) Amravati and Nagpur
(b) Mumbai and Ahmedabad
(c) Hyderabad and Bangalore
(d) Gautam Buddha Nagar and Ghaziabad

[ESE Prelims : 2024]

3.25 Which one of the following Institutions launched Centre of Data for Public Good (CDPG) for multidisciplinary research, bringing together experts from academia, industry, and Government to harness the power of data to benefit the public?

(a) IISc (b) IIT Madras
(c) DRDO (d) NITI Aayog

[ESE Prelims : 2024]

3.26 Which Union Ministry announced 5G & Beyond Hackathon 2023' aimed at shortlisting India-focused cutting-edge ideas workable beyond products and solutions?

(a) Ministry of Science and Technology
(b) Ministry of Communication
(c) Ministry of Micro, Small and Medium Enterprises
(d) Ministry of Electronics and Information Technology

[ESE Prelims : 2024]

3.27 "Scheme for Expansion and Modernization of Fire Services in the States" from the allocation of preparedness and Capacity Building Funding Window under the National Disaster Response Fund for strengthening fire services in the States was introduced by which Union Ministry?

(a) Ministry of Family and Health Affairs
(b) Ministry of Youth Affairs and Sports
(c) Ministry of Defence
(d) Ministry of Home Affairs

[ESE Prelims : 2024]

3.28 To increase the transparency and consumer awareness and handle the customer complaint a 'Centralised Receipt and Processing Centre' and 'Integrated Ombudsman Scheme has been set up, these two schemes are related to which one of the following institutions?

(a) NITI Aayog (b) DPIIT
(c) ISRO (d) RBI

[ESE Prelims : 2024]

4. INTERNATIONAL AFFAIRS

4.1 HDI is a better index of development because

(a) GDP growth may not consider personal growth situations.
(b) It takes into consideration reduction of poverty.
(c) It covers income, health and education aspects of development.
(d) It covers promotion of growth.

[ESE Prelims : 2018]

4.2 Which of the following are the main functions of WTO?

1. To organize meetings of member countries to arrive at trade agreements covering international trade.
2. To ensure that member countries conduct trade practices as per agreements agreed upon and signed by the member countries.
3. To provide a platform to negotiate and settle disputes related to international trade between and among member countries.

(a) 1 and 2 only (b) 1 and 3 only
(c) 2 and 3 only (d) 1, 2 and 3

[ESE Prelims : 2020]

4.3 How many Indian companies have found a place in 2021 Fortune's Global 500 list?

(a) Three (b) Five
(c) Seven (d) Nine

[ESE Prelims : 2022]

4.4 Which one of the following cities has been named as the world's safest city from among 60 global cities, in Safe Cities Index 2021, released by the Economist Intelligence Unit (EIU)?

(a) Yangon (b) Copenhagen
(c) New York (d) Toronto

[ESE Prelims : 2022]

4.5 The First-ever G20 Ministerial Conference on Women's Empowerment was held at

(a) Vienna, Austria
(b) Hamburg, Berlin
(c) Geneva, Switzerland
(d) Santa Margherita Ligure, Italy

[ESE Prelims : 2022]

4.6 What is 'The Pandora Papers'?

(a) It is the document related to the top 100 highest tax payers of the world.
(b) It is the project of investigation which leaked almost 12 million documents that reveals hidden wealth, money laundering by some of the world's rich and powerful.
(c) It is the record of total revenue collected at the world level.
(d) It is the document containing record of the top young talented entrepreneurs under the age 30.

[ESE Prelims : 2023]

4.7 The Wasseanaar Arrangement is

(a) an elite club of countries which subscribe to arms export controls.
(b) a group of countries concerned with unconventional energy sources in the world.
(c) concerned with the preservation of extinct animal species.
(d) an arrangement which seeks to study recurring cyclone patterns.

[ESE Prelims : 2023]

4.8 Which one of the following is NOT include in the 12 areas of "Doing Business 2020"?

(a) Getting credit
(b) Paying taxes
(c) Promoting small scale industries
(d) Getting electricity

[ESE Prelims : 2023]

4.9 Government of India has moved a resolution in UN General Assembly to declare the year 2023 as the International Year of Millets for which of the following reasons?

1. Support will be provided for post-harvest value addition, enhancing domestic consumption.

2. Support will be provided for branding millet products nationally and internationally.

Select the correct answer using the code given below:

(a) 1 only (b) Neither 1 nor 2
(c) Both 1 and 2 (d) 2 only

[ESE Prelims : 2023]

4.10 Consider the following statements regarding the Global Peace Index 2023:

1. Iceland has retained its position as the most peaceful country since the inaugural study in 2008.
2. Five out of the top 10 most peaceful countries in the world are located in Europe.

Which of the above statements is/are NOT correct?

(a) Both 1 and 2 (b) 1 only
(c) 2 only (d) Neither1 nor 2

[ESE Prelims : 2024]

4.11 Consider the following statements regarding Henley Passport Index 2023:

1. Japan holds the title of the world's most powerful passport, granting visa-free access to 192 out of 227 global travel destinations.
2. Three European countries, namely Germany, Italy, and Spain, share the second position, with visa-free access to 190 destinations.

Which of the above statements are NOT correct?

(a) Both 1 and 2 (b) 1 only
(c) 2 only (d) Neither1 nor 2

[ESE Prelims : 2024]

5. DEFENCE IN NEWS

5.1 The Defence Research and Development Organization (DRDO) has successfully test-fired medium-range subsonic cruise missile Nirbhay from the Integrated Test Range (ITR) at

(a) Chandipur, Odisha
(b) Cuddalore, Tamil Nadu
(c) Nellore, Andhra Pradesh
(d) Digha, West Bengal

[ESE Prelims : 2022]

5.2 Ministry of Defence signed contract with which one of the following organizations for Upgraded Super Rapid Gun Mount (SRGM) and other equipment for around 3000 crores?

(a) DRDO (b) BHEL
(c) ISRO (d) BEL

[ESE Prelims : 2024]

5.3 Which one of the following countries did the Indian Navy participate in the U.S. Navy-led Southeast Asia Cooperation and Training (SEACAT) military exercise, to demonstrate its a maritime manoeuvres?

(a) Malaysia (b) Australia
(c) Singapore (d) New Zealand

[ESE Prelims : 2022]

5.4 Which one of the following ships does NOT come under Indian Nayy's eight ASW Shallow Water Craft project?

(a) Mahanav (b) Mahe
(c) Malvan (d) Mangrol

[ESE Prelims : 2024]

5.5 Aim of exercise 'Nomadic Elephant' is to build positive military relations, exchange best practices, develop inter- operability, bonhornie, camaraderie and friendship between India and which one of the following countries?

(a) Bangladesh (b) Mongolia
(c) Botswana (d) South Africa

[ESE Prelims : 2024]

5.6 Which one of the following is associated with 'SPRINT Challenges' aimed at giving a boost to the usage of 75 new indigenous technologies/ products in collaboration with Innovations for Defence Excellence, NIIO and Technology Development Acceleration Cell?

(a) Indian Coast Guard
(b) Indian Air Force
(c) Indian Army
(d) Indian Navy

[ESE Prelims : 2024]

6. SCIENCE AND TECHNOLOGY

6.1 Consider the following statements about ingenuity:

1. It is man's decades-long quest to fly a helicopter on Mars.
2. It is 0.6 metres tall and weigh less than 1.8 kg.
3. It aims to look for habitability.

Which of the above statements is/are correct?

(a) 3 only (b) 1, 2 and 3
(c) 1 and 2 only (d) 2 only

[ESE Prelims : 2021]

6.2 Name the NASA astronaut who after setting the record of 328-day stay on the International Space Station (ISS), returned through Earth's atmosphere and landed on the Kazakhstani desert on 6th Feb 2020.

(a) Josh Cassada (b) Jeanette Epps
(c) Christina Koch (d) Peggy Whitson

[ESE Prelims : 2021]

6.4 Consider the following statements regarding Cybersecurity Multi-Donor Trust Fund:

1. The World bank has launched a new 'Cybersecurity Multi-Donor Trust Fund' to better roll out cybersecurity development agenda in a systematic manner.
2. The new fund has been developed as an associated trust fund under the broader Digital Development Partnership (DDP) umbrella program.
3. World Bank has partnered with four countries, namely Estonia, Japan, Germany, and the Netherlands, to launch the fund.

Which of the above statements are correct?

(a) 1 and 2 only (b) 2 and 3 only
(c) 1 and 3 only (d) 1, 2 and 3

[ESE Prelims : 2022]

6.5 Consider the following statements regarding Hydrogen Breakthrough Iron-making Technology:

1. Swedish green steel venture HYBRIT, which had made the 'world's first' customer delivery of steel produced with using coal.
2. The steel was made using Hydrogen Breakthrough Iron-making Technology, which uses 100% fossil-free hydrogen instead of coal and coke.
3. The venture has started delivering the fossil-free steel to the Volvo Group as part of its trial run.

Which of the above statements is/are correct?

(a) 1 only (b) 2 and 3 only
(c) 1 and 3 only (d) 2 only

[ESE Prelims : 2022]

6.6 US-based Ohmium International has started India's first green hydrogen electrolyzer manufacturing unit at

(a) Pune, Maharashtra
(b) Hyderabad, Telangana
(c) Bengaluru, Karnataka
(d) Noida, Uttar Pradesh

[ESE Prelims : 2022]

6.7 AstroSat space telescope has crossed a major milestone by detecting 600th Gamma-Ray Burst launched by which one of the following countries?

(a) USA (b) Russia
(c) China (d) India

[ESE Prelims : 2024]

6.8 These days V-CIP is simple, safe and secure. You can complete your V-CIP from wherever you are in India, you only need your PAN card and Aadhaar card. Then, what is the full form of the term "V-CIP"?

(a) Venture Capital Identification Process
(b) Venture Process Capital Investment
(c) Voice based Customer Identification Process
(d) Video based Customer Identification Process

[ESE Prelims : 2024]

7. MISCELLANEOUS

7.1 The words 'Satyamave Jayate' inscribed below the base plate of the Emblem of India have been taken from which one of the following?

(a) Mundaka Upanishad
(b) The Rig Veda
(c) The Bhagavad Gita
(d) Vaalmiki's Ramaayana

[ESE Prelims : 2017]

7.2 The Olympic Flame symbolizes

(a) Unity among various nations of the world
(b) Speed, perfection and strength
(c) The development of sportsmanship
(d) Continuity between ancient and modern games

[ESE Prelims : 2018]

7.3 The Ministry of Sports and Youth Affairs has recently approved the inclusion of four indigenous games to be part of Khelo India Youth

Games 2021. Which one of the following is NOT included?

(a) Thang-Ta (b) Lagori
(c) Kalaripayattu (d) Gatka

[ESE Prelims : 2021]

7.4 Which one of the following is NOT correct pair of Author-Book published in the year 2020?

(a) Arundhati Roy : Azadi
(b) Jairam Ramesh : A Chequered Brilliance
(c) Zadie Smith : Intimations
(d) Diane Cook : One Arranged Murder

[ESE Prelims : 2021]

7.5 Which iconic figure set a Guinness World Record in 2020 for receiving 1 million followers for debut on instagram in just 4 hours and 44 minutes?

(a) Bong Joon-ho
(b) Amy Coney Barrett
(c) David Attenborough
(d) Sanna Marin

[ESE Prelims : 2021]

7.6 Who received the prestigious Abel Prize for the year 2020?

(a) Eric Adelberger and Blayne Heckel.
(b) Hillel Furstenberg and Gregory Margulis.
(c) Yvonne Farrell and Shelley McNamara.
(d) Nina Holden and Lisa Piccirillo.

[ESE Prelims : 2021]

7.7 Select the incorrect pair of the 2020 Nobel Prize Winners with their respective areas of contribution:

(a) Louise Gluck - Literature
(b) Andrea Ghez - Physics
(c) Jennifer A. Doudna - Chemistry
(d) Harvey J. Alter - Economic Sciences

[ESE Prelims : 2021]

7.8 Which one of the following is NOT correct regarding the Khel Ratna Award?

(a) The award was inaugurated in 1991-92
(b) The award comprises a medallion, a certificate, and a cash prize of Rs. 15 lakh.
(c) The first recipient of the Khel Ratna was chess legend, Viswanathan Anand
(d) The award renamed as Major Dhyan Chand Khel Ratna Award

[ESE Prelims : 2022]

7.9 Which one of the following is NOT correctly matched?

(a) The Last Queen : Chitra Banerjee Divakaruni
(b) Inseparable : Simone de Beauvior
(c) Great Circle : Rumaan Alam
(d) Jungle Nama : Amitav Ghosh

[ESE Prelims : 2023]

7.10 Match the following:

List I	List II
P. V Shanta	1. Film Director
Q. Akhtar Ali	2. Former Davis Cup Coach
R. Anil Dharkar	3. Noted Journalist
S. Sumitra Bhave	4. Renowned Indian Oncologist

Select the correct pair using the code given below:

	P	Q	R	S
(a)	4	2	3	1
(b)	2	4	1	3
(c)	3	1	2	4
(d)	1	3	4	2

[ESE Prelims : 2023]

7.11 Which one of the following pairs is NOT correct under women achievers?

(a) Megha Rajagopalan : Winner of Pulitzer prize in feature writing
(b) Anvee Bhutani : Indian-origin student elected as the President of Oxford Student Union
(c) Delisha Davis : 24 year old female heavy vehicle driver carrying hazardous goods.
(d) Bela M Trivedi : Took oath as the Judge of the Supreme Court of India

[ESE Prelims : 2023]

7.12 'Bahujan Hitaya : Bahujan Sukhaya' is the motto of

(a) Central Board of Film Certification
(b) Indian Railways
(c) Doordarshan
(d) All India Radio

[ESE Prelims : 2023]

7.13 Who among the following is India's first Space Tourist?

(a) Santosh George Kulangara
(b) Sirisha Bandla
(c) Raja J V Chari
(d) Pankaj Lokhani

[ESE Prelims : 2023]

Current Issues of National & International Importance : Answer Key

1.1	(a)	**1.2.**	(a)	**1.3**	(d)	**1.4**	(c)	**1.5**	(c)	**1.6**	(c)	**1.7**	(a)	**1.8**	(b)
1.9	(c)	**1.10**	(c)	**1.11**	(c)	**1.12**	(b)	**1.13**	(c)	**1.14**	(d)	**1.15**	(d)	**1.16**	(c)
1.17	(b)	**1.18**	(b)	**1.19**	(b)	**1.20**	(d)	**1.21**	(a)	**1.22**	**(a, b)**	**1.23**	(a)	**1.24**	(d)
1.25	(d)	**1.26**	(d)	**2.1**	(b)	**2.2**	**(a, d)**	**2.3**	(d)	**3.1**	(d)	**3.2**	(a)	**3.3**	(b)
3.4	(c)	**3.5**	(a)	**3.6**	(c)	**3.7**	(c)	**3.8**	(c)	**3.9**	**(*)**	**3.10**	(c)	**3.11**	(a)
3.12	(a)	**3.13**	**(a, d)**	**3.14**	(a)	**3.15**	(a)	**3.16**	(d)	**3.17**	(b)	**3.18**	(d)	**3.19**	(d)
3.20	(b)	**3.21**	(b)	**3.22**	(c)	**3.23**	(d)	**3.24**	(a)	**3.25**	(a)	**3.26**	(b)	**3.27**	(d)
3.28	(d)	**4.1**	(c)	**4.2**	(d)	**4.3**	(c)	**4.4**	(b)	**4.5**	(d)	**4.6**	(b)	**4.7**	(a)
4.8	(c)	**4.9**	(b)	**4.10**	(d)	**4.11**	(b)	**5.1**	(a)	**5.2**	(b)	**5.3**	(c)	**5.4**	(a)
5.5	(b)	**5.6**	(d)	**6.1**	(c)	**6.3**	(c)	**6.4**	(d)	**6.5**	(b)	**6.6**	(c)	**6.7**	(d)
6.8	(d)	**7.1**	(a)	**7.2**	(d)	**7.3**	(b)	**7.4**	(d)	**7.5**	(c)	**7.6**	(b)	**7.7**	(d)
7.8	(b)	**7.9**	(c)	**7.10**	(a)	**7.11**	(c)	**7.12**	(d)	**7.13**	(a)				

EXPLANATIONS

1.4 (c)

Capacity utilization: It is the extent or level to which the productive capacity of a plant, firm, or country is being used in generation of goods and services.

- It can be expressed as a percentage. It is computed by dividing the total capacity with the portion being utilized.
- NPA decline in value when capacity utilization increases.

1.5 (c)

Crowdfunding: It is the practice of funding a project or venture by raising small amounts of capital from a large number of individuals to finance a new business venture, typically via the internet.

1.6 (c)

The components of Capital Receipts are:

(i) Recovery of loans and advances
(ii) Disinvestment
(iii) Borrowing (domestic and external)
(iv) Small savings

1.7 (a)

The said option is a normal sequence of behaviour becoming a custom.

1.8 **(b)**

The chosen option is one of the defining dictums of the convergence theory on social change.

1.11 **(c)**

In economics, fiscal policy is the use of government revenue collection (mainly taxes) and expenditure (spending) to influence the economy.

1.12 **(b)**

MUDRA Bank has been established to refinance Micro Finance Institutions (MFIs) through the Pradhan Mantri Mudra Yojana. MUDRA Bank will work with coordinators at the national, state and regional levels. Additionally, it will also work with last mile financers (LMFs) of micro businesses at the regional levels.

1.13 **(c)**

The key reasons for encouraging start up entrepreneurship are innovations and bringing the values of proactivity into the society.

1.14 **(d)**

The purpose of Gold Monetization Scheme (GMS) is to mobilize unutilized gold from individuals, households and institutions and make them available to gold-based industries including jewellers.
Its primary objectives are to reduce dependence on imported gold and recycle the unutilized gold in the country.

1.23 **(a)**

The Pension Fund Regulatory and Development Authority (PFRDA) has increased the entry age for the National Pension System (NPS) from 65 years to 70 years. Earlier the eligible age to invest in NPS was 18-65 years which has now been revised to 18-70 years.

1.26 **(d)**

- 6.0%
- 8.7% (projected)

For the world, growth will slow down from 6% in 2021 to 3.2% 2022 and 2.7% in 2023.

2.1 **(b)**

CMS COP-13 was held in Gandhinagar, Gujrat from 17th to 22nd February 2020. The theme of CMS COP-13 was " Migratory species connect the planet and we welcome them home"

2.2 **(a, d)**

Energy services are directly linked to poverty and women. Energy services also directly have the impacts on the lifestyles, living conditions, income and livelihood, especially of poor people. So the preferred option shall be (a). However, sometimes we have seen UPSC people do have some interpretations of their own.

2.3 **(d)**

- NITI Aayog and World Resources Institute(WRI), India, jointly launched the 'Forum for Decarbonizing Transport' in India as part of the NDC-Transport Initiative for Asia (NDC-TIA) project.
- The launch was held virtually and inaugurated by NITI Aayog.
- The project aims at bringing down the peak level of GHG emissions (transport sector) in Asia (in line with a well below 2-degree pathway), resulting in problems like congestion and air pollution.

3.1 **(d)**

Vision area of digital India includes:

(i) Digital infrastructure as a utility to every citizen.
(ii) Governance and services on demand.
(iii) Digital empowerment of every citizen.

3.3 **(b)**

PRAGATI stands for Pro Active Governance and Timely Implementation.

3.4 **(c)**

- The Bill replaces the Consumer Protection Act, 1986. The Bill enforces consumer rights, and provides a mechanism for redressal of complaints regarding defects in goods and deficiency in services.

- The bill proposes to establish an investigating, prosecuting, reviewing and recommending body-the Central Consumer Protection Authority.
- It also provides right to consumer to file a complaint for overcharging.

3.5 (a)

Features of Technology Promotion, Development and Utilization Programme (TSDP) in the Department of Scientific and Industrial Research include: "... R&D over existing technologies ... for support, however "Only full-fledged complete for development of technology will be considered under TSDP", mainly "Projects related to design and development of Software/IT as required for products and processes", however "Pure Software Development does not fall in the scope of the programme".

Its two main features are: Development and integration of technologies in identified areas; and to promote application of advanced technology for improving the performance, value addition and exportability of various products".

Its specific objectives include: develop and integrate technologies with holistic approach; application of modern / advanced technology for socio-economic problem solving; development and application of R&D activities, and promoting activities for improving technologies and other related issues.

3.6 (c)

Integrated Power Development Scheme (IPDS): Launched by the Government of India (GoI) in November 2014 it improves distribution network in urban areas by strengthening of sub-transmission and distribution network, metering of distribution transformers / feeders / consumers, and Information Technology (IT) enablement of Distribution Network. It is a modification on Restructured Accelerated Power Development and Reforms Programme (RAPDRP). **Deendayal Upadhyaya Gram Jyoti Yojana (DDUGJY)**: Launched on 15th July 2015 in Patna, it goes for feeder separation in household and agricultural sector, strengthening of sub-transmission and distribution network, metering to all, and 24 X 7 Power Supply in Rural Areas. Similar such Scheme was already there in Gujarat, and the Union Government's erstwhile scheme Rajiv Gandhi Grameen Vidyutikaran Yojana (RGGVY) is subsumed in DDUGJY. **Domestic Efficient Lighting Programme (DELP)**: Launched in January 2015, renamed as **'Unnat Jyoti by Affordable LEDs for All (UJALA)'** in March 2016, it focuses on 'LED-bulbs/tube lights' and energy efficient fans, and it is being implemented by Energy Efficiency Services Limited (EESL).

3.7 (c)

National Agriculture Market is a pan India electronic trading portal which networks the existing Agriculture Produce Market Committees (APMC) to create a unified national market for agricultural commodities.

- Farmers can showcase their produce online from their nearest market and traders can quote price from anywhere.
- It also ensures open price discovery and better returns to farmers.

3.8 (c)

- Utkarsh Bangla Scheme: West Bengal
- Placement Linked Skill Training Programme: Assam
- SURYA Scheme: Haryana
- Employment Linked Skill Training Programme: Rajasthan

3.9 (*)

21st Hunar Haat was organised in Feb., 2020 at Ranchi whereas 23rd was organised in December 2020 at Rampur. From 23rd October to 1st November 2020, Hunar Haat was proposed for Jaipur.

3.10 (c)

The National Education Policy 2020 has fixed target of :

- Universalization of education from preschool to secondary level with 100% Gross Enrolment Ratio (GER) in school education by 2030.
- Gross Enrolment Ratio in higher education to be raised to 50% by 2035.

3.11 (a)

Among educational institutions in the country, IIT Madras ranked 1st in the National Institutional Ranking Framework 2020, followed by IISc Bangalore.

The National Institutional Ranking Framework or NIRF is an annual ranking and was launched in 2015 and the first rankings were released in 2016. NIRF 2020 has ranked institutions across 10 categories: Overall, university, engineering, management, pharmacy, college, medical, law, architecture and dental.

3.12 (a)

- PM SVANidhi stands for Prime Minister Street Vendor's AtmaNirbhar Nidhi. It is a central sector scheme launched in June 2020. It aims to provide micro-credit facilities to street vendors affected due to COVID-19 pandemic.
- The "SATYABHAMA (Science and Technology Yojana for AtmaNirbhar Bharat in Mining Advancement)" is a portal launched by the Ministry of Mines with an aim to promote research and development in the mining and minerals sector.

3.13 (a, d)

- The Atal Tunnel in Rohtang, Himachal Pradesh, India is the World's longest highway tunnel. It is constructed at an altitude of 3000 meters and is about 9.02 kilometers long, connecting Solang Valley near Manali to Sissu in Lahaul and Spiti district.
- The Atal Tunnel is constructed with a traffic density of approximately 1500 trucks and 3000 cars per day.

3.14 (a)

- Agri India Hackathon is the largest virtual gathering to create dialogues and accelerate innovations.
- Virtual Agri-hackathon 2020 is organised by the Department of Agriculture and Cooperation and Farmers' Welfare in association with IARI, Pusa, New Delhi.

3.15 (a)

Kaziranga National Park (KNP) of Assam state has become the first national park in India to be equipped with satellite phones.

This first-of-its-kind move was taken to provide assistance to prevent poaching and ensure coordinated functioning.

3.16 (d)

EASE 4.0 reforms look after four key initiatives to be adopted by public sector banks:

1. **Smart lending backed by analytics:**
   - Dial-a-loan for doorstep lending
   - Credit@Click: End to End digital retail and MSME lending
   - Analytics based credit offers
2. **24×7 banking with resilient technology and cloud based IT systems**
   - Deeper penetration of Mobile & internet banking
   - Cloud-based IT Systems and improved cyber resilience
   - Process automation
3. **Data enabled agriculture financing**
   - Dial-a-loan for agri loans
   - Partnership with AgriTechs for data exchange
   - Automated Processing & sanctioning
4. **Collaborating with the financial ecosystem**
   - Digital Payments in semi-urban and rural areas
   - At scale delivery of doorstep banking services
   - Co-lending with NBFCs

3.19 (d)

Mahila Kisan Sashaktikaran Pariyojana (MKSP):

- Implemented by Department of Agriculture and farmer welfare.
- Atleast 30% of the expenditure incurred for women to bring them into mainstream agriculture.
- Started in 2011
- It seeks to improve the present status of women in agriculture and to enhance the opportunities available to empower her.

3.20 **(b)**

Jal Jeevan Mission:
Slogan: HAR GHAR JAL
Objective: Safe and adequate drinking water to all household in rural India.

3.21 **(b)**

SDG: State score in zero Hunger quality education and affordable and clean energy.
Kerala : 75 Points
Chandigarh : 79 points

3.22 **(c)**

Malvika Krishna is the new CEO of Coffee Day Enterprises Ltd.

3.24 **(a)**

Nagpur Metro has created a Guinness World Record by constructing the longest Double Decker Viaduct (metro) of 3.14 km.
It has highway flyover and metro rail supported on a single column.
It is located on Wardha Road, Nagpur.
Length of bituminous concrete road : 42.2 km.

3.25 **(a)**

IISc launched Centre of Data for Public Good (CDPG) with a first of its kind symposium. Its initiative is to leverage data for social good.
Centre of Data for Public Good will serve as a hub for multidisciplinary research, bringing together experts from academia industry and government to harness the power of data to benefit the public.

4.1 **(c)**

Human Development Index (HDI), say used by United Nations Development Programme (UNDP) in its Reports, is a better Index of Development as it conceives and perceives development not merely as economic growth but rather as broader Human Development and Human Capital formation taking care of income, health and education aspects.

4.3 **(c)**

Seven Indian companies have found a place in 2021 Fortune's Global 500 list. Indian Companies in the list are:

1. Reliance Industries (Rank: 155)
2. State Bank of India (205)
3. Indian Oil Corporation (212)
4. Oil & Natural Gas Corporation (243)
5. Rajesh Exports (348)
6. Tata Motors (357)
7. Bharat Petroleum (394)

4.4 **(b)**

Copenhagen, the capital city of Denmark, has been named as the world's safest city from among 60 global cities, in Safe Cities Index 2021, released by the Economist Intelligence Unit (EIU). Safe Cities Index 2021 report ranks 60 cities across 76 indicators covering digital, health, infrastructure, personal and environmental security.
Delhi ranks 48 according to the index.

4.5 **(d)**

First ever G20 Ministerial Conference on Women's Empowerment Conference was held at Santa Margherita Ligure, Italy in a hybrid format.

4.6 **(b)**

Pandora papers are a release of nearly 12 million leaked documents that reveal the hidden and sometimes unethical or corrupt dealings of global wealthy and elite including prominent world leaders, politicians, corporate executives, celebrities and billionaires.

4.7 **(a)**

Wassenaar Arrangement on export control for conventional arrows and dual– use goods and Technologies is a multilateral export control region with 42 participating states.

4.8 **(c)**

12 areas of doing business

1. Starting a business.
2. Dealing with construction permit.
3. Getting electricity.
4. Registering property.
5. Getting credit.
6. Protecting minority investors.
7. Paying taxes.
8. Trading across borders.
9. Inforcing contracts.

10. Resolving insolvency.
11. Employing workers.
12. Contracting with the government.

4.9 (b)

- "Indian Wealth, Millets for Health"
- Millets included in PDS.

4.10 (d)

Global Peace Index experienced a total decline of 0.42%.

Iceland maintains its status as the most peaceful country in the world for the 17th consecutive year.

Top 10 countries are:

1. Iceland
2. Denmark
3. Ireland
4. New Zealand
5. Austria
6. Singapore
7. Portugal
8. Slovenia
9. Japan
10. Switzerland

4.11 (b)

Henley Passport Index is a ranking system that evaluates all the world's passport based on the number of travel destinations their holders can access without the need for a prior visa.

Indian passport secured 80th rank, in 2022 rank was 87th that gives visa free access to or countries.

Singapore holds the most powerful passport in the world.

Previously Japan held the top sport on the Henley passport for 5 years.

Germany, Italy and Spain shared the 2nd position.

Japan, Austria, Finland, France, Luxembourg, South Korea and Sweden shared 3rd spot.

5.2 (b)

Ministry of Defence signed contract with Bharat Heavy Electricals Limited, Haridwar for procurement of 16 upgraded Super Rapid Gun Mount (SRGM) along with associated equipment for Indian Navy. SRGM is a light-weight, rapid-fire naval gun providing unrivalled performances and flexibility in any air-defence and anti-surface role, particularly in anti-missile role.

5.6 (d)

SPRINT: Supporting Pole-valuting in R & D through Innovation for Defence Excellence.

6.1 (c)

Ingenuity is the first robot helicopter to fly on Mars.

The helicopter weighs 1.8 kilograms. It is a solar powered helicopter.

It aims to collect samples from the locations on Mars where the rover cannot reach.

6.5 (b)

Swedish green steel venture 'HYBRIT' had made the 'world's first' customer delivery of steel produced without using coal.

The steel was made using Hydrogen Breakthrough Ironmaking Technology, which uses 100% fossil-free hydrogen instead of coal and coke.

The venture has started delivering the fossil-free steel to the Volvo Group as part of its trial run.

6.7 (d)

AstroSat is the first dedicated Indian astronomy mission aimed of studying celestial sources in X-Ray, optical and UV spectral bands. Astrosat achieved a significant milestone by detecting more than 600 Gamma Ray burst marking the death of a massive star.

6.8 (d)

V-CIP is alternate method of customer identification with facial recognition.

7.2 (d)

- The olympic flame represents link between the ancient and modern games and underlines the profound connection between these two events.
- There was no olympic flame used in first olympic games, held in 1896 (Athens, Greece). 1st time it was used in 1928 olympics in Amsterdam (Netherlands).

7.3 (b)

The Sports Ministry included four indigenous games to be a part of the Khelo India Youth Games 2021, which was conducted in Haryana. The games include Gatka, Kalaripayattu, Thang-Ta, and Mallakhamba.

7.4 **(d)**

One Arranged Murder book is written by the Indian author Chetan Bhagat.

7.7 **(d)**

Harvey James Alter is an American medical researcher, virologist, physician and Nobel Prize laureate, who is best known for his work that led to the discovery of the hepatitis C virus.

7.8 **(b)**

The Khel Ratna Award, renamed Major Dhyan Chand Khel Ratna award, is the highest sporting award given by the Ministry of Youth Affairs and Sports.

It is awarded for the spectacular and most outstanding performance in the field of sports by a sportsperson over a period of four years. It comes with a cash prize of Rs 25 lakh.

The Khel Ratna award was instituted in 1991-1992 and the first recipient was Chess legend Viswanathan Anand.

7.9 **(c)**

- **The last Queen:** Chitra Banerjee Divakaruni.
- **Inseparable :** Simone de Beavour.
- **Great Circle :** Maggie Shipsteal.
- **Jungle Nama** : Amitav Ghos.

7.10 **(a)**

V Shanta : Renowned Indian Oncologist
Akhtar Ali : Former Davis Cup Coach
Anil Dharkar: Journalist and Writer
Sumitra Bhave: Film Director

7.11 **(c)**

- **Megha Rajgopalan**: American journalist of Indian descent who works at the New York Times. She won 2021 Pulitzer Prize.
- **Anvee Bhutani**: President of oxford university student union.
- **Delisha Davis**: First Indian woman to drive a fuel tanker.
- **Bela M Trivedi**: Took oaths as the judge of the supreme court of India.

7.13 **(a)**

Santosh George Kulangara was the first Indian space tourist after paying a whopping $ 2,00, 000 and booked a seat in Virgin Atlantic. Santosh George is a resident of Kerala.

■■■■